Francis Ysidro Edgeworth

In memory of

Ernest Lluch,
Lluís Argemí
and
A.W. 'Bob' Coats

Francis Ysidro Edgeworth

A Portrait with Family and Friends

Lluís Barbé

Professor of Economic Theory, Universitat Autònoma de Barcelona, Spain

Translated by Mary C. Black

Edward Elgar

Cheltenham, UK • Northampton, MA, USA

This is a substantially revised version of the book originally published in Catalan as: *Francis Ysidro Edgeworth, Crònica Familiar* by Universitat Autònoma de Barcelona, Servei de Publicacions, Bellaterra, 2006

Published by
Edward Elgar Publishing Limited
The Lypiatts
15 Lansdown Road
Cheltenham
Glos GL50 2JA
UK

Edward Elgar Publishing, Inc.
William Pratt House
9 Dewey Court
Northampton
Massachusetts 01060
USA

A catalogue record for this book
is available from the British Library

Library of Congress Control Number: 2009937924

FSC
Mixed Sources
Product group from well-managed
forests and other controlled sources
www.fsc.org Cert no. SA-COC-1565
© 1996 Forest Stewardship Council

ISBN 978 1 84844 716 5

Printed and bound by MPG Books Group, UK

Contents

Illustrations

Genealogical branches

Abbreviations

Family

C.E. = Christina Edgeworth (née MacPherson)
C.S.E. = Charles Sneyd Edgeworth
F.A.E. = Frances Anne Edgeworth (née Beaufort)
F.B.E. = Francis Beaufort Edgeworth
F.W. = Fanny Wilson (née Edgeworth)
F.Y.E. = Francis Ysidro Edgeworth
H.B. = Harriet Butler (née Edgeworth)
H.E. = Honora Edgeworth (after 1838, Honora Beaufort)
H.J.B. = Harriet Jessie Butler, 'Harrie' (née Edgeworth)
M.E. = Maria Edgeworth
M.P.E. = Michael Pakenham Edgeworth
R.F.E. = Rosa Florentina Edgeworth (née Eroles)
R.L.E. = Richard Lovell Edgeworth

Manuscripts (Ms)

CP TRINITY CLD = Crampton Papers from Trinity College Library, Dublin
EP BODLEIAN UO = Edgeworth Papers from the Bodleian Library, University of Oxford
EP LSE = Edgeworth Papers from the British Library of Political and Economic Science of the London School of Economics and Political Science
EP NL IRELAND = Edgeworth Papers from the National Library of Ireland
EP NUFFIELD C UO = Edgeworth Papers from Nuffield College, University of Oxford
GP & PP UC LONDON = Galton Papers and Pearson Papers from University College London
KP KING'S C CAMBRIDGE MA= John Maynard Keynes Papers from King's College Cambridge, Modern Archives

Journals

EJ = *Economic Journal*
JRSS = *Journal of the Royal Statistical Society*
QJE = *Quarterly Journal of Economics*

Preface

John Creedy

Francis Ysidro Edgeworth was a leading figure in the rapid development of economics during the last quarter of the nineteenth century and the first quarter of the twentieth century, by which time it was firmly established as an academic subject. He held the Drummond Chair at Oxford and was regarded as second only to the great Cambridge economist Alfred Marshall. He was a prolific and highly original author who, in a cosmopolitan age, had probably the widest correspondence with economists all over the world. He was a man of enormously wide reading and considerable linguistic skills. He was the first editor of the *Economic Journal*, published by the newly formed Royal Economic Society. He was President of Section F of the British Association in 1889 and 1922. He achieved eminence as a statistician as well as an economist, becoming a Guy Medallist (Gold) of the Royal Statistical Society in 1907, and was President of the Society in 1912–14.

His name is familiar to all students of economics, if only from learning about the 'Edgeworth box', one of the most widely used analytical devices in the subject. This diagrammatic tool was first introduced by Edgeworth in 1881 in his first publication in economics, *Mathematical Psychics: An Essay on the Application of Mathematics to the Moral Sciences*. This small book is remarkable for its highly original and far-reaching contributions to economics; indeed Marshall began his review with the statement: 'This book shows clear signs of genius'. However, it was written in such a terse and unique style that it took many years before its contributions were fully appreciated, despite the fact that Edgeworth became one of the most prominent economists of his age. The title itself does not clearly signal a book on economics, and his use of mathematics put it well beyond the reach of most of the economists of the period. The technical difficulty of much of his published output contributed to its slow assimilation into textbooks and he continues to remain relatively neglected in texts on the history of economic analysis.

Mathematical Psychics, written right at the start of Edgeworth's career as an economist, also provides the key to all his later work and his lasting importance to economists. He wrote extensively on a wide range of topics, but the central importance of his *Mathematical Psychics* means that some discussion and clarification is perhaps warranted here. Before discussing the book, it is worth remembering that the period, the last quarter of the nineteenth century, marks a distinct change of emphasis in the study of economics, in the transition to neoclassical economics from the classical economics associated with Adam Smith. Instead of concentration on the dynamic themes of growth and development and the important monetary debates associated with the numerous banking crises of the first half of the nineteenth century, the emphasis of the neoclassical economists was on the nature of exchange. This was seen as the 'central' problem. As Hicks stressed when discussing the neoclassical economists, 'It was possible, they found, to construct a "vision" of economic life out of the theory of exchange, as the classics had done out of the social product. It was quite a different vision.'[1] Edgeworth himself was later to remark that 'in pure economics there is only one fundamental theorem, but that is a very difficult one: the theory of bargain in a wide sense'.[2] The period in question was also one of intense debate in moral philosophy, with utilitarianism as the dominant principle. There were also significant developments in experimental psychology, and of course Darwin's theory of evolution was being hotly debated, with extensions being made towards forms of social Darwinism. Edgeworth's book reflects all these strong influences. Many of his allusions are perhaps unclear to a modern reader, but they would have been familiar to his contemporaries.

Taking Jevons's basic analysis of exchange of two goods between two traders, also examined by Walras, as his starting point, Edgeworth supposed that the objective of each trader is to maximise utility, considered to be a general function of the quantities of the goods held and consumed after trade is concluded. The utility maximising approach was immediately congenial to Edgeworth, who was steeped in utilitiarian moral philosophy. He first concentrated on the nature of barter, instead of describing the characteristics of an equilibrium set of prices, that is, one which ensures that the individuals' responses are mutually consistent. If the traders in barter are allowed freely to vary the terms of provisional 'contracts', Edgeworth showed that there is a range of 'final settlements', from which no further 'recontracting' would take place. In a rectangular box where the base and height are determined by the initial stocks of the two goods, these final settlements define what Edgeworth called the 'contract curve'. These settlements are also efficient trades, in the sense that if a settlement is not on the contract curve, movement to it can make one person better off without the other being worse off: this original idea of efficiency later came to be called Pareto efficiency. Movement along the contract curve involves one trader becoming worse off while the other gains.

Edgeworth then defined indifference curves for a trader as showing combinations of amounts consumed for which utility is constant. Using several approaches, he demonstrated that the contract curve is the locus of points of tangency between traders' indifference curves, between limits given by their pre-trade curves (those going through the initial endowment point in the box). The existence of a range of final settlements has important implications. First, without introducing further structure to the barter framework, it is not possible to say what the implied rate of exchange is, given only information about the preferences and endowments of individuals. It results in 'indeterminacy' whereby all that can be said is that the actual trade depends on the relative bargaining strength of the traders.

On the argument that such higgling is widespread, Edgeworth stated in his unique style: 'The whole creation groans and yearns, desiderating a principle of arbitration, and end of strifes' (1881, p. 51). His next argument involved two steps. First, he showed that the utilitarian principle of maximising total utility places individuals on the contract curve, because the mathematical conditions are equivalent to the tangency of indifference curves. Indeed, if it is possible to make someone better off without someone else being worse off, total utility cannot be at maximum and individuals cannot be on the contract curve. While this may seem a small step, to Edgeworth it was of great significance:

> It is a circumstance of momentous interest that one of the in general indefinitely numerous settlements between contractors is the utilitarian arrangement . . . the contract tending to the greatest possible total utility of the contractors. (1881, p. 53)

However, he recognised that this result is not sufficient to justify the use of utilitarianism as a principle of arbitration; it is only a necessary condition. Edgeworth's justification for utilitarianism as a principle of justice, comparing points along the contract curve, was as follows:

> Now these positions lie in a reverse order of desirability for each party; and it may seem to each that as he cannot have his own way, in the absence of any definite principle of selection, he has about as good a chance of one of the arrangements as another . . . both parties may agree to commute their chance of any of the arrangements for . . . the utilitarian arrangement. (1881, p. 55)

The important point to stress about this statement is that Edgeworth clearly considered willingness to accept the utilitarian arbitration in terms of choice under uncertainty. His argument is that the contractors, faced with uncertainty about their prospects but viewing alternatives along the contract curve as equally likely, would choose to accept an arrangement along utilitarian lines. Thus a crucial component of this argument is the use of equal a priori probabilities, something that was later important to Edgeworth in his statistical work. In taking this second step, Edgeworth believed that he had provided an answer to an age-old question, stating, 'by what mechanism the force of self-love can be applied so as to support the structure of utilitarian politics, neither Helvetius, nor Bentham, nor any deductive egoist has made clear' (1881, p. 128).

The importance to him of this new justification of utilitarianism cannot be exaggerated. Indeed the whole of *Mathematical Psychics* seems to be imbued with a feeling of excitement generated by his discovery of this justification based on a 'social contract'. This provided the crucial link between 'impure' and 'pure' utilitarianism in a more satisfactory way than his earlier appeal to evolutionary forces, made in his book entitled *New and Old Methods of Ethics*, written before turning to economics.

The nature of price-taking behaviour – involving an equimarginal principle whereby the ratio of prices must be equal, for both traders, to the ratio of their marginal utilities for each of the relevant goods – had been explored with great originality by Jevons with his 'equations of exchange'. Edgeworth made important extensions to this analysis, as well as providing his succinct diagrammatical synthesis (which included showing how Marshall's 'offer curves' can be derived from indifference curves). He showed how his box diagram can be used to illustrate a price-taking equilibrium. This arises where one or more of the mutual tangency positions of indifference curves along the contract curve also corresponds to tangency with a straight line going through the endowment point. This line represents a common budget constraint for the choices of the individuals, whereby the slope represents the exchange ratio and hence the relative price. In equilibrium, individuals acting in isolation and taking prices as given (in contrast to those engaged in barter) have mutually consistent demands and supplies. A price-taking equilibrium, as such a tangency position, must therefore correspond to a point on the contract curve.

Edgeworth was thus able to clarify the sense in which a price-taking (often called competitive) equilibrium is 'optimal', fully recognising that it is just one of many Pareto optimal points. This gives rise to what is now referred to as the 'first fundamental theorem' of welfare economics – that a price-taking equilibrium is Pareto efficient. The use of price taking also provides a considerable reduction in the amount of information required by traders compared with barter. Individuals only need to know the equilibrium prices, whereas in barter they have to learn a considerable amount of information about other individuals' preferences and endowments. Of course, this merely describes the properties of an equilibrium and does not, as Edgeworth was fully aware, explain how it might be achieved in practice. However, he later showed that a sequence of price adjustments, where trading – at the minimum of demand and supply – takes place at disequilibrium prices, leads to a point on the contract curve, although precisely where is indeterminate.

Edgeworth then returned to indeterminacy in barter, asking whether this indeterminacy results from the absence of competition in the simple two-person market. He quickly moved on to examine the implications of introducing further pairs of traders. The analysis of barter with numerous traders again involves Edgeworth's stylised description of the recontracting process of barter mentioned above. With more traders, the importance of the recontracting process, apart from allowing the dissemination of information, lies in the fact that it makes it possible to analyse the use of collusion among some of the traders. Individuals are allowed to form coalitions in order to improve bargaining strength. Recontracting enables the coalitions to be broken up by outsiders who may attract members of a group away with more favourable terms of exchange.

The analysis of many traders, where coalitions can be temporarily formed and broken up by the offer of improved terms from other traders, would appear to present formidable difficulties. Yet Edgeworth rapidly demonstrated, again using his famous box diagram, that the introduction of further similar pairs of traders gradually reduces the range of indeterminacy; that is, the length of the contract curve shrinks. With a sufficiently large number of traders, the range of indeterminacy shrinks to the finite number of price-taking equilibria. Barter thus replicates price-taking behaviour. Given that coalitions among traders are allowed in the recontracting process, a price-taking equilibrium cannot be 'blocked' by a coalition of traders. In this sense the competitive equilibrium is robust.

The argument that a complex process of bargaining among a large number of individuals produces a result which is identical to a price-taking equilibrium is an important result that is far from intuitively obvious. The recontracting process can be said to represent a competitive process, and the contract curve shrinks essentially because of the competition between suppliers of the same good, although it is carried out in a barter framework in which explicit prices are not used. The price-taking equilibrium, in contrast, does not actually involve a competitive process. Individuals simply believe that they must take market prices as given and outside their control. They respond to those prices without any reference to other individuals. But the result is that the price-taking equilibrium looks just like a situation in which all activity is perfectly coordinated.

Great stress was placed by Edgeworth on comparison with Lagrange's 'principle of least action' in examining the overall effects produced by the interactions among many particles. The connection with Edgeworth's analysis of competition, involving interaction among a large number of competitors to produce a determinate rate of exchange, is clear. The fact that in the natural sciences so much could be derived from a single principle was important for both Jevons and Edgeworth. But Edgeworth took this to its ultimate limit in arguing that the comparable single principle in social sciences, that of maximum utility, would produce results of comparable value. Referring to Laplace's massive work, *Mécanique Celeste*, he suggested that:

> 'Mécanique Sociale' may one day take her place along with 'Mécanique Celeste', throned each upon the double-sided height of one maximum principle, the supreme pinnacle of moral as of physical science . . . the movements of each soul, whether selfishly isolated or linked sympathetically, may continually be realising the maximum energy of pleasure, the Divine love of the universe. (1881, p. 12)

A strong belief in the value of mathematical analysis in economics, even where the precise numerical form of the relevant relationships cannot be known, imbues all of Edgeworth's work. When this is combined with his strong adherence to utilitarianism, it is not difficult to see how Edgeworth was excited to be showing not only why this principle may be accepted in the form of a 'social contract', but how the actions of many utility maximising individuals in a market can lead to a determinate solution. Thus, while the comparison with Laplace may seem fanciful to some readers, it was far from fanciful to Edgeworth. These elements provide the 'plan' with which virtually all his work in economics could be viewed. It is no wonder that Alexander Pope's statement, in his *Essay on Man*, that it presents 'A mighty maze, but not without a plan' was borrowed by Edgeworth to describe the competitive barter process. It also nicely fits Edgeworth's own oeuvre. Although he went on to write on a wide range of economic topics, and to make original contributions to mathematical statistics which alone would guarantee a lasting reputation, an appreciation of the preoccupations leading towards, and nature of, this first work is important in placing everything else in perspective.

It is clear from even a small sample of Edgeworth's work that the writer brings to it not just a deep and fertile originality, but also a vast range of knowledge covering natural sciences and literature. His writing is highly allusive and contains quotations from Greek and Latin classics as well as a range of English poets. It displays a sharp wit of a kind found in no other writing in the subject. It cannot fail to raise curiosity about the background, training and character of the man which went into generating such a prolific and ingenious author. Yet, very little has previously been known about Edgeworth, despite the fact mentioned above that he held prominent positions as editor of the *Economic Journal* and as the Drummond Professor at Oxford, and corresponded extensively with a wide range of economists. Furthermore, Edgeworth's aunt was the famous novelist Maria Edgeworth, his grandfather was the highly colourful inventor Richard Lovell Edgeworth, and other family connections link him with a large number of other eminent families. Much has been written about these other relatives and indeed several members of the family have produced extensive historical records of their own. Yet very little indeed was known about his mother's background and family – Edgeworth combines Spanish ancestry with one featuring prominent Anglo-Irish intellectuals. Edgeworth turned to economics at a relatively late age, but his activities during the period between his university education and his first publication in economics have previously only partially been known. Earlier writers have even had to speculate about the ways in which Edgeworth might have acquired his proficiency in mathematics.

The present biography is therefore especially welcome in providing a considerable amount of fresh information about Edgeworth's life and background. The author of this book has carried out a vast amount of exhausting primary research in a range of archives in several countries. He has managed to assemble much interesting detail which provides a fresh view of Edgeworth's life and his relationships with many of his contemporaries.

In writing a book of this kind a difficult decision is required about the mixture of biographical and technical material – a tricky balancing act has to be performed. The author places greater emphasis on the biographical detail, which indeed represents the fruits of his extensive scholarship and will be new even to those with considerable familiarity with the development of the subject and its main authors during the relevant period. Of course, in some respects such a deep and complex subject must always remain an enigma, and Edgeworth himself left no introspective writing. But we can be grateful to the author for bringing us closer to one of the truly great and creative figures in economics.

NOTES

1. See Hicks (1984a) p. 250.
2. 'On the Application of Mathematics to Political Economy', in Edgeworth (1925), II, p. 288.

Acknowledgements

I began this research fifteen years ago. Through its various stages there have been many friends – both old and new – who have guided, encouraged and led me through the entire process that has made this book possible. Professors Fabian Estapé, Josep Fontana, Jordi Pascual, Salvador Almenar, Antoin E. Murphy, José Luis Cardoso, Alfonso Sánchez Hormigo, Muriel Casals, Luci Nussbaum, Amparo Tusón, Carlos Lomas, Howard Petith, and the late Professors Ernest Lluch, Lluís Argemí and A.W. 'Bob' Coats are just a few names that come to me when I try to recall the people who in one way or another have contributed to this biography. Many thanks to them and also to Francesc Prats, Amadeu Rocamora, Anita Puigcerver-Rumbold, Xavier López Andrés, and to the countless others who have remained unnamed.

The following have all kindly given permission to publish the quotations from manuscripts in their collections: The Boards of the National Library of Ireland, Trinity College Dublin, the London School of Economics and Political Science, and the Bodleian Library of the University of Oxford; the Director of Library Services, Special Collections, of University College London; the Provost and Fellows of King's College, Cambridge; the Warden and Fellows of Nuffield College, Oxford; and Dr David Edgeworth Butler. Dr Butler has also permitted the use of Edgeworth's pictures.

The librarians of these institutions and the personnel in charge of the Oriental and India Office Collections at the British Library in London, the Diocesan Archives in La Seu d'Urgell and the Historical Archives of the Universitat de Barcelona (UB) have most open-handedly assisted me in finding the material that I was looking for. I would especially like to express appreciation for the helpfulness of Elizabeth Martin of Nuffield College and Josep Casals of the UB Historical Archives. Their efforts went far beyond mere professional duty. I am also indebted to Carlos Alonso and the entire team of the Servei de Publicacions de la Universitat Autònoma de Barcelona (UAB), which has always believed in my projects.

For this English version, I have to acknowledge the crucial academic and editorial support of Professor John Creedy, the main expert on the economics of Edgeworth, who personally fought to have this book published in times of technological and financial upheaval, generously contributed a preface, restrained my speculative penchant and tried to redress my misapprehensions. I should also mention the efficient help of Felicity Plester, Commissioning Editor of Edward Elgar Publishing Ltd and the support of the Institut Ramon Llull in rewarding the diligent work of Mary C. Black with a grant for the translation of Catalan literary and scholarly works.

Given the nature of this undertaking, I have considered it necessary to reproduce brief extracts from the books and articles which are listed as references at the end. Consequently, using the convention of inserting short quotations without explicit permission, I gladly pay their authors the academic tribute they deserve.

In any case, all blameable errors, misjudgments and omissions fall under my exclusive responsibility.

The translation of this work was supported by a grant from the Institut Ramon Llull.

institut ramon llull
Llengua i cultura catalanes

Introduction

WHY A BIOGRAPHY OF FRANCIS YSIDRO EDGEWORTH?

A book-sized biography of the economist and statistician Francis Ysidro Edgeworth (1845–1926) is a worthy endeavour for three reasons.

First, there is a series of articles portraying Edgeworth: Bonar (1926), Keynes (1926), Price (1926), Bowley (1934), Hildreth (1968), Kendall (1968), Stigler (1978), Hicks (1984b), Creedy (1986), Newman (1987) and Mirowski (1994). However, there is no complete book about his life.

Second, over time historians of economic thought have come to pay greater attention to Edgeworth's work. As just a foretaste of this interest, I could mention that though in Edgeworth's times his scientific work was relegated to the background due to Alfred Marshall's predominance and Karl Pearson's and Maynard Keynes's lack of generosity, his work is now indispensable for grasping the genesis of current thinking in economics. Moreover, knowledge of Edgeworth's life supplies a vivid portrait of the personal and intellectual relationships amongst the marginalist and non-marginalist economists of those times, including Jevons, Marshall, Foxwell, Walras, Pantaleoni, Pareto, Wicksell and Keynes.

Third, a more personal reason is that Edgeworth's mother was Catalan: Rosa Florentina Eroles. The only thing known about her, when I started my research, was that she was the daughter of a Catalan refugee in London (Keynes, 1926) whose name was Antonio. According to a note from Lord Holland in December 1831, he was a general who participated in an expedition against Ferdinand VII of Spain and was very likely a close relative of the Baron of Eroles, the well-known proponent of absolutism (Hicks, 1984b). As a Catalan economist myself, it was a challenge for me to discover who the Eroles were and where they came from. Moreover, the Nobel prize-winning John Hicks accepted Lord Holland's assertion and conjectured that this Antonio Eroles was a brother of the Baron. Obviously, the possibility of proving this thesis wrong added appeal to the research.

The Four Stages in the Research

Throughout the fifteen years of research, I can distinguish four stages.

In the first stage, my goal was to write an article to prove the falsity of Hicks's conjecture about Antonio Eroles, since both Fontana and Lluch were convinced that as he was an exile in London and had participated in the 1830 expedition against Ferdinand VII, he could never have been an absolutist.[1] The files in the Arxiu Nacional de Catalunya about Joaquín Ibáñez-Cuevas y de Valonga, Baron of Eroles prove that he had only sisters and helped me to conclude that Lord Holland had erroneously confused a surname, Eroles, with a title, Baron of Eroles. However, since there are no documents about General Antonio Eroles in the Archivo General Militar of Segovia I did not know how to pursue the research, so I temporarily set it aside.

Without knowing that I was entering a second stage in the research, in the summer of 1998 I travelled to Ireland as a tourist. While there, I took advantage of the fact that the coach was going through Edgeworthstown, to visit the church and the manor house of the Edgeworth family, now a senior citizens' residence, to ask the nun who greeted me if there were any family letters or documents. The nun replied that there had been a huge amount, which was ultimately deposited in the National Library of Ireland. During the summers of 1999 and 2000, I stayed in Dublin for three weeks and in Oxford for two, where I examined the Edgeworth manuscripts kept at the National Library and the Bodleian Library. It is an endless task because Francis Ysidro Edgeworth's Irish grandfather, Richard Lovell Edgeworth, married four times and had twenty-two children, including Maria, the famous writer, and Francis Beaufort Edgeworth, child number twenty-one, who at the age of twenty-two married Rosa Florentina Eroles, then sixteen. They went on to have seven children, the youngest of whom was the economist and statistician. The heap of correspondence is spectacular – more than two thousand letters, among them one maintaining that Antonio Eroles was endorsed by the generals Mina (Espoz y Mina) and Àlava, and therefore that he was a liberal constitutionalist. These initial findings were published as an article in the *Revista Econòmica de Catalunya*. Meanwhile, I realised that the Edgeworths were an Irish family that had experienced noteworthy historical events in Britain during the end of the eighteenth and first half of the nineteenth centuries. As a by-product of this research, I wrote a fictional narrative on the most relevant events – both good and bad – that they had witnessed. This novel, written in Catalan and called *Retrat de família sobre fons de trèvols*, won a literary prize, the Premi Sant Joan de Literatura Catalana 2001, that led to more than 50 000 copies being printed.

Immediately after winning the prize, I decided to concentrate on Francis Ysidro Edgeworth, whom I had left in the novel at the age of nineteen on his way to Oxford. This ushered in a third stage of research that has driven me to consult a variety of manuscripts and documents once again at the National Library of Ireland, the Library of Trinity College Dublin, the Bodleian Library and Nuffield College at Oxford, the London School of Economics, the British Library, University College London, and King's College Cambridge. I got a good deal of information, and among all the documents consulted I noticed a letter dating from 1832 from Francis Beaufort Edgeworth to his mother, in which he tells her that his father-in-law, Antonio Eroles, appears in a French book as the commander of the 'Organce' militia that in 1822 had seized a pair of smugglers in Andorra following orders from the staff of general Mina. There was no reference for the French book in the letter, but I found it at the Biblioteca de Catalunya: F. Galli (1831), *Campagne du Général Mina en Catalogne*. Since 'Organce' did not match any existing place, with a map I deduced that it meant Organyà.

With this location pinpointed, a fourth stage of research, which ran parallel to the third, got under way. Via the Internet, I contacted an economist from Organyà, Amadeu Rocamora, who kindly searched the registers of marriages, burials and baptisms in Alt Urgell at the Diocesan Archives of La Seu. Unfortunately, the baptismal record in Organyà has disappeared, and we have been unable to confirm the christening of Francis Ysidro Edgeworth's mother, Rosa Florentina Eroles, in around 1815. Nevertheless, Rocamora found in the books of El Pla de Sant Tirs the baptismal and marriage records for Edgeworth's maternal grandparents, as well as the baptisms of his aunt and uncles Eroles. All these findings were confirmed with the help of a letter written in 1844 by his aunt Maria Edgeworth, where she explains that Isidro Eroles had become a lawyer. This induced me to look for his student file in the Historical Archives of the Universitat de Barcelona, where all the information dovetails with Rocamora's findings in the books of El Pla de Sant Tirs. This search is explained in my articles 'Els orígens alturgellencs d'un economista famós: Francis Ysidro Edgeworth' (2004a) and 'Francis Ysidro Edgeworth's Catalan grandfather' (2004b).

Nature of the Portrait

This book does not pretend to offer a profound intellectual portrait of Francis Ysidro Edgeworth; rather it is simply a personal portrait that can help us grasp his temperament and his feelings in order to better understand his development as an individual and as a social scientist. His works in both economics and statistics are simply described; the analysis and exegesis of them are not attempted as there are recent texts that satisfactorily cover both fields, namely those mentioned above and particularly Stigler (1978) on statistics and Creedy (1986, 1998) on economics.

I have tried to focus on the challenge of building a personal portrait, in spite of the fact that the biographical sources were mostly family letters which were dispersed in several archives in Dublin and Oxford. I have especially worked with manuscripts – with handwriting that is often difficult to decipher – which overwhelm us with apparently insignificant details when examining each document. However, along with Seurat I believe that by joining thousands of tiny specks of colour we may manage to paint a vivid image, and from the thousands of details contained in the family letters and poems we may yield a chronicle through which we can paint a quite precise personal portrait. As Maria Edgeworth said:

> We cannot judge either the feelings or of the characters of men with perfect accuracy, from their actions or their appearance in public; it is from their careless conversations, their half-finished sentences, that we may hope with the greatest probability of success to discover their real characters. The life of a great or a little man written by himself, the familiar letters, the diary of any individual published by his friends or by his enemies, after his decease, are esteemed important literary curiosities.[2]

NOTES

1. Professor Josep Fontana and Professor Ernest Lluch have been two of the most distinguished scholars on eighteenth and nineteenth-century Spanish economic history. Josep Fontana is now Emeritus Professor of the Universitat Pompeu i Fabra. Ernest Lluch was assassinated by ETA in November 2000.
2. Maria Edgeworth (1800), Preface.

F.Y. EDGEWORTH'S RELATIVES

The Edgeworths

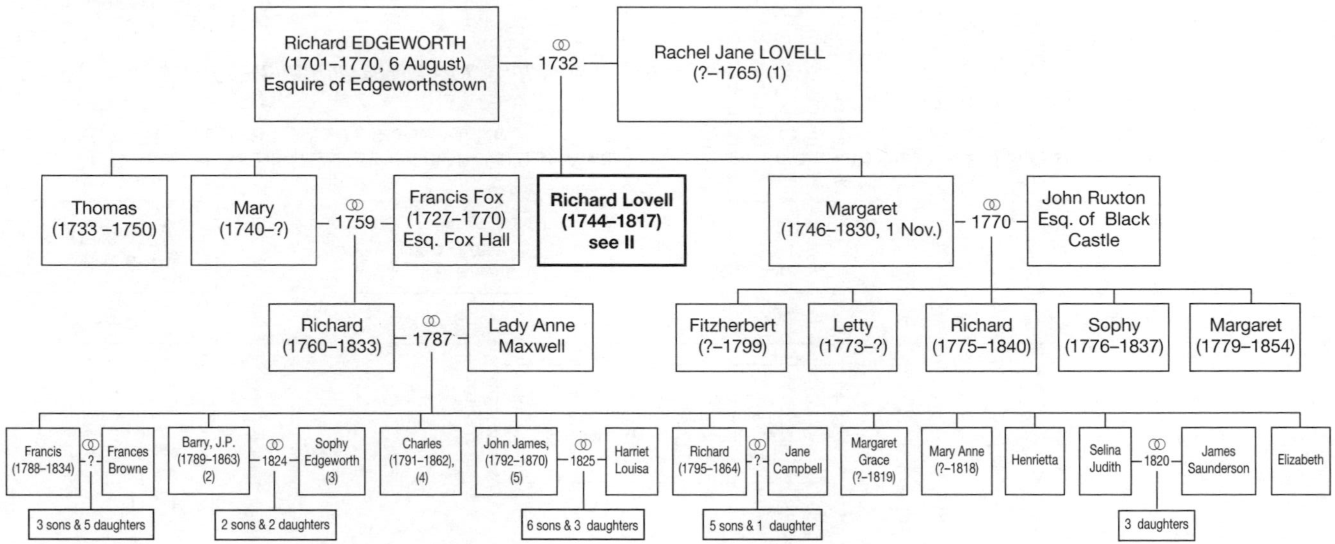

Notes: 1. Daughter of a Welsh judge. 2. Army officer. 3. See branch IId. 4. Lawyer. 5. Reverend.

I The family of Great-Grandfather Richard Edgeworth

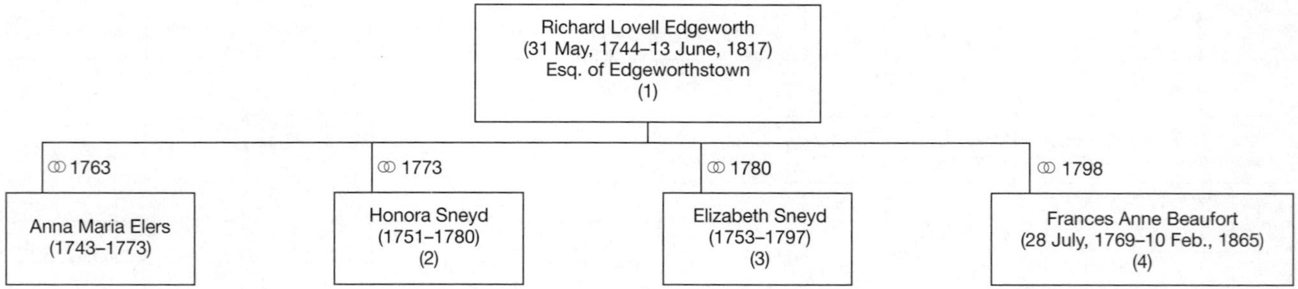

Notes: 1. Founder of the Lunar Society and the Royal Irish Academy. 2. Daughter of Edward Sneyd. 3. Honora's sister. 4. Daughter of Rev. Daniel Augustus Beaufort (1739–1821). See branch III.

II The families of Grandfather Richard Lovell Edgeworth

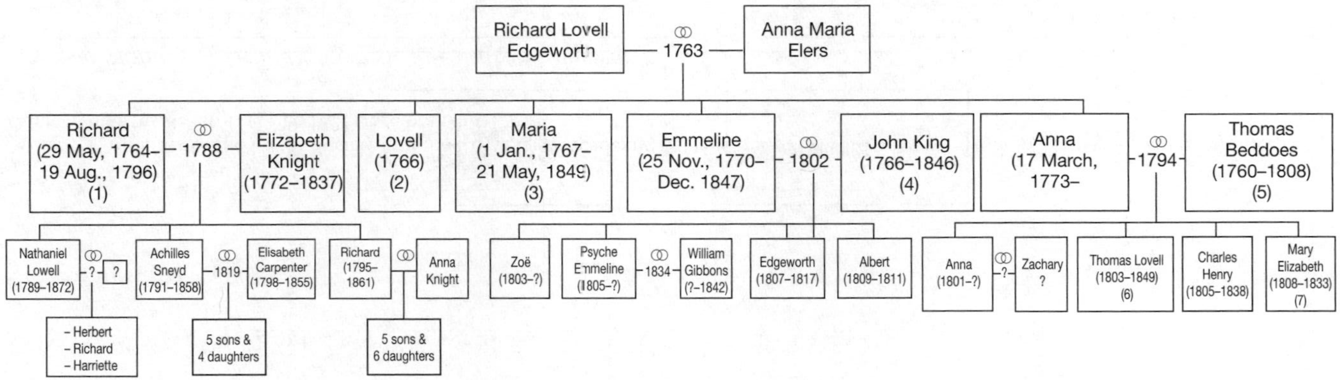

Notes: 1. Emigrated to America (Virginia & the Carolinas) around 1785–87. 2. Lived a few weeks. 3. Writer, Honorary member of the Royal Irish Academy. 4. Surgeon at Clifton. 5. Professor of Oxford & physician at Clifton. 6. Dramatist & poet. 7. Illustrator.

IIa First marriage

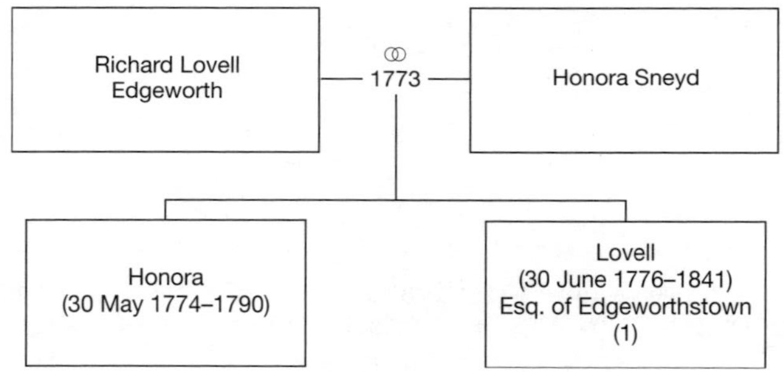

Notes: 1. School director. Expelled from Edgeworthstown in 1833, went to live in the Liverpool area.

IIb Second marriage

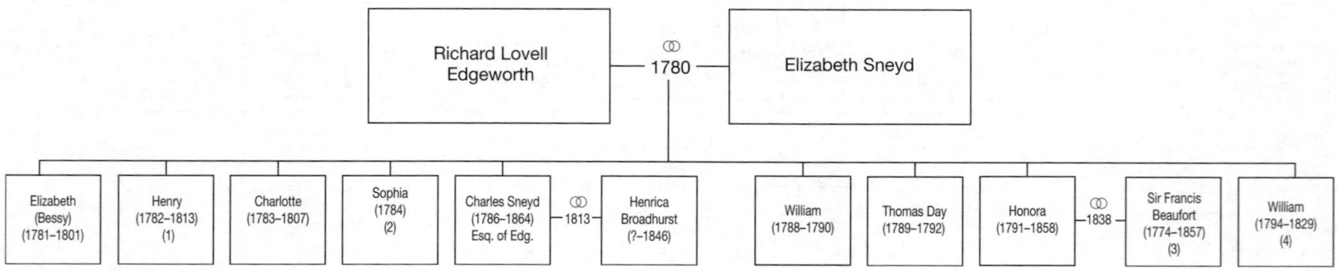

Notes: 1. Physician. 2. Lived only a few days. 3. See branch III. 4. Engineer.

IIc Third marriage

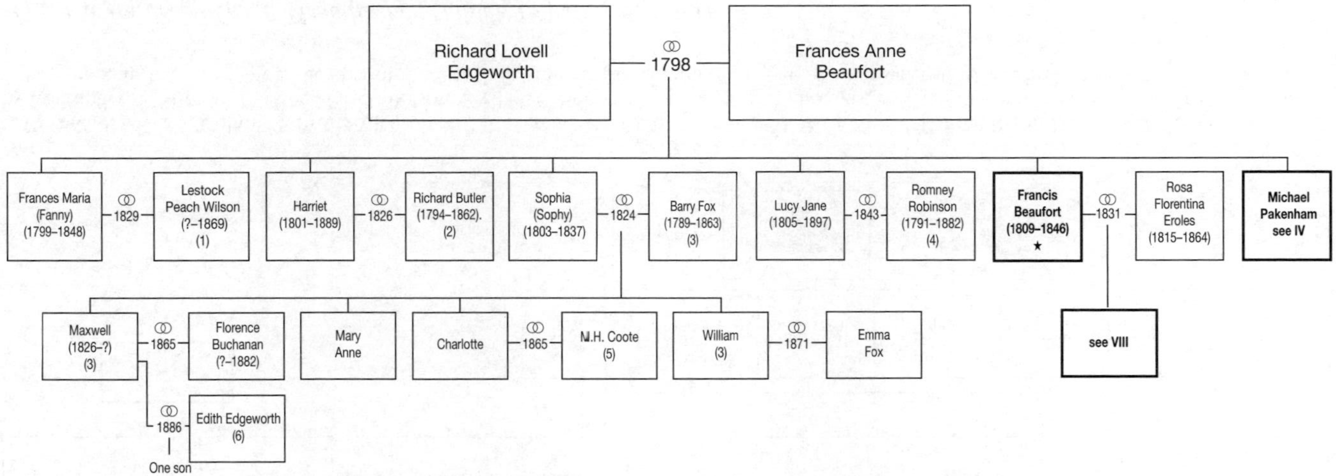

Notes:
* F.Y.E.'s father
1. London's merchant, brother of Alicia Beaufort. See branch III. 2. Vicar & Director of the Diocesan School of Trim. 3. Army officer. 4. Astronomer, President of the Royal Irish Academy. Doctor Honoris Causa for Cambridge University. 5. Reverend. 6. Daughter of Rev. Essex Edgeworth of Kilshrewly.

IId Fourth marriage

The Beauforts

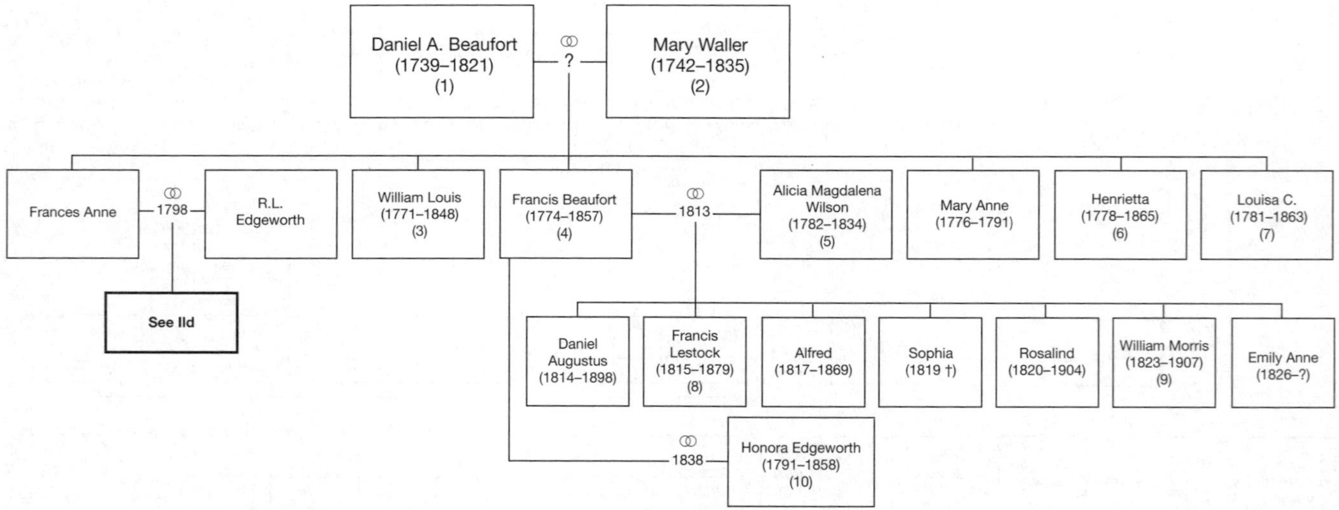

Notes:
1. Reverend, Vicar of Collon. 2. From Allenstown. 3. Reverend, Rector of Glanmere. 4. Rear Admiral of the Royal Navy, meteorologist & hydrographer. 5. Sister of Laestock Peach Wilson. See IId. 6. Botanist & writer. 7. Entomologist & writer & member of the Royal Irish Academy. 8. Lawyer & author. 9. Founder member of the British Economic Association (Royal Economic Society). 10. See branch IIc.

III The family of Great-Grandfather Daniel Augustus Beaufort

The Edgeworths

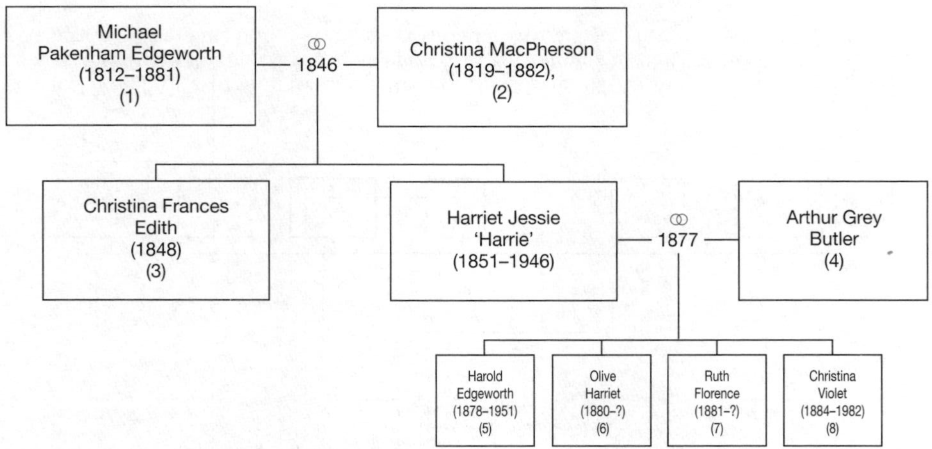

Notes: 1. Biologist, East India Co. employee & writer. 2. Daughter of Hugh MacPherson, surgeon & professor of Aberdeen University. 3. Lived a few months. 4. Reverend, director of East India College & tutor at Oriel College Oxford. 5. Latin Professor at University College, London. 6. Responsible for the Lady Margaret Hall Settlement of Lambeth. 7. Fellow of St Anne's College of Oxford. 8. Sociologist.

IV The family of Uncle M. Pakenham Edgeworth

The Eroles

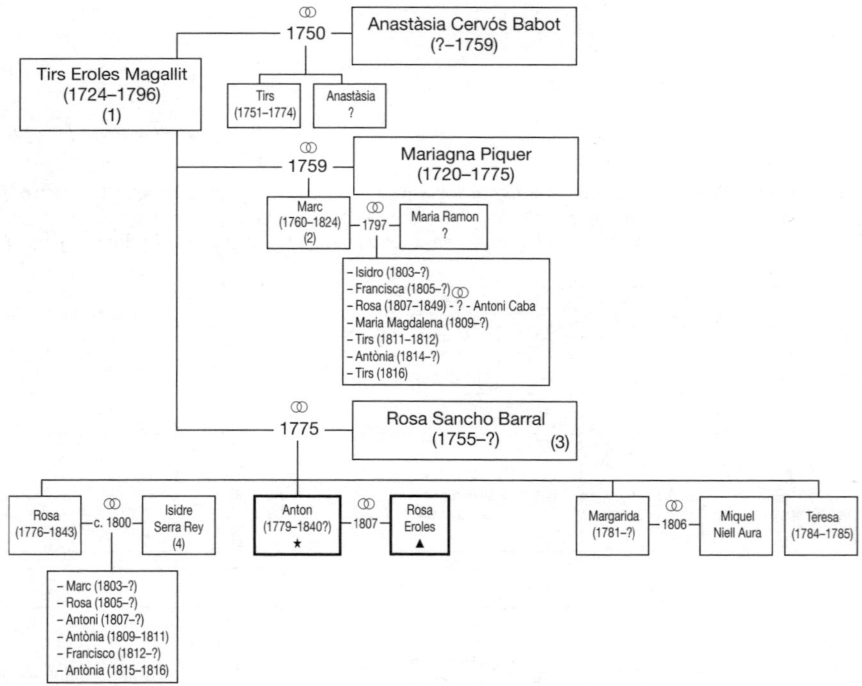

Notes:
★ F.Y.E.'s grandfather, see VII. ▲ F.Y.E.'s grandfather, see VI & VII
1. Farmer. 2. Farmer & stonecutter. 3. Daughter of Salvador & Agnès. 4. Livestock veterinarian

V The family of Great-Grandfather Tirs Eroles i Magallit

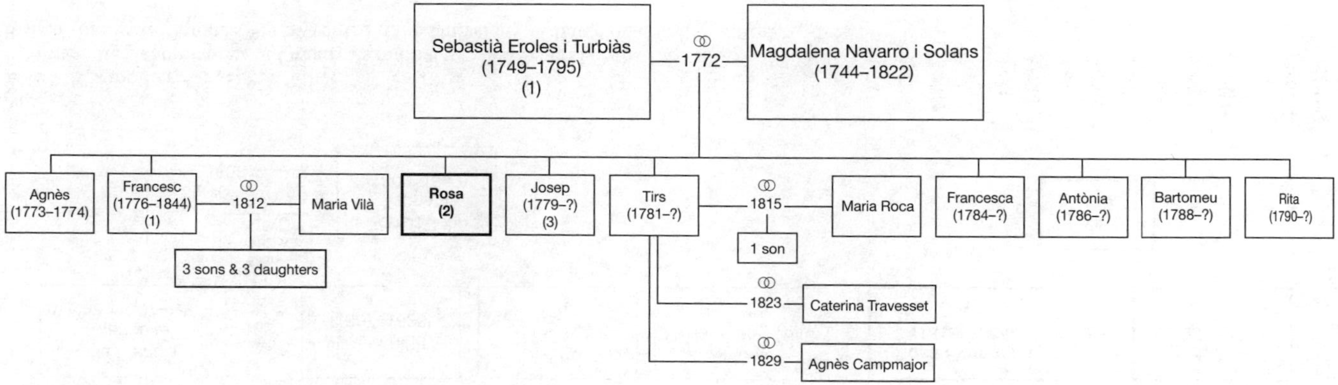

Notes: 1. Farmer. 2. F.Y.E.'s grandmother, see branch VII. 3. Priest.

VI The family of Great-Grandfather Sebastià Eroles i Turbiàs ('Marquet')

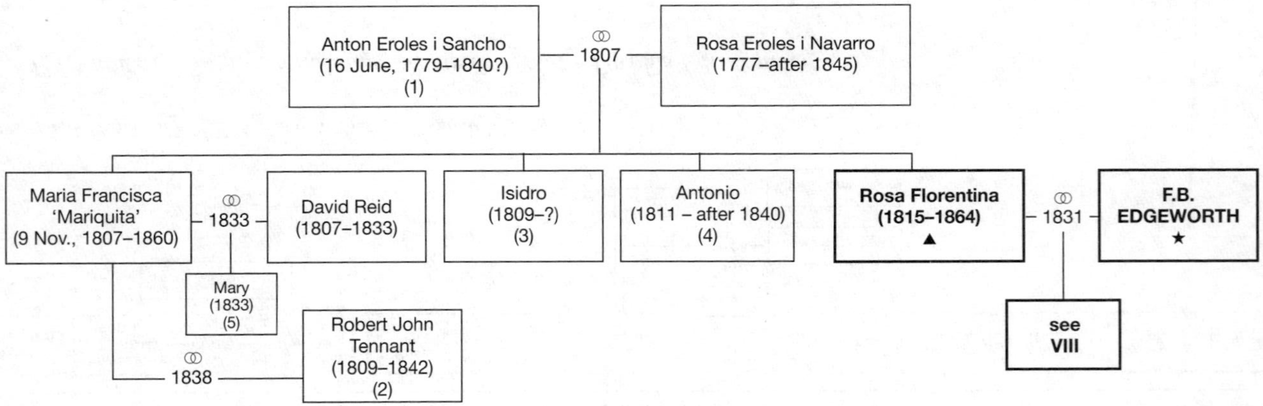

Notes:
▲ F.Y.E.'s mother ★ F.Y.E.'s father
1. Stonecutter, commander of Organyà's militia & exiled general of the Liberal Army. F.Y.E.'s grandfather. 2. Reverend, Anglican pastor of the British colony at Florence. 3. Lawyer from the University of Barcelona. 4. Illustrator. 5. Lived 30 days.

VII The family of Grandfather Anton Eroles i Sancho

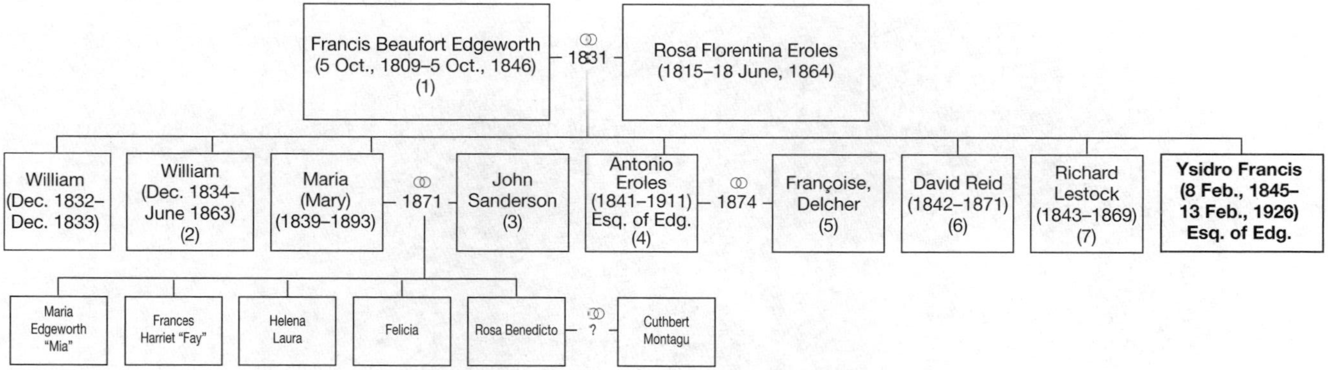

Notes: 1. Philosopher, poet & Edgeworthstown's administrator. 2. British army officer. 3. Reverend, rector of Winchfield. 4. Master of Arts of Trinity College Dublin, Justice of the Peace & founder member of the British Economic Association. 5. Daughter of Colonel Delcher, of the French Army. 6. Engineer. 7. Physician & poet.

Source: Burke's Genealogic and Heraldic Dictionary, 7th ed., 1886, with some addenda and corrections from: family letters kept at the National Library of Ireland and the Bodleian Library of Oxford; the books of Butler and Butler (1972) and of Halsey ed. (1976); from the website www.houseofedgeworthsc.com and www.genforum.com and from registers of the Arxiu Diocesà de la Seu d'Urgell.

VIII The family of Francis Beaufort Edgeworth

Richard Lovell Edgeworth

Engraving by Anthony Cardon, 1812

1. Edgeworth's background

Blest be thou, Triunal Deity!
That hast such favour in my seed vouchsafed.

(Paradise, Canto XV, vv. 47–48).
Dante Alighieri, *The Divine Comedy*,
English translation by Henry F. Cary.

Francis Ysidro Edgeworth, economist and statistician, was born on 8 February 1845 in Edgeworthstown, County Longford, Ireland. His name was actually Ysidro Francis Edgeworth, but his family and friends called him Francis or Frank, and when he started publishing in 1876 he transposed his Christian names. He was the sixth son and seventh child of Francis Beaufort Edgeworth and Rosa Florentina Eroles.[1] His father came from an Anglo-Irish family of Protestant descent and his mother from a Catalan family with a liberal and constitutionalist penchant.

Both families played an important role in moulding Edgeworth's personality: the Edgeworth ancestry accounts for most of his intellectual development and his scientific leanings, while some of the peculiarities of his character can be traced to the Eroles side. One of these peculiarities, which became more prominent as he grew older, was his passion for learning about his own ancestors and deceased relatives. Therefore, we shall pay special attention to both families, the Edgeworths and the Eroles.

1.1 EDGEWORTH ANCESTORS

The story of the Edgeworths[2] in Ireland starts in the sixteenth century during the reign of Elizabeth I, when the two sons of Roger Edgeworth, a former monk who had married, emigrated from England and came to be landowners in central Ireland. One of the brothers, Edward, became Bishop of Down and Connor and died without heirs. As a result, the land passed to the descendants of his brother, who was named Francis. Therefore, this Francis was the first Irish ancestor of Francis Ysidro Edgeworth.

Three generations later we find another Francis Edgeworth, who raised a regiment for King William III and was known as Protestant Frank. He was an avid gambler. One night when he lost all his money, his wife offered him her diamond earrings. He won back the money he had lost that night, but in exchange she made him promise not to gamble again with dice or cards. Nevertheless, he persisted in gambling with friends by drawing straws out of a hayrick. This gambling habit reappeared in successive generations of Edgeworths.

Protestant Frank had two sons. The elder became the owner of Fairy Mount or Firmont, and his descendants are known as the Edgeworths of Firmont or the Essex Edgeworth branch. The second son of Protestant Frank, Richard (1701–1770) was the great-grandfather of Francis Ysidro Edgeworth. Richard had inherited the part of the land known as Edgeworthstown, which he enlarged with some new acquisitions, substantially improving the revenues. He was a member of the Irish Parliament for 25 years and wrote *The Black Book of Edgeworthstown* about his ancestors. In 1732 Richard Edgeworth married Rachel Jane Lovell, daughter of a Welsh judge famous for his fits of rage and foul temper. They had four children, Thomas Edgeworth, who died in 1750 when he was not yet eighteen years old; Mary, who in 1759 married her neighbour Francis Fox, Esquire of Fox Hall, and had one child; Richard Lovell Edgeworth, the grandfather of Francis Ysidro, who married four times, had 22 children and eventually inherited Edgeworthstown; and Margaret (1746–1830), who married John Ruxton, Esquire of Black Castle, near Collon, and had five children.

1.2 RICHARD LOVELL EDGEWORTH'S FOUR FAMILIES

Francis Ysidro Edgeworth's grandfather, Richard Lovell Edgeworth, was born in Bath on 31 May 1744. He was educated at Warwick, Trinity College, Dublin and Corpus Christi College, Oxford. During his time at Oxford he lived in Black Bourton, Oxfordshire, at the house of Mr Elers, a friend of his father.

First Family

At the end of 1763, before finishing his studies and not yet twenty years old, Richard Lovell Edgeworth eloped to Gretna Green, on the Scottish side of the border, to marry one of Mr Elers's daughters, Anna Maria, who was pregnant by him. On 29 May 1764, a first son, Richard ('Dick'), was born in Black Bourton. Five or six months after the birth, the young couple travelled to Edgeworthstown to receive the blessings of Richard's dying mother, and they remained there. In the autumn of 1765, they went back to England, where they found that Mr Elers had been imprisoned for debt. In consequence, they stayed at Black Bourton, where Richard Lovell Edgeworth helped to educate the children. He had to travel often to London to attend lectures at the Lincoln's Inn Temple in order to finish his studies in law, although his main interests were inventing mechanical devices, literature and educational methods.[3]

In 1766, Anna Maria gave birth to a second boy, Lovell, in Black Bourton; he died a short while later. He was buried on 26 May of that year. She then gave birth to a daughter, Maria, on the first day of 1767. Though she was born in Black Bourton, Richard Lovell Edgeworth had moved his family to a house in Hare Hatch, between Reading and Maidenhead in Berkshire, not very far from Black Bourton. There, he met Thomas Day, who became his best friend, much to the dismay of Anna Maria, who despised Day because of his dreadful table manners. Day was quite a well-read man who was also interested in disciplinary educational methods, and he introduced Richard to a group of people in Lichfield who were fond of science and literature, including Dr Erasmus Darwin, whose grandchildren Charles Darwin and Francis Galton would become famous. In fact, Galton was a champion of Francis Ysidro Edgeworth in his work as a statistician.

Under the auspices of Dr Darwin, Day and Edgeworth became members of the Lunar Society of Birmingham, which assembled a group of eminent men of the Enlightenment who were interested in science and its applications. The members included Matthew Boulton, James Watt, James Keir, William Small, Joseph Priestley, Josiah Wedgwood, John Whitehurst, William Withering, Thomas Bentley and Samuel Galton. The Lunar Society was thus named because the members gathered at each other's houses on the Mondays nearest the full moon in order to have light for the ride back home.[4]

Thomas Day and Richard Lovell Edgeworth also participated in Lichfield's cultural life, which led them to meet the enticing Miss Honora Sneyd and her sisters. At one of those gatherings in Lichfield or Birmingham, Richard made the acquaintance of William Roscoe, a banker from Liverpool, and the two became friends. In the years to come, Richard visited Roscoe, who in turn entertained him several times. More than a century later, one of Roscoe's grandchildren, William Stanley Jevons, met Francis Ysidro Edgeworth at Hampstead, London, and exerted a strong academic influence on him.

In the summer of 1768, Richard Lovell Edgeworth's father died, whereupon he inherited Edgeworthstown. After becoming an estate owner, he lost interest in being called to the Bar, where Thomas Day entered in 1769.

Instead, Richard occupied himself with chemistry and mechanical experiments – such as the design of carts – with such zeal that his wife Anna felt neglected. On 25 November 1770 she gave birth to a second daughter, Emmeline, born at Hare Hatch.

Meanwhile, Richard Lovell Edgeworth discovered that he had fallen in love with Miss Honora Sneyd and, to try to get her out of his mind, went to Lyons (France) in 1772 with his son Dick and his friend Thomas Day. He had been educating Dick according to Rousseau's schemes of permissiveness and he was very interested in meeting Rousseau in Paris. The encounter proved to be a fiasco and Richard left Dick's education in the hands of an English preceptor who spoke no French at all and depended on Dick to be understood. This meant that Dick became wildly independent at the tender age of seven. Once in Lyons, Richard took an active role in efforts to divert the courses of the Rhône and Saône rivers. In the summer, his wife Anna Maria came to visit him, but she could not adapt herself to Lyons society and went back to London, escorted by Mr Thomas Day.

On 17 March 1773, Anna Maria gave birth in London to a third daughter, Anna, and died of puerperal fever a few days later. When the news of Anna Maria's death reached Lyons, Richard decided to go back to England and went to Lichfield to meet Honora Sneyd. He lost no time proposing marriage to her.[5]

Second Family

Before the end of 1773, the widower Richard Lovell Edgeworth had married Honora Sneyd and they settled with his four children at Edgeworthstown, where they renovated the manor house. They lived there for three years. During this time, on 30 May 1774 Honora gave birth to her first daughter, also named Honora, and on 30 June 1776 to her first son and last child, Lovell. All the children in the family were supposed to be educated by their parents, but Dick and Maria felt neglected, showed mischievous behaviour and were sent to boarding schools: Charterhouse in London for him and Mrs Lattufière's boarding school in Derby and, after 1778, Mrs Devis's at Upper Walpole Street, London, for Maria.

At the end of 1776, the whole family left Ireland and moved to Hertfordshire, England, settling at Northchurch, where Richard Lovell Edgeworth spent a lot of time experimenting with clocks and mechanical pursuits in the mornings. These earned him prizes from the Society for Encouragement of Arts and Manufactures. Meanwhile, he dedicated the evenings to reading the best literature and books on a variety of subjects, wisely chosen thanks to the 'clearness of Honora's judgement'.[6]

In December 1778, Dick, who was 13 years old, escaped from Charterhouse and boarded a ship to pursue the sailor's life. Later on, Dick joined the crew of the *Monmouth* at Portsmouth in February 1781 and served there until February 1783, when he deserted in Goa and his father had to pay £10 to the Royal Navy for his expenses. Richard Lovell Edgeworth resented this misbehaviour and revised his will, naming his second son, Lovell, heir to the family estate and leaving Dick an annuity of £300 per year from the time of his return to Ireland. Between 1785 and 1787, Dick was in Virginia, and then he migrated to the Upper PeeDee, where he became a tutor in Chesterfield District, South Carolina. He married Miss Elizabeth Knight and established himself at the Anson Company, North Carolina, thus starting a branch of American Edgeworths.[7]

Honora had always had a tendency towards consumption, a disease from which she escaped with her life when she was fifteen. However, in 1779 she became infected again, and she finally died in April 1780 after suggesting her sister Elizabeth as a wife for her husband Richard.

Third Family

On Christmas Day, 1780, Richard Lovell Edgeworth and Elizabeth Sneyd were married in St Andrew's Church, Holborn. During this third marriage, Elizabeth added nine new children to the family, five boys and four girls. In 1782, Richard decided to go back to live at Edgeworthstown; therefore, most of these children, seven of them to be precise, were born there. Three of them died as infants and four died from consumption when they were young: two daughters in their twenties and two sons in their thirties. Only two lived long enough to be part of Ysidro Francis Edgeworth's life: Charles Sneyd and Honora.

During this period, Maria Edgeworth, who went back to Edgeworthstown in 1782 to join the rest of the family, became her father's main collaborator, in both his literary and educational texts and the home schooling of her half brothers and sisters, as well as in the administration of the Edgeworthstown estate. She also started a literary career by writing tales for the children in the family, and in 1795 published her first book, *Letters for Literary Ladies*.

Meanwhile, Richard Lovell Edgeworth participated in the 1785 founding of the Royal Irish Academy. He also engaged in political activities, much along the lines of Henry Grattan, detaching himself from the Anglo-Irish gentry's positions and standing up in favour of restoring Catholic rights.[8] He also continued his experiments with a kind of semaphore-telegraph which he familiarly called the 'logograph', and in 1795 he published 'A letter on the Telegraph, and on the Defence of Ireland', suggesting military applications for the device during those times of political unrest after the French Revolution in 1789 and the founding of the Society of United Irishmen in 1791. In fact, in 1795 Richard unsuccessfully campaigned for a seat in the Irish Parliament as an upholder of Catholic relief.

In those years, the menace of hereditary predisposition to consumption was confirmed by the death of Richard Lovell Edgeworth's favourite daughter, Honora, who died in 1790 when she was sixteen years old. Since her brother Lovell, Richard Lovell Edgeworth's heir, had also displayed alarming symptoms, the whole family went in two stages to live for a couple of years in Clifton, a health resort near Bristol, where Elizabeth gave birth to a girl who was also named Honora. There, Anna and Emmeline met their future husbands, Dr Thomas Beddoes and Surgeon John King at Clifton. Dick Edgeworth visited them in 1792, and convinced a few members of the family – including the wealthy Sneyd aunts – to let him invest in American land on their behalf. The investment was a fiasco, as was a similar investment in 1795 that followed a second visit, this time to Edgeworthstown. During this second trip, Dick came down with pleurisy and later died in his North Carolina home in August 1796, leaving a widow and three children.

In November 1797, Elizabeth Edgeworth died from consumption, just like her sister Honora. Despite the fact that Richard Lovell was 53 years old when his third wife died, he lost no time getting married for the fourth time.

Fourth Family

His new wife was Frances Anne Beaufort, born on 28 July 1769 in Flower Hill, near Navan, County Meath, and therefore two years younger than Maria. She was the eldest offspring of Daniel Augustus Beaufort (1739–1821) of Huguenot ancestry and Vicar of Collon from 1790 to 1821, and Mary Waller of Allenstown (1742–1835). After Frances Anne, they had two sons, Rev. William Louis Beaufort (1771–1848), Rector of Glanmere; and Royal Navy Captain – and later Rear Admiral — Francis Beaufort (1774–1857), cartographer and hydrographer. After them came three more daughters, Mary Anne (1776–1791), Harriet (1778–1865), author of *Dialogues on Botany*, and Louisa Catherine (1781–1863), author of *Dialogues on Entomology* and *Essay on the Round Towers of Ireland*, and an elected member of the Royal Irish Academy.

Frances Anne Beaufort's marriage to Richard Lovell Edgeworth took place at St Anne's Church in Dublin on 31 May 1798. In the same year, Richard was elected to a seat in the Irish Parliament and when, at the request of the Society of United Irishmen, General Humbert landed at

Killala in August at the head of French troops and invaded northwest and central Ireland, Edgeworth organised a mixed force of Protestants and Catholics to defend Longford County. On 5 September, the entire family was forced to leave Edgeworthstown for five days, and they took refuge in an inn in Longford. In the move from the Manor House to Longford, they were kindly invited to travel with two officers and four dragoons of the Anglo-Irish army. However, Richard declined so as not to delay the soldiers. This was a wise decision since the munitions cart exploded when it was halfway to Longford, killing several soldiers and horses.

Richard Lovell had initially been in favour of the Act of Union, by which the Irish parliamentary institutions would disappear and an Irish representation would go to the House of Commons and the House of Lords. However, later on in Parliament, he changed his mind when he became aware of the non-democratic pressures that were being exerted from London. After the Act of Union was approved, he left politics for the rest of his life.

Through this fourth marriage, the Edgeworth family grew with six new children. Four girls came in a row: Frances Maria, known as 'Fanny' in 1799, Harriet in 1801, Sophy in 1803 and Lucy Jane in 1805. The last two children were boys: Francis Beaufort Edgeworth, born on 5 October 1809, who would become the father of our Francis Ysidro Edgeworth 35 years later, and Michael Pakenham Edgeworth, 'Pakenham', born in 1812, when his father was sixty-eight years old. All these Edgeworths of the Beaufort branch lived to be adults and married. One of them – Sophy – would die of consumption at 34 when she had four children, who would later have a fine relationship with their cousin Ysidro Francis. However, his most important influences were his aunt Harriet and his uncle Pakenham.

Richard Lovell Edgeworth and his Daughter Maria

Meanwhile, the intellectual and administrative father and daughter team continued working as smoothly as ever. In 1798 they jointly authored and published *A Treatise on Practical Education*, and in 1802 they came out with their *Essay on Irish Bulls* – 'bulls' in the sense of foolish talk – which tried to prove that 'bulls' were neither exclusive to nor more frequent in Ireland than in the rest of the United Kingdom.

The subsequent years were ones of abundant literary output for Maria Edgeworth, this time independently of her father. She published her first stories *Castle Rackrent* (1800) and *Belinda* (1801) to great acclaim, as well as the more pedagogical texts *Moral Tales* and *Early Lessons* (1802). In 1802 as well, Richard Lovell Edgeworth published *Poetry Explained for the Use of Young People*.

In this same year, during the Peace of Amiens, Richard travelled to London and Paris with his wife Frances Anne and daughters Maria and Charlotte. Emmeline accompanied them as far as Clifton, where she was married to the surgeon John King, although somewhat surprisingly, the four travellers did not stay for the wedding. In Paris they were met by their son and brother Lovell, and Richard Lovell Edgeworth made the acquaintance of Étienne Dumont, the man who had popularised Bentham and Adam Smith in France and Switzerland. Afterwards, Dumont spent some time tutoring Lord Henry Petty, son of the Marquis of Lansdowne, in Bowood, near London, where William Petty had lived. All of them would become friends of Maria Edgeworth in the years to come. Also in Paris, Richard Lovell Edgeworth convinced a Swedish diplomat, Chevalier Edelcrantz, to propose marriage to Maria. But she refused him because the prospect of living in Sweden, far away from her father, was not appealing to her.

In France, they met *la crème* of French society, and Richard Lovell Edgeworth was appointed a member of the *Société d'Encouragement pour l'Industrie Nationale*, in recognition of his work in Lyons. However, the political situation shifted suddenly and they had to escape back to England in March 1803, just before war was again declared between France and England. The young Lovell Edgeworth was careless enough to be taken prisoner, and he remained in captivity until 1814, that is, for eleven years, despite all his father's diplomatic efforts.

Back in Edgeworthstown, Maria Edgeworth published her most acclaimed works between 1804 and 1809: *Popular Tales* and *The Modern Griselda, a Tale*, mostly for children and their parents as educators; and the moral stories *Leonora*, *Adelaide* and *Tales of Fashionable Life*, containing 'Almeria', 'Manœuvering', 'Ennui' and 'The Absentee'. Much later, in 1894, when *Palgrave's Dictionary of Political Economy* was published, her nephew Francis Ysidro Edgeworth wrote the entry on 'Absentee' in homage to Maria and left the entry 'Edgeworth, Maria' to his colleague Clara E. Collet, who stated that 'Ennui' is considered to be 'the nearest approach to definite exposition of economic theories'. Maria Edgeworth's interest in everyday economic problems prompted her to seek the acquaintance of the greatest British economists of her day. Meantime, the father's and daughter's joint text entitled *Essays on Professional Education* appeared in 1809 and won instant recognition. In 1810, Maria's *The Wife* and Richard's *An Essay on the Construction of Roads and Carriages* and the article 'On Telegraphic Communications' were published.

In 1813, Frances Anne, Maria and Richard Lovell Edgeworth visited London, where Maria became a resounding social success. She met Lord Byron, who admitted that she, along with Madame de Stäel, was one of the 'exhibitions of the year'. As we shall detail further on, she also befriended the economists Ricardo, Malthus, Jones and Mrs Marcet, the politician Samuel Romilly and the writer Joanna Baillie. Back in Edgeworthstown, Maria Edgeworth wrote *Patronage*, one of her best social satires. Meanwhile, the first symptoms of the virulent internal inflammation that would cause Richard Lovell Edgeworth's death flared up in April 1814. However, he still had time to publish his pamphlet *Observations upon Wheel Carriage Experiments* and *On Aërostation*, and to write the prologue to Maria's *Harrington and Ormond*, which was published in 1817, one year after her *Readings on Poetry*.

Richard Lovell Edgeworth died on the 13 June 1817. He was comforted by the presence of his heir Lovell, who upon his return from France in 1814 had been the perfect image of the most devoted son. On his deathbed, Richard asked his son to curtail any possible financial extravagance by Maria, who tended to be moved by her generosity.

1.3 SOME EDGEWORTH UNCLES AND AUNTS

When Richard Lovell Edgeworth died, his youngest sons Francis Beaufort and Michael Pakenham were only seven and five years old respectively. His older children tried to retain his influence in Edgeworthstown by further developing some of their father's preferred projects.

Uncles

In 1818 Richard Lovell Edgeworth's heir, Lovell Edgeworth, started a school for day boys and boarders of all classes and creeds – most unusual in Ireland. Like his father, Lovell was an excellent teacher, but he was also a heavy gambler and wine drinker. In December 1825, the Edgeworths realised that they were on the verge of financial ruin: Lovell had run up a debt of £26 000 and was obliged to put the management of Edgeworthstown in Maria Edgeworth's hands. Within a few years, thanks to her half brothers Charles Sneyd Edgeworth and William Edgeworth, the latter a civil engineer, as well as her own private resources and the cooperation of the tenants, who followed Maria's request for early payment of some rent, the debt was reduced to £12 000.[9] After 1826, Lovell Edgeworth continued teaching at the school, but the acting head of the family was Maria, with remote support from her half brother Sneyd.

Charles Sneyd Edgeworth, 'Sneyd', who in 1813 had married a rich heiress from Derby, Henrica Broadhurst, went to live in the county of Kent, southeast of London, and became a *bon vivant*, with occasional journeys to France. He edited the *Memoirs of the Abbé Edgeworth; Containing its Narrative of the Last Hours of Louis XVI*, written by his second cousin Henry Essex Edgeworth. The latter was the well-known Abbé Edgeworth of Firmont who had been *grand vicaire* in charge of the Archbishopric of Paris during the French Revolution and had accompanied the deposed king to the scaffold. The book was published in 1813, when Sneyd was just twenty-seven years old, but this publishing work would go no further. In 1835 he tried to enter politics, running for the Longford County seat in the general election, but he was defeated. After Lovell's death in 1841, Sneyd became the official landlord of Edgeworthstown and, though still living in Kent, he kept this position during Francis Ysidro Edgeworth's childhood and adolescence.

Aunts

Much more important to Francis Ysidro than these uncles was the role played – mostly posthumously for him, yet central to his intellectual development and financial ease – by his aunt Maria Edgeworth. Before detailing – in the next section – her vital contributions, we proceed with the rest of Francis Ysidro Edgeworth's aunts.

The job of finding suitable husbands for Honora, Fanny, Harriet, Sophy and Lucy Jane was left in Maria's hands. This task became more desperate after Lovell Edgeworth's bankruptcy, as their dowries were drastically reduced. Maria fostered a social and cultural atmosphere at Edgeworthstown, trying to marry her half sisters to intellectual men. The first to leave Edgeworthstown was Sophy, who in 1824, that is before Lovell's financial crisis, married the son of a nearby cousin and former Army officer, Barry Fox. However, in this case Maria had no hand in the match.

With Harriet, Maria Edgeworth did get to play matchmaker. In 1826, Harriet married the Rev. Richard Butler, vicar of Trim, an Oxford-educated man who ran the Diocesan School of Meath with the help of James Hamilton. It was one of the best schools in Ireland, housed in what remained of Talbot Castle, where Jonathan Swift had lived in 1718 and where Arthur Wellesley, Duke of Wellington, had studied. A former pupil of this school and nephew of James Hamilton, William Rowan Hamilton, was introduced to Maria by Butler in 1824 and invited to Edgeworthstown when he was just nineteen. Richard Butler and Harriet, who later became a very important family link for Francis Ysidro Edgeworth, will come to play a prominent role in this story.

Another aunt, Fanny, Maria Edgeworth's most beloved half sister, married Mr Lestock Peach Wilson, a London-based businessman. The wedding took place on the first day of 1829, and the couple lived happily until her sudden death in 1848. Despite the fact that they had no children, Lestock Wilson remained very close to the Edgeworth family.

Honora, from Richard Lovell Edgeworth's third marriage, and Lucy, from the fourth, were still spinsters when the first act of our story begins in 1831, and they play only minor roles.

1.4 AUNT MARIA EDGEWORTH: FRIENDSHIPS WITH RICARDO AND HAMILTON

After her father's death, Maria Edgeworth fulfilled her promise to publish *Memoirs of Richard Lovell Edgeworth; Begun by Himself and Concluded by his Daughter*. This book was issued in 1820. It garnered scathing reviews from puritanical critics and accolades from her London friends. Robert Malthus, the famous political economist, 'spoke most highly of it' and Mrs Marcet, also a fashionable political economist, claimed that the critical report in the *Quarterly Review* 'makes my blood boil with indignation and rouses every feeling of contempt and abhorrence'.[10]

Maria Edgeworth continued her travels to England, Scotland, France and Switzerland. In 1823, she and her half sisters Sophy and Harriet visited Walter Scott in Abbotsford, and they spent some time then and in the following years with Fanny and Harriet in Edinburgh, London, Paris and Geneva, where she maintained friendships with men of letters, politicians, philosophers and scientists, as well as a select group of political economists such as Dugald Stewart, James Mill, Mrs Marcet, Robert Malthus, Richard Jones and David Ricardo. In Switzerland, she visited M. Sismondi and again met Étienne Dumont. Through Dumont, Maria Edgeworth also became very good friends with Lady Lansdowne and was often her guest at Bowood.

The friendships with Ricardo and Jones exceeded pure formality, and their correspondence is full of confidences that we shall detail soon. However, these confidences were even more plentiful between Maria Edgeworth and Mrs Marcet, who received her 'with only herself and children'. Maria admired her public discretion in the field of political economy:

> It has become high fashion with blue ladies to talk Political Economy and make a great jabbering on the subject, while others who have more sense, like Mrs Marcet hold their tongues and listen. Meantime, fine ladies require that their daughters' governesses should teach Political Economy.[11]

Besides all this travelling, Maria Edgeworth spent a great deal of energy keeping Edgeworthstown as one of the top social hubs in Ireland. Walter

Scott was there in 1825, and William Wordsworth in 1829. A long list of men of high intellect, such as the astronomer John Frederick Herschel and the chemist Humphry Davy, were invited with the hopes of making a match with the Edgeworths' girls. The plan was successful because Maria, who inherited the Lunar Society side of her father, was truly interested in all scientific advances and knew how to create a pleasant atmosphere for scientists in Edgeworthstown.

However, all this activity had its price. After publishing her father's *Memoirs*, Maria Edgeworth wrote only one novel, *Helen*, which kept her busy from 1830 to 1833 and was published in 1834. Nevertheless, she felt that keeping Edgeworthstown as an enlightened *citadelle* was a worthwhile effort.

Some of Maria Edgeworth's scientific friends exerted an influence on the Edgeworthstown inhabitants of those days; however, a couple of them transcended time and won the admiration of her youngest nephew, Francis Ysidro Edgeworth. Such influential friendships that deserve more detailed attention include Maria's relationship with David Ricardo and, last but not least, William Rowan Hamilton's friendship first with Maria, and later with Francis Beaufort Edgeworth.

Maria Edgeworth and David Ricardo

Francis Ysidro Edgeworth was impressed by Maria's closeness to Ricardo. In 1888, he asked his aunt Harriet Butler, who was a living archive of the Edgeworth family history, about the prominent political economist and his children. In her last letter to him she reported that Ricardo 'was undersized but his face handsome, his manners delightful. He died comparatively young, an imposthume in his ear carried him off in two days.'[12]

Francis also asked her about Ricardo's living descendants, since he wanted to inquire about Maria Edgeworth's correspondence with Ricardo. In fact, he eventually found the letters and published them in the *Economic Journal* in 1907. But let us return to the facts.

After Maria Edgeworth's acquaintance with David Ricardo in 1813, we have to wait until her stay with her half sisters Fanny and Harriet in England from November 1821 until June 1822 to see how their friendship flourished. Ricardo had invited them to his residence at Gatcomb Park, Gloucestershire, in November. Maria reported to her stepmother Frances Anne:

> Mr Ricardo, with a very composed manner has a continual life of mind that starts perpetually new game in conversation, I never argued or discussed a question with any person who argues more fairly or less for victory and more for truth. . . . It seems quite indifferent to him whether you find the truth, or

> whether he finds it, provided it be found. . . . Mr. Ricardo is altogether one of the most agreeable persons, as well as the best informed and most clever that I ever knew.

Afterwards in London, contact with Ricardo continued. They were invited to breakfast at his place several times. The first time, Maria Edgeworth described the macabre post-breakfast spectacle of Ricardo's brother-in-law, Mr Wilkinson, who brought a skull and tried to prove that it had contained the brain of Oliver Cromwell.[13] At another breakfast, as Fanny would explain to her brother Pakenham some years hence, Mr Ricardo 'told us the manner in which he was cured of personal vanity'. He was referring to 'the only time in his life when he remembered to have cared about his dress':

> He was about nine years old when he was staying in Holland with his uncle. His fancy was struck with a pair of red shoes nicely edged with fur; he could not be happy without them and many days he begged his uncle, till at last the shoes were bought and he thought himself quite happy. He put them on and went out to walk, but he had soon cause to repent of his desire to have these shoes. For the soles were made of wood and as he walked along the streets everybody looked at him and laughed at his clacking shoes; in a very short time he went to his uncle to get another pair of shoes for him but his uncle told him that he must be satisfied with his own choice. For that he should not have other shoes till those were come out.[14]

Once Maria Edgeworth was back in Edgeworthstown in July 1822, she kept up a regular correspondence with Ricardo that ended with his sudden death in September 1823. In these letters, they started an interesting discussion on the troubles raised by Ireland's dependence on potatoes. However, Ricardo apparently got tired of the Irish problems, since in his last letter he wrote:

> Your restless nation gives us a great deal of trouble in Parliament. The best amongst us do not know how to manage you, nor what course to take to give you the blessings of peace, order and good government. You have been so long subjected to misrule as hardly to be in a fit state to be reclaimed by common means. Coercion and severity have proved of little use, and I hope the system of indulgence, kindness, and conciliation will now be tried. If that system will not succeed I hope we shall get rid of you altogether; – we could do very well without you – you are a great expence to us, and prevent us from making any great improvements in our own government, as all our time is taken up in attending to yours.[15]

Maria Edgeworth and William Rowan Hamilton

Another remarkable man of science befriended by Maria Edgeworth, and whose work would become an inspiration for Francis Ysidro Edgeworth many years later, was the astronomer and mathematician W.R. Hamilton. He had been a precocious child under the tutelage of his uncle James in Trim from his second year of life, and he had received an extraordinary education. He could read Latin, Greek and Hebrew by the time he was four years and five months old; French and Italian at seven years and nine months; and Sanskrit at nine years and six months. When he was thirteen, he developed a *Compendious Treatise of Algebra*, a *Grammar of the Sanskrit Language*, an *Arabic Praxis* and an *Analysis of a Passage in Syriac*.[16] At the age of seventeen, he described the 'raptures' of mathematical research to his sister Eliza.[17]

During the long vacation of 1824, Maria Edgeworth invited the Rev. Richard Butler, then a 30-year-old bachelor, and Hamilton, who was just nineteen and in his first year at Trinity College, Dublin, to spend some days at Edgeworthstown.[18] Hamilton's impressions of Edgeworthstown were immediately described to his sister Grace: 'I do not know any place so pleasant as Edgeworthstown in the extensive circle of my acquaintance.'[19]

What Hamilton does not tell Grace is that he had succeeded in gaining greater intimacy with Maria Edgeworth by asking her: 'Do I squint?' The reason behind this question is that Hamilton suffered from double vision.[20]

After this August 1824 stay in Edgeworthstown, Maria Edgeworth and Hamilton maintained constant correspondence. The letters from April 1825, in which Maria thanked Hamilton for sending her his *Novum Organum* papers on optics, 'On Caustics', and the ones from April and May 1828, may be a representative sample. Maria thanked Hamilton for sending Fanny the second volume of *Système du Monde*.[21] The intellectual influence was reciprocal. As Hankins reports, 'when Hamilton began to study philosophy on his own in 1826, it was the works of Dugald Stewart, an old Edgeworths' friend that he read, and from a copy of Stewart's *Philosohical Essays* in the Edgeworth library'.[22]

During his years at Trinity College, Dublin, Hamilton went back to Edgeworthstown in October 1825. Later, in 1828, when he was the Royal Astronomer living at the Dunsink Observatory, he repeated the visit and was impressed by William and Fanny Edgeworth's interest in practical astronomy.[23]

Despite the fact that Francis Beaufort Edgeworth was only fourteen years old when Hamilton visited Edgeworthstown for the first time, they became great friends. Curiously enough, throughout their friendship, which spanned many years, William Rowan Hamilton's intellectual debt to Francis, despite the fact that the former was a great mathematician and astronomer, was much greater than the other way around.

1.5 FRANCIS B. EDGEWORTH AND M. PAKENHAM EDGEWORTH

Due to Maria Edgeworth's unease about Lovell's school, her two youngest half brothers, Francis Beaufort and Michael Pakenham, were sent to Charterhouse as boarders, following the tradition of the Edgeworth boys in years past. Francis entered in 1819 and reported: 'I find school the most odious place.' In spite of heavy fagging,[24] he got used to the place and soon started writing poetry and theatrical dramas. His tutor was Dr Russell.[25] A few years later his brother Pakenham followed him to Charterhouse, and in September 1822 they surprised the family by performing the play *Catiline*, written by Francis, with their sisters. On another occasion they also performed *Remus: A Tragedy*, written by one or both of them; the roles of Romulus and Remus were, of course, played by Francis and Pakenham. In 1824, Francis won Charterhouse's second prize for English poems for *Carthage*, while Thomas Mozley won third prize. Mozley describes him in the following way:

> Francis Beaufort Edgeworth was a little fair-haired, blue eyed, pale-faced fellow, ready and smooth of utterance, always with something in his head and on his tongue and very much loved in a small circle at Charterhouse. With a fertile imagination and with infinite good-nature he would fall in with any idea for the time and help you on with it.[26]

The following year, in November 1825, Francis's poem *Saül* won a medal and Pakenham's *Jacob* won a miniature Horace.

At Charterhouse, Francis Beaufort Edgeworth's best friend was David Reid, the son of a Scottish brewer who 'looks anything but a brewer, poor D. Reid himself has not got a very prepossessing appearance but I am sure his sentiments and his actions are truly gentlemanlike'.[27] Pakenham Edgeworth's best friend was William MacPherson, who would come to exert a major influence in his life. He was also a Scot, and at that time his father was a professor at King's Collage, Aberdeen University, after a career as a surgeon in the Indian Medical Service. Pakenham followed the path laid out for his friend and studied first at Aberdeen and afterwards, from the first semester of 1829 to December 1830, at the East Indies College of Haileybury, where Robert Malthus was teaching. He was ranked at the top of his class and was offered an employment contract by the United Company of Merchants of England, which traded in the East Indies. In May 1831 he set sail from London to India with a pair of pistols provided by his aunts Maria and Fanny, as well as locks of hair as mementos of his mother and sisters.

Meanwhile, after finishing school, Francis Beaufort Edgeworth went to

Trinity College, Cambridge, in the academic year of 1827–28. There, he met John Sterling[28] and visited Thomas Mozley in Oxford. However, he profoundly disliked the type of mathematics taught at Cambridge, and he left. Maria Edgeworth tried to overcome this aversion in May 1828 by hiring William Rowan Hamilton as a private teacher for Francis, but Hamilton declined the offer.

Francis Beaufort Edgeworth's Poems and Travels

Advised by his nephew, the Gothic poet Thomas Lovell Beddoes, who was six years his senior, Francis Beaufort Edgeworth was planning to go to Germany at the end of January 1829 to study philosophy. He left Edgeworthstown with his half brother, William, a successful civil engineer, to collect his baggage, which had been left at Cambridge. Yet after William left him in London, when in the Library of the British Museum, he realised that he would save time and money by reading philosophy there to prepare for the German courses. So he took a room at a cheap, cosy inn in Hampstead and stayed in London. On 26 May, after being informed of his dear half brother William's death from consumption, Francis went back to Edgeworthstown but arrived too late for the funeral.[29] At St John's cemetery, he dedicated a lovely, mournful poem two hundred verses long, to his brother. This poem was praised by the poet William Wordsworth, who had become a friend of Hamilton: 'The specimens of your young friend's genius [Francis Beaufort Edgeworth] are very promising. His poetical powers are there strikingly exhibited.'[30] In September 1829, Wordsworth visited Edgeworthstown and both Hamilton and Francis enjoyed what Maria qualified as the 'tiresome lengthiness' of most of his conversation.[31]

Despite all his efforts, Hamilton's poetry was rather crude. Francis teased him about it by writing that 'after all, Hamilton, your poetry will not disgrace you'.[32] That same summer of 1829, Hamilton and Francis had been speculating about philosophical questions such as the one proposed by Francis: 'Is our knowledge of the Idea of Beauty improved by knowing the real solar system?'[33] As Hankins points out:

> it was in his arguments with Francis that Hamilton worked out his ideas on the relationship between science and poetry. Francis was a thoroughgoing Platonist, even rejecting Christianity and modern science in favour of the idealistic vision of Timæus. He argued that truth and beauty could be found only in the unity of an ideal world and not in the inductive process of modern science, which focuses on bits of observed nature but never on the whole.[34]

The next year, Francis and David Reid travelled to France, Switzerland and Italy. Francis had previously checked with Mozley, who thought that

Reid was completely mad, and 'remonstrated against the plan, on the ground of its dangers, the inevitable excitement, the chance of paroxysms far from help, the little good a maniac could derive from new scenes and experiences'.[35] Despite Mozley's admonitions, the trip was a resounding success, and they returned to England and Ireland in August 1831.

In Italy, Francis Beaufort Edgeworth had reached the conclusion that he could live there happily married with the annual income earned from the amounts inherited from his father and other relatives, and he could therefore spend his time reading philosophy and literature and writing poetry. He had the example of his niece, Anna Beddoes, who lived with her companion Zachary in Florence. Francis was thus in the mood for marriage, and he convinced his aunt Maria to invite Hamilton and his two sisters Sydney and Elizabeth to Edgeworthstown. The visit took place in September 1831. Maria described Sydney as 'intellect' and Elizabeth as 'sensibility',[36] but in a letter she confessed that she did not like 'those sisters'.

A few weeks after the visit, Francis once again took up his plan to study philosophy in Germany, and on his way to London he visited Dublin, where he came within 'one quarter of an inch' of proposing to Elizabeth. But just after arriving in London he happened to meet Rosa.[37]

1.6 NOVEMBER AND DECEMBER 1831: THE WEDDING

In London, Francis Beaufort Edgeworth headed to the inn in Hampstead where he had stayed and, as his niece Harriet Jessie Butler would report many years later, 'the maid by mistake showed him into a room full of ladies with beautiful eyes and charming manners' who 'proved to be the wife and daughters of General Eroles, a Spanish exile'. The next day he came back and 'he found himself in a room full of silent, gesticulating young men; the Eroles brother was deaf and dumb, but took likeness to support his family. It soon became evident that Francis was hopelessly entangled, and before November was at an end he had proposed to Rosa Florentina (aged 16).'[38]

Once Francis Beaufort Edgeworth had proposed marriage to Rosa, he informed his half brother Charles Sneyd Edgeworth of the tidings and introduced Rosa and her mother, sister and brother to him. Sneyd did not like her appearance. Rosa was not physically attractive, as Francis himself admitted when he described her to his brother Pakenham:

> [Rosa is] about my height – and fat, shortnecked, olive complexioned, oval faced, black haired – an emigrant – speaks English, though very unenglish

> English, she makes it so beautiful by the distinct pronunciation and such deep voice – black lustrous eyes with a kind expression, very decided oriental nose and large [beeved?] lips, and very fine hand and arm but bad ankles, very soft skin, very mild and kind temper, with a freedom in acting, character . . . constancy, tranquillity and feeling deep and unexpressable because unexpressable.[39]

Francis tried to express the unexpressable through a poem dedicated to Rosa:

> The love of thy sweet beauty is gone down
> Into my heart and its soft lustre there
> Glows, mingled with the essence of that thought
> In which we live and move and have our being.
> When I behold those splendid rolling orbs
> Flush'd with new youth, inspir'd with hope, elate
> With joy, look through
> And smile, my spirit amaz'd at the sweet sight
> . . .
> I am up loose
> High over earth, high over all mortal toils
> Uplifted to the universal heav'n
> Where the great heart of the whole world pants and flames
> Increasingly forever; there absorb´d
> Quite to forget this solitary life.[40]

It is curious to observe how sometimes poetry anticipates science, since the lines, 'Uplifted to the universal heav'n / where the great heart of the whole world pants and flames / increasingly forever' is a striking poetical metaphor of cosmological theories which are very fashionable nowadays. Francis also wrote to his mother, Honora and Maria about his determination to marry Rosa immediately.

The Edgeworth women were very much against the wedding. Honora was even cruel in the way she stated her position to Charles Sneyd Edgeworth, by writing disparagingly about Francis and Rosa: 'To tie himself thus early to a mere piece of humanity – without even beauty sufficient to plead his own forgiveness'.[41] However, Maria was more diplomatic in a letter to her favourite half sister, Fanny Wilson:

> *If* it be for his happiness -no matter whether she be ugly or pretty, poking or graceful, ignorant or cultivated-, no matter comfort is being penniless, *if* they can both be happy on the small means he possesses, *if* . . . But how much hangs on these *ifs* -and when I look back and consider how his mind has changed or matured in those love affairs in the course of the last six months, I tremble.[42]

Not all the Edgeworth women were against Rosa. Fanny expressed her conviction that 'she is a person I could grow very fond of'. Meanwhile,

Sneyd tried simply to buy some time by proposing 'to Donna Eroles [the mother] and her brilliant oldest daughter [Mariquita], to defer the publication of bans till the girl's father who is at Limoges has been heard from on the subject'.[43] Moreover, Sneyd, acting as *de facto* head of the Edgeworth family, asked the Hispanist Lord Holland for any references known about the Eroles family. He replied thus:

> I have some acquaintance and took great interest in the fortunes of Madame Eroles and her daughters at the time her husband somewhat imprudently engaged in the expedition of his countrymen against Ferdinand's government a year and a half ago. Mrs. Eroles is a lady of great merit and fortitude and most respectable character. . . . They are Catalans and I am much mistaken if either her father or uncle-in-law was not the Baron L'Erolles [sic] who commanded a very considerable corps in Catalonia during the revolutionary war. . . . I will inquire of General Alava and Don Augustin [sic] Argüelles more particulars of the family which though exiled and poor I believe to be in any way respectable.[44]

Lord Holland, whose family name was Fox, was indeed mistaken when he conflated General Eroles with the Baron of Eroles, since he was confusing a surname with a title. In fact, the name of the Baron was Joaquín de Ibáñez-Cuevas,[45] and he had been a war enemy of General Eroles. Nevertheless, the Edgeworth family accepted the report[46] and their attitude towards the wedding changed. A few days later, General Alava informed Lord Holland that 'General Eroles had been in a high command under Mina in Catalonia'. This information dissipated all remnants of distrust since General Mina was a highly esteemed and popular figure in the United Kingdom.[47]

Meanwhile, Charles Sneyd Edgeworth had received a dramatic letter from General Eroles saying that 'he was willing to give his daughter to Francis, and that he will give his life in concern of her provision to him'. As a consequence, Sneyd abandoned all hope that the marriage could be discouraged from the bride's side, and the wedding date was set for 19 December.[48]

Hamilton was personally informed by Francis Beaufort Edgeworth, who described Rosa's simplicity to him as 'a simplicity like nothing but that of a clear simple truth in science'. Francis also sent him a copy of the poem dedicated to Rosa.[49] The wedding took place in Saint Pancras on 19 December 1831, under the curate E.P. Hannam and with Charles Sneyd Edgeworth and Antonio Eroles acting as witnesses.[50] This Antonio Eroles must have been Rosa's deaf-mute brother, because her father Antonio was detained at Limoges with his eldest son Isidro.

The newly-weds stayed at Sneyd's place several days before departing for Florence. Francis wrote to his mother trying to explain his refusal to take Rosa immediately to Edgeworthstown, due not only to the expenses '20 £!', but also out of 'consideration to her timidity and utter helplessness of manner'. He went on to promise that 'when I find myself more familiar with her and more a friend and a confidant for her to lean upon, I will bring her to you . . . give me confidence and freedom which complete the full sphere of my happiness'.[51]

1.7 GENERAL EROLES'S STORY

Who was General Eroles, the Catalan grandfather of Francis Ysidro Edgeworth? All that was known about him before the present research was that around 1831 he had been in London as 'a political refugee from Catalonia' (Keynes, 1926), that his name was Antonio Eroles and he held the rank of general (Hicks, 1984).[52] Moreover, as mentioned in note 46, Sir John Hicks, under Lord Holland's influence, incorrectly speculated that since the General was a close relative of the Baron of Eroles, he had also upheld an absolutist and anti-liberal position, popularly known as the 'royalist' position. Both speculations are wrong: Antonio Eroles was no relative of the Baron of Eroles, and he always fought in favour of the liberal and constitutionalist cause, as General Mina himself recorded in his *Memoirs*.

In Mina's *Memoirs*, Antonio Eroles is mentioned three times. The first two are in January 1823, when Mina, then Captain General of Catalonia, was campaigning in Alt Urgell against the 'royalists' led by Joaquín de Ibáñez-Cuevas, Baron of Eroles. He refers to Edgeworth's Catalan grandfather as the 'commander of the militia of Organyà, Don Antonio Eroles'.[53]

The third of Mina's references to Antonio Eroles is in November 1830, when Mina, then in exile, was campaigning in the Pyrenees against Ferdinand VII, the expedition mentioned by Lord Holland. The political shift in France with the rise of Louis Philippe (July 1830) and his hostile attitude towards Ferdinand VII of Spain encouraged the liberal exiles in London to organise several expeditions against the Spanish regime. In October 1830, General Mina travelled from London to Bayonne (France). However, through an *entente* with Ferdinand VII, the attitude of the French government then changed from encouraging the expeditions to a neutrality which, in reality, discouraged them. In spite of this change, Mina tried to unite and coordinate forces with several other military chiefs in exile and decided to enter Spain with a small number of troops through the Basque country (Valdés and Mina), Aragon (Gurrea) and Andorra-

Alt Urgell (Miranda) with the idea of garnering popular support. Their reception in the villages was not hostile, but the royalist troops caused serious trouble for them. Mina sent Eroles and his eldest son on the Alt Urgell expedition as guides for Colonel Miranda, who reported that on the third day he 'sent another man that Eroles had introduced to me to La Seu in order to obtain some news' (translated from Espoz y Mina 1852/1962, II, pp. 181–2).

The expeditions lasted one week. Back in France, the French authorities, having altered their policy in favour of Ferdinand VII, kept them under arrest for more than a year. This is the reason General Eroles and his son Isidro were detained in Limoges at the time the wedding took place.[54]

The Parish Registers

After being informed that Antonio Eroles was the 'commander of the militia of Organyà' in 1823, we located our research around Organyà[55] in an attempt to find the corresponding parish registers. The results are as follows.[56]

Antonio Eroles was born in Pla de Sant Tirs[57] on 16 June 1779, the son of Tirs Eroles (born 1724, died 1796), a farmer, and Rosa Sancho, who had been married in 1775, he as a widower twice over from two previous marriages. Antonio had three sisters, two half brothers and one half sister.[58]

Antonio Eroles married Rosa Eroles[59] (born 30 December 1777), who also lived in Pla de Sant Tirs, the daughter of Sebastià Eroles (born 1749, died 1795), a farmer, and Magdalena Navarro (born 1744, died 1822). Rosa had four sisters and four brothers.

Antonio and Rosa's wedding took place on 13 March 1807 in the church in Pla de Sant Tirs, and even though the bridegroom was described as a *picapedrer* (stonecutter and builder), marriage articles were read and signed before Mr Bartolomé Domínguez, notary public of La Seu. Moreover, there is further proof indicating that the Eroles family was middle class: in the parish books the registers show the presence of from six to eight priests at the funerals of the head of the family, his wife and his heir. Wealthy families paid for the presence of twelve to thirty priests, while common people could afford two to three and the poor, unable to pay anything, had just one.

The christening books of Pla de Sant Tirs show the baptisms of Antonio and Rosa's three elder children: Maria Francisca (called familiarly 'Mariquita'), born on 9 November 1807; Isidro, born on 28 October 1809, and Antonio, born on 8 January 1811 and christened at his home – Cal Ton del Tirs – by the midwife because 'before being completely born she thought that he was dying'.[60]

These parish registers of Pla de Sant Tirs do not show the christening of Rosa Florentina Eroles. However, since her father Antonio Eroles was mentioned as commander of the Organyà militia in 1823, we can surmise that the family moved to this nearby village sometime after 1811. Unfortunately, the parish christening books of Organyà have disappeared, and therefore we cannot confirm her precise date of birth. We only know that Rosa was born in 1815 because she was sixteen when she married Francis Beaufort Edgeworth in December 1831.

Learning About the Father-in-Law

On his wedding day, Francis Beaufort Edgeworth would have known even less about his father-in-law than we have discovered, although he was to learn more details later. There is a letter that he wrote to his mother Frances Anne Edgeworth, dated July 1832, where he copied a passage about Rosa's father contained in 'a French history of the war in Catalonia':[61]

> At Organce [*sic:* Organyà] – village at a certain distance from Urgel [La Seu d'Urgell], there was a partisans' chief named Eroles (he is not a relative of the Baron Eroles) who was considered the most courageous and enterprising man of that part of Catalonia. General Zorrequin [in fact, Zorraquín[62]] called him and asked him if he was capable of seizing Blasi and take him to Urgel. . . . Eroles accepted the mission eagerly.

The mission was accomplished valiantly with the help of 'the point of his sword'. And F.B. Edgeworth added:

> Rosa says this affair was what first made her father join the revolutionists [in fact, the government's militia] whom before he had assisted only with money and mules. He would never wear any uniform when he formed the army except an épaulette. He said he was not a soldier. But they [Mina and Àlava] made him a colonel [and before 1831, a general], his eldest son only was made a lieutenant which turned to his advantage as he received the pay as an officer.

At the beginning of this same letter, F.B. Edgeworth explained several interesting details about his father-in-law that Rosa had told him:

> I had often wondered how Rosa's father being banished and having to fly away suddenly from Spain, could have afforded to give his family so much education when they arrived in England. He sent his elder son [Isidro] to an English school – to have a music master for Rosa – and all the expense of drawing and [?] masters for the [deaf and] dum[b] brother [Antonio], besides supporting an old servant . . .; well, the father had been saving up money for a long time. He had been a West Indies merchant dealing in indigo chiefly [and had saved] to

build a hospital in his native place in the Pyrenees; and this money he was able to carry off with him in the hour of need. [63]

So we learn that Rosa's father, Antonio Eroles, had risen from being a stonecutter and builder in 1811 to a rich West Indies merchant who in 1823 was also a reputed militia chief. How should we interpret this information?

First, we should keep in mind that Rosa was nine years old when she left Organyà for London. Therefore, what she told her husband was a childish version of the facts that her father had brought home.[64]

Second, we should be aware that Alt Urgell is a territory that borders on Andorra and France, where the smuggling of Andorran tobacco, French mules and Catalan textiles was a widespread occupation during those years.[65] The type of textile most common in this trade was made of cotton printed on one side and called *indiana*, roughly similar to calico. F.B. Edgeworth translated *indiana* incorrectly as indigo from the West Indies. He also alluded to a gift from Antonio Eroles of some mules as an aid to the liberal cause. Therefore, we should not discard the possibility that part of his income came from illegal trade or from the mere conveyance of smuggled goods.[66] Still, there is no evidence of any wealth being built up by the people of Alt Urgell through smuggling. They simply smuggled as a supplementary source of income, mostly in hard times. Thus, to find out how Antonio Eroles amassed his wealth, we would also have to inquire into his activities as a militia commander in around 1823, when he was described by Mina's aide-de-camp, Galli, as 'the most courageous and enterprising man of that part of Catalonia'. Yet what kind of honourable enterprises may a courageous militia commander undertake which allow him to get rich very rapidly?

Whatever the answer to this disturbing question may be, there is another reputable way to explain the source of the money that Antonio Eroles took to London. In Mina's *Memoirs* he writes:

I established [in January 1823] an Economic-administrative 'Junta' for the villages of Alt Urgell, to collect income and taxes for the State, since there was nobody in charge, and these resources would serve for the subsistence of the army. This 'Junta' was composed of residents of those villages and was presided over by a 'war deputy'[67]

There were 1200 soldiers stationed at La Seu. If Mina had appointed Eroles to preside over that 'Junta', he collected a lot of money that he administered but did not spend completely, since a lot of soldiers deserted when the royalists invaded Spain in 1823. At that point, he took all this money with him and gave it to Mina in London for the liberal cause (in fact, this was public money raised by the liberal regime). Mina thanked him by promoting him to colonel and afterwards to general with the purpose of warranting him a better subsidy from the British government, in spite of the fact that, during the 1830 campaign in the Pyrenees, Mina placed General Eroles under the orders of an officer of lower rank, Colonel Miranda.

Before leaving Francis Beaufort Edgeworth's report, let us make a final point: in most villages in Alt Urgell there was a tradition, which still persists today in some of them, that rich people would contribute to the upkeep of the rooms that the town offered to poor people when they were sick. Therefore, the fact that Antonio Eroles saved 'to build a hospital in his native place in the Pyrenees' can be viewed as Rosa's naïve interpretation of her father's words.[68]

1.8 ROSA AND F.B. EDGEWORTH: LIFE IN FLORENCE, LONDON AND IRELAND

News from Edgeworthstown

In the year after the wedding, a series of unfavourable events took place in Edgeworthstown. At the end of November, news of the death of Sir Walter Scott was a blow for Maria Edgeworth. Scott had been her most avid supporter in Great Britain and had declared himself to be her disciple in the realm of the 'regional' novel. At that time, Maria was trying to finish her new novel, which would be the last, entitled *Helen*. In May 1833 she completed it and read it to the family.[69] By then she was involved with the latest chapter in the drama of her half brother Lovell.

Lovell Edgeworth had been gambling again and had been obliged to borrow £3000. To deal with the situation, his half brother Sneyd came to Edgeworthstown for a family council. Lovell was forced to leave Ireland and settle in the Liverpool area, and the school was closed, fifteen years after being founded.[70] Lovell never visited London or Edgeworthstown again even though he was repeatedly invited by Maria and Sneyd. From Liverpool, Lovell moved to the nearby village of Ruthin, where he died in 1841.

A Sojourn in Florence

When Rosa Florentina and Francis Beaufort Edgeworth and their Irish maid left London in January 1832, just a fortnight after their marriage, they first stopped in Paris,[71] and then they most likely went to see Rosa's father and brother in Limoges, as Francis had promised her. In March

1832 they were reported to be in Florence with Francis's niece, Anna Beddoes,[72] and with Zachary, Anna's partner and possibly her husband.

For a year and a half, Rosa and Francis were indeed very happy; he proudly reported to his family how cheap life was in Florence.[73] In December, they had their first child, a boy:

> His voice at any rate is good and better than his father's. . . . His name is to be William. . . . David [Reid, who had been with them] is going tomorrow or day after back to England and in case he finds a *convenevole* wife, he is to come back in three months and we are to live together. This would be very convenient.[74]

The *convenevole* wife was found. Mariquita Eroles, Rosa's sister, agreed to marry David Reid and the wedding took place in London at Saint Pancras on 16 February 1833. They had probably met at Rosa and Francis's wedding, and Rosa had written favourably about him to her sister. When they married, Mariquita was 25 and David was 26.

By the time the Reids arrived in Florence, Mariquita was pregnant. Everything went fine at the Edgeworths' and Reids' home until early November, when the situation unexpectedly deteriorated, as Frances Anne Edgeworth reports:

> Poor David Reid who was seized with brain fever . . . and in one of the paroxysms had thrown himself out of the window in spite of the efforts made by his wife Mariquita to hold him back. . . . Poor Francis says he devoted all his time to the endeavour and solace the last moments of his friend. The poor wife is within a short time of her confinement and has shown more than ordinary strength both of mind and body.[75]

David Reid died in Florence on 19 November 1833. Ten days after Reid's death, his widow Mariquita had a daughter who was named Mary and lived only one month. Moreover, the 'poor Irish maid they had worn out with fatigue' also died around this time.[76] Yet that was not all, since William, Francis and Rosa's infant son, was afflicted with 'the setting of eleven teeth and some operation' and became very ill – 'pulse 130 and feet swelled', according to Maria Edgeworth, who added with a touch of frankness: 'I was content that he [Francis] should be chastened by some touch of sorrow but this is too much'.[77] William Edgeworth died on 20 December 1833.[78]

Visiting Edgeworthstown

After the death of both children, Francis and Rosa Edgeworth accompanied Mariquita to the Reids' house in London, and they continued afterwards to Clifton and Edgeworthstown. In London, they met Charles

Sneyd Edgeworth and visited the Eroles, who were once again reunited when father and son were permitted to leave Limoges.[79]

At the end of March, they arrived in Edgeworthstown, where Rosa finally met Maria and Honora Edgeworth. By then, Maria had published *Helen* and her last children's story, *Orlandino, a tale*. Both of them were immediate successes. However, she was not tired of literary writing, since in a letter to Pakenham Edgeworth, who had been placed in February in Amballa (Punjab), she gave a full account of Rosa, which was peppered with a few common stereotypes, as we can see in this excerpt:

> [She is] not very handsome, scarcely handsome at all, yet picturesque and fixing attention. Her face varies very much as of a Murillo seen in a good or bad light. But the light comes from within, not from without in this picture. . . . The more I see of Rosa the more I like her. Francis is very happy.[80]

During their stay in Ireland, Francis and Rosa visited William Rowan Hamilton at his observatory in Dublin. Hamilton reported that 'Francis's foreign wife (whom by the way I like) has conquered (it would seem) his dislike to learning modern languages'.[81]

Settling Down in London

During the month of May, Rosa and Francis went back to London, and upon hearing that Mariquita had left the house of her father-in-law, Andrew Reid, who had been very generous with her, they decided to live together and settled at number 8, Portland Terrace, [St] John's Wood, opposite Regent's Park. This put an end to Francis Beaufort Edgeworth's plans of becoming a poet, since life in London was much more expensive than in Florence and his income would not allow him to devote himself full-time to literature and poetry.[82]

During this summer of 1834, the Eroles, Antonio and Rosa with their sons Isidro and Antonio, went back to Spain, financed by Mariquita. Their ten years of exile had come to an end.

Rosa and Francis's married life in England and Ireland can be divided in two periods. The first runs from May 1834 until the spring of 1839, when they were mostly in London. The second period spans spring 1839 until October 1846, when they were both mainly at Edgeworthstown. However, in this second period there was another sojourn in Florence for Rosa, from April 1840 to June 1841. During this period of more than twelve years, 1834 to 1846, Rosa and Francis had six more children: one daughter and five sons.

The first child to arrive, born in London on 14 December 1834, was a boy named after the son who had died in Florence, William.[83] Because the name was repeated, which was quite usual in those times, all of Edgeworth's biographers have been unaware of the Florentine William.

Throughout the London period (1834–1839), Francis B. Edgeworth organised a school in Eltham where he taught English, the classics and possibly philosophy and history. These were the times when Francis's intellectual potential was best exhibited, and we have several testimonies in this regard. During the years 1835 to 1837, we find the pinnacle of his discussions on philosophy, science and poetry in his correspondence with Hamilton. During the summer of 1837, Rosa and Francis were in Edgeworthstown with their son William, and in September they visited Hamilton at the observatory.[84]

In London, Francis got back in contact with his Charterhouse friend, Thomas Mozley, and with his Cambridge companion, John Sterling, who introduced him to Thomas Carlyle. As Keynes quoted in his 1926 article, Carlyle described Francis in his *Life of John Sterling*:

> Frank was a short neat man, of sleek, square, colourless face (resembling the portraits of his father), with small blue eyes in which twinkled curiously a joyless smile; his voice was croaky and shrill, with a tone of shrewish obstinacy in it, and perhaps of sarcasm withal. A composed, dogmatic, speculative, exact and not melodious man. He was learned in Plato and likewise in Kant; well read in philosophies and literatures; entertained no creeds, but the Platonic or Kantean ghosts of creeds; coldly sneering away from him, in the joyless twinkle of those eyes, in the inexorable jingle of that shrill voice, all manner of Toryisms, superstitions; for the rest, a man of perfect veracity, great diligence and other worth. . . . He now hoped to find support in preparing young men for the university, in taking pupils to board; and with this view was endeavouring to form an establishment somewhere in the environs. . . . Poor Edgeworth tried this business for a while, but found no success at all.[85]

Mozley, however, differs from Carlyle in his description of Francis:

> Carlyle felt a singular and even affectionate interest in Edgeworth. Yet some expressions are so harsh and so unmerited, they jar on my memories very much. My ear still testifies that there was sweetness in his voice and gentleness in his manner and tone. My eye still recalls his soft and steady gaze. I felt sure then, and I feel sure now, that he wished to be a Christian. . . . Francis Edgeworth was torn by conflicting systems and I may add conflicting sensibilities from childhood. He was a most sympathetic, self-sacrificing being.[86]

Maria Edgeworth's Plan

In March 1837, Sophy Fox died of consumption. Her half sister Maria, now past her seventieth birthday, realised that her stepmother Frances Anne was 68 and her half brother Sneyd, the absentee head of the family, was already 51 and had no children. If Sneyd died unexpectedly, she might possibly have problems again with Lovell, now 61 and exiled in Liverpool.

After Charles Sneyd Edgeworth, the next in the line of succession to inherit Edgeworthstown was Francis Beaufort Edgeworth. Maria clearly saw that the future of Edgeworthstown was in Francis's hands and that Francis had to be involved as soon as possible in its administration, which she was once again in charge of after Lovell's crisis in 1826. Therefore, the first thing she did was to encourage her stepmother Frances Anne's plan to invite Francis, Rosa and William to Edgeworthstown, and to have Mariquita come along as well.[87] As mentioned above, during the summer of 1837, Rosa, Francis and William were in Ireland with Frances Anne and Maria, who was very happy with them and reported that 'Rosa is teaching me Spanish, and Francis is laughing at me for learning a new language at seventy!'[88]

Maria Edgeworth's plan was bolstered in 1838 by the marriages of Mariquita Reid and Honora Edgeworth. Mariquita was going to leave Francis and Rosa to settle down with her new husband in Fiesole, again in the Florence region. Additionally, Honora got married and left Edgeworthstown to settle in London. Therefore, Edgeworthstown became emptier, and Mariquita had vanished as a pretext for keeping Francis and Rosa in London.

Mariquita married again in July 1838 and her second husband, just like her first, was a friend of Francis, this time from Trinity College Cambridge, Robert John Tennant. Tennant, a friend of the poets Tennyson and Hallan, was just one year younger than Mariquita: they were 29 and 30 when they married. He seems to have accepted the prospect of becoming the chaplain at legation in Florence for the English expatriate colony just to please her. The first thing Mariquita and Robert Tennant did after marrying was to renounce her rights within the Reid family. Moreover, to perform the role of the chaplain's wife, Mariquita embraced Anglicanism.[89]

Honora's marriage took place in September 1838. She married Captain Francis Beaufort, the brother of her living stepmother Frances Anne and the best friend of her brother Sneyd. Beaufort was a distinguished naval officer, hydrographer, cartographer of the Greek islands and the southern coasts of Anatolia, and meteorologist. In 1806 he had invented the wind scale that was finally adopted in 1926 by the International Meteorological Committee and is still in use today under the name of the Beaufort Wind Scale. His first wife, Alice Wilson, the sister of Lestock Wilson, Fanny's husband, had died in 1834 after giving birth to three sons and four daughters. Beaufort was living in a big house in Gloucester Place, London, and was rich enough to pay for his whims.[90] Honora was certainly not a whim, since he had been in love with her during his youth, although in those early days she had not reciprocated the feeling.[91]

Captain Beaufort, who was promoted to Rear Admiral of the Royal Navy in 1846, became one of Charles Sneyd Edgeworth's main helpers in solving the problems that would affect the Edgeworth family. Both Frances Anne and Maria Edgeworth were exultant about this marriage. Now Maria's only unmarried half sister was Lucy Jane, who spent much time visiting her sisters and relatives in Clifton and London.

Maria's effort to attract Francis and Rosa to Edgeworthstown was finally rewarded when in the spring of 1839 they moved there to live. Maria was enthusiastic about their presence, and she described William as 'the finest boy perhaps you ever beheld'. Also in Edgeworthstown, Rosa and Francis had a daughter who was named Maria after her aunt on 25 August 1839. Henceforth, she would be known as Mary.

Despite Carlyle's opinion: 'He was the last man of whom one could expect would become the manager of an Irish estate', in 1839 Francis had taken the administration of Edgeworthstown from Maria, who was now 72 years old.[92] Consequently, she could finally savour a happily leisured old age surrounded by children, and with the future of the Edgeworthstown estate ensured through Francis Beaufort Edgeworth.

A Florentine Eroles

However, the soothing state of Maria Edgeworth's mind was endangered in the spring of 1840 when Mariquita's health required serious attention and her husband asked Francis to send Rosa to be with her. Francis, Rosa and their son William – they left the baby girl in Edgeworthstown – arrived in Florence in May. Mariquita was recovering slowly, despite recurring nervous collapses, and at the beginning of September Francis left Rosa and William there with the Tennants to go back to Edgeworthstown, where he arrived at the beginning of October. Rosa could not go with him because she became pregnant in June. Therefore, she and William stayed with Mariquita and Robert Tennant, awaiting the newcomer.

In the letters addressed to Rosa dating from February and early March, Francis pondered an entire list of girls' names and just one for a boy: 'Antonio Eroles could be a nice name for a boy'.[93] We do not know whether Rosa's father was alive or dead at the beginning of 1841; after leaving London in 1834, there is no direct information precisely where in Catalonia the Eroles settled. On 25 October 1835, Mina was named again 'Capitán General de Catalunya' and organized a battalion made up of former exiles to fight against the Carlists (supporters of Ferdinand VII's brother Carlos's claim to the Spanish throne in opposition to his eldest daughter, Isabel II, Queen of Spain). It is conceivable that General Antonio Eroles was part of that battalion, although his name is not mentioned in any war chronicle. An important part of the action took place in the Alt Urgell (and in Berguedà and Cerdanya). In fact, from 1836 to July 1840, the date marking the end of the first Carlist war, Organyà was in Carlist hands, despite the fact that La Seu was never Carlist. Mina died almost at the beginning of this war, on 24 December 1836.[94]

Despite Mina's death, we find General Eroles's eldest son, Isidro, fighting against the Carlists in the governmental militia under Captain D. Pablo Massanet during the period 1835–1840. Afterwards, in October 1840, he applied to be the beneficiary of a governmental decree commuting his studies in law, which at that time was a seven-year university degree in Spain, to two years, by virtue of his five years in the militia. The commutation involved, by a formality provided for in the decree, an examination on all the subjects contained in the first five years. Isidro surmounted this 'formality', and in the following academic year, 1840 to 1841, he undertook the sixth year of law, which he passed with excellent overall marks of *notablemente aprovechado*. The next academic year he succeeded in passing the seventh year of law with the same marks, and in June 1842 he unanimously passed (*nemine discrepante*) his final exam and became a full-fledged lawyer admitted to the Spanish bar.[95]

Let us go back now to Florence, where on 19 March 1841, the anniversary of the Spanish Constitution of Cádiz so dear to General Eroles, his daughter Rosa gave birth to a boy, who was christened Antonio Eroles. As a child he was called 'Erolino' due to his Italian birth. Upon the birth, Francis Beaufort Edgeworth expressed his happiness and mentioned Rosa's need to return. In the next letter, dated 10 April, Francis proposed leaving William with Mariquita. They did so and William grew very attached to his aunt Mariquita Tennant and came to be quite spoiled, too, in the process.[96]

Edgeworthstown Forever

After settling down again in Edgeworthstown, life was placid for Francis and Rosa Edgeworth. In March 1842, Pakenham had returned to Great Britain from India on a four-year leave. He visited his mother in April and amazed Maria with his botanical drawings and sketches of the country.[97]

After a long confinement, on 14 April 1842, David Reid Edgeworth, Francis and Rosa's fifth child, named after Francis's best friend and Mariquita's first husband, was happily born at Edgeworthstown. Mariquita was also there to return William to his own enlarged family. Then, in July 1842, Mariquita was again in Fiesole when her second husband, Robert John Tennant, died unexpectedly on 24 July at the young age of 33. Thereupon, Mariquita Tennant decided to go back to England the following spring.

Meanwhile, in June 1842, Maria Edgeworth greatly enjoyed the distinction of being named 'honorary member of the Royal Irish Academy', the scientific institution founded in 1785 by her father, amongst others. According to Maria, this high honour was exclusively due to her friendship with the president of the Academy, Sir William Rowan Hamilton – he had been knighted in 1835 – since she humbly considered that, despite her texts on education, she had no scientific merits whatsoever.

In 1843, Rosa became pregnant again, but this time she felt well. At the end of May, she had a boy, Richard Lestock, named after his grandfather Richard Lovell Edgeworth and his uncle Lestock Peach Wilson, Fanny's husband.

A few days later, on 8 June 1843, Lucy Jane Edgeworth married Dr T. Romney Robinson, a widower since December 1839. Robinson, an astronomer at the Armagh observatory, became President of the Royal Irish Academy some years after William Rowan Hamilton. Maria wrote about this union: 'Never was a marriage hailed with a more family acclaim of universal joy.'[98] In fact, Maria was exultant because she had successfully accomplished her goal of marrying off her half sisters.

Despite his frustration as a poet, Francis Beaufort Edgeworth proved to be an excellent administrator. However, he did not enjoy the job and was anxious about the overall economic situation. In early May 1844, he felt sick and Rosa felt miserable because she could not give her son William, then ten years old, his Latin lessons. Maria Edgeworth explained to Fanny Wilson that 'Dr. Marsh says that there is nothing the matter with Francis's liver or spleen, but the fault is in the colon. He advises warm baths.'[99]

Ysidro Francis Edgeworth's Birth

Rosa had not felt well either, and according to Maria, she was trying to regain her strength by having 'a delightful cordial from her brother Ysidro [*sic*] and her mother'.[100] The combined effect of the cordial – a Catalan brandy, perhaps – on Rosa and the warm baths on Francis was unexpected but of capital importance for our story: Rosa became pregnant for the seventh time, and on 8 February 1845 she gave birth to their sixth son, who was named after both the donor of the cordial, uncle Isidro, misspelled as Ysidro, and his father, Francis. According to Maria Edgeworth, Rosa's confinement was rather problematic: 'How ill Rosa had been. It is all well now – thank God – but it makes me shudder – I am sure you will feel as I did when I read Francis's note. How beautiful his description and quotation from Dante.'[101] It is a pity that F.B. Edgeworth's note has not been preserved so that we could know precisely which of Dante's original verses had been associated with Ysidro Francis's birth. Was it the one extolling the seed which was used in an English translation on the opening page of this chapter? The Italian original reads:

> Benedetto sia tu' fu 'trino e uno
> che nel mio seme se' tanto cortese![102]

Whether or not these were the particular verses chosen, it is anyway rather flattering for any child to merit a quotation from Dante, perhaps as a portent of future greatness.

Death in Dublin

Ysidro Francis's first five years were hard for his family. Let us trace in detail the events that might have affected Edgeworth's later life.

At the beginning of 1846, Pakenham Edgeworth was engaged to Christina MacPherson, sister of his best friend William. The wedding took place in Aberdeen in February, and the couple went to Armagh, Trim and Edgeworthstown, where Christina met Maria, Frances Anne and Rosa's children. Neither Rosa nor Francis Beaufort Edgeworth were there; rather, they were in Dublin because of his health. He had experienced intestinal troubles again and was being treated there by Sir Henry Marsh, a reputed physician. The prognoses were optimistic when Christina and Pakenham visited them in Dublin. Francis responded favourably to the treatment, and by the end of March the couple was back in Edgeworthstown. The day after their return, Francis attended a meeting on actions to alleviate the early effects of the potato crisis in County Longford, as he wrote to his brother Pakenham:

> The meeting today has certainly raised my spirits with regard to our prospects for the summer, at least as far as the account of the low price of Indian corn [imported by the Government] and its nutritional properties. . . . Our potatoes are certainly very bad, and I do not know how people live who have to buy them; but as the population, thank heaven, has not increased as suddenly as the food has diminished, there is not increase of work . . . I am going on with my restorative course.[103]

In fact, his recovery was illusory. Francis and Rosa went back to Dublin at the end of May.[104] In spite of several optimistic medical opinions, Francis's intestinal crises only worsened during the summer of 1846. Back in Edgeworthstown, Maria took care of the children and wrote letters to them every two or three days and sent them money to pay for their stay in Dublin. 'I can do you more good by money than by words'. She also agreed with Rosa 'in never saying anything to Francis about expenses. It would be the most anxious and dangerous topic.'[105] This means that Francis and Rosa had to be very careful with money, and any unexpected source of expenditures, such as Francis's illness, posed a major problem for them.

The summer was also inauspicious for other members of the Edgeworth family. In August, surgeon John King, husband of Emmeline, 'died poor' according to an obituary in Clifton, 'where he practised with eminent success for almost fifty years'.[106] At the very end of September, Henrica, the wealthy wife of Charles Sneyd Edgeworth, died in London. As a widower, Sneyd, then nearly sixty years old, stayed at his residence in Kent, and from there, as we shall see, he would be in contact with Edgeworthstown's Manor House to manage the problems of the family and the estate.

Meanwhile, Pakenham and Christina Edgeworth were travelling to India aboard the steamer *Madrid.* They were accompanied by Jessie MacPherson, Christina's sister.[107] Before reaching Gibraltar, Pakenham wrote to his mother that he hoped 'that the letter will arrive at Francis's birthday, may it be an auspicious [. . .sign?] for him'.[108] In fact, on 5 October, the very day of Francis's 37th birthday, he died in Dublin.

One of the people who remained very close to Rosa during the last days of her husband was Dr Crampton. He had been dismissed by Francis in order to be treated by Dr Marsh, and he only intervened as a family friend when nothing could be done, and he refused to accept any payment from Rosa. An autopsy was conducted, the brain was examined, and the physicians concluded that they could not find any hereditary disease. The corpse was finally cremated.

After the funerals in Dublin and Edgeworthstown, Rosa was comforted by the family members' appreciation of her virtues. The most consolatory gesture was the one she received from Charles Sneyd Edgeworth, who proposed a plan of providing the sum of £300 per annum to each Edgeworth child through a fund.[109] This was more than a mere gesture; it was a token of hope for Rosa. If Sneyd had instead planned a new marriage in order to have children of his own, her financial situation would have been quite precarious. But as long as Sneyd had no male descendants, Rosa's eldest son would inherit Edgeworthstown.[110]

1.9 THE POTATO FAMINE

After Francis Beaufort Edgeworth's death, Maria once again had to rely upon an agent to manage the estate and the trust. She turned eighty on

New Year's Day of 1847, but she was still able to campaign to alleviate the suffering of the poor of County Longford at the peak of the potato famine. The Irish peasants had already suffered from the effects of the low prices in 1842, and in County Longford the first attack of the fungus *Phytophtora infestans* spread in 1845 when the prevailing opinion was still that the trouble was due to the excessively humid weather. The successive waves of the fungus reduced the yields per acre to 66 per cent in 1845 and 20 per cent in 1846 compared with the yields of 1844.

In the Longford area, the potato crop in the summer of 1846 was also poor, and on 1 January 1847, Maria, her stepmother Frances Anne and their neighbour Mr Parnell filled in the questionnaire from the Central Relief Committee. This document gives a reliable description of the dramatic situation in Edgeworthstown when Francis was two years old; it is reproduced in its entirety at the end of the book as Appendix A.[111]

After this initial contact, Maria exchanged a dozen letters with the Central Relief Committee. In February 1847, she received £40 from them – '£30 for soup and £10 for female employment'. In April and May of the same year, the Committee granted Maria half a ton of rice valued £6 and announced the shipment of '4 sacks of Indian meal and half ton of rice to distribute in a cooked state. No funds may be sent for emigration purposes.' At that time, three vessels were taking passengers to America at the price of seven pounds ten shillings per person.[112]

Later, Maria exploited her popularity in America amongst the children who had read her tales and republished *Orlandino*, her last story, giving her royalties to the cause of Irish relief and asking for food contributions to mitigate the famine from which the Irish children were suffering. Some of this food was sent directly from Boston children 'to Maria Edgeworth, for the Poor' with no address. The largest shipments came through the Irish Relief Committees established in the United States of America and were sent to the Central Relief Comité (CRC). In October, Maria informed the CRC that 'the Irish Relief Committee from Cincinnati is sending to her $180 in corn meal'. The last letters date from March and May 1848, and they prove that in the spring of 1848 Maria was still busy trying to help the people of County Longford. She always bore in mind that most of these people were the children of those tenants who had helped her in 1826 during the first Lovell crisis.[113]

Maria Edgeworth and Richard Jones

Notwithstanding all her activity, we also hear from Maria Edgeworth in a letter to the political economist Richard Jones, Malthus's successor in the East India College of Haileybury, that the plague did not affect Longford crops of 1845 as heavily as in other parts of Ireland. The following year, Maria discussed some economic aspects of this plague with Richard Jones, in a letter dated 14 June 1847:

> Our harvest promises well even the potatoes. But we must not brag or hope too soon . . . I have letters of Ricardo in which all the requisites for a safe national food are ably stated and the potato the lowest in his scale. . . . I do not consider it as an evil in itself but a good that has been abused. The introduction of other food or vegetables [that] require more care and cultivation will be an improvement no doubt.

She followed with some criticism of the British policy:

> The quantity of provisions of corn that have been sent over to Ireland to be locked up in stores till certain time and certain prices prevailed and the quantity that has been wasted and spoiled of these stores when opened is lamentable! Also the quantity of money that has been wasted and is now wasting in paying officers watching officers.

And she added some words in defence of the Irish landlords:

> I could name at least ten or twelve great landed proprietors who have, this season and last, lost their lives from over exertion and from fevers caught in attending their tenants and the poor and protestant clergymen in great numbers have so zealously exhausted themselves that they have won the affection of the poor Catholics.

Finally, she closed the letter with a question: 'How shall we get the people who have been fed gratis to believe that the government and their landlords are not bound to feed them always?'[114]

In these letters to Richard Jones we grasp that Maria was in favour of the middle ground, as she stated several times, even in the postscript of her last letter to him dated April 1849:

> Potatoes are coming again! Tell me whether Ireland ought to be glad? or sorry? or between both?.
> Post-Scriptum: I think it would be folly to give potatoes up altogether or to depend upon them as the principal food of the people. Truth often falls between two stoves.[115]

Maria Edgeworth's combative mood with regard to the consequences of the famine contrasts with Hamilton's attitude. The astronomer wrote to his friend Aubrey De Vere that he was donating through various channels, but he dedicated his time to abstract reasoning even though this reasoning might seem impractical. Nevertheless, he hoped that through his impractical effort 'the fame of our country shall be in any degree raised thereby and . . . the industry of a particular kind thus shown shall tend to remove the prejudice which supposes Irishmen to be incapable of perseverance'.[116] However, in August 1849, when the Queen travelled to Ireland, Hamilton rushed to dedicate a sonnet to her: 'To the Queen in Her First Visit to Ireland'.

1.10 FRANCIS'S FIRST AND MARIA EDGEWORTH'S LAST YEARS

The first specific reference to Ysidro Francis Edgeworth after his birth is in a letter written by his mother Rosa to his uncle Michael Pakenham Edgeworth. Pakenham and Christina had arrived in Calcutta in November 1846, and the East India Company stationed him in Bandar, on the Bengal Gulf, 215 miles north of Madras.[117] Once they were installed there, Rosa kept up a prolific correspondence with Pakenham, who became her main family informant. Ysidro Francis lost his first name after the death of her father and became Francis to Rosa and Frank to the rest of the family. Rosa wrote to Pakenham that 'Francis baby is magnificent, so good, so clever, so solid'.[118]

Francis was two years old when he received these flattering epithets from his mother. In addition to her friendship with Dr Crampton, Rosa also kept in contact with her late husband's friend, Sir William Rowan Hamilton.[119] Hamilton saw Maria Edgeworth for the last time in January 1848 at the Butlers' vicarage in Trim, where Maria appeared to be 'in remarkable health, spirits and vigour, for a lady in her eighty-second year!'[120] But her spirits and those of her stepmother Frances Anne waned a month later with the terrible blow of Fanny Wilson's death after a very brief illness. As Harriet Butler wrote, for Maria 'the life of her life is gone; Fanny had been for forty-seven years the object and sharer of every thought'.[121]

In June, Pakenham reported from Bandar on the birth of his daughter Christina Frances Edith Edgeworth, who lived only a few months.[122] Nearly one year later, in May 1849, Pakenham was appointed to the Punjab Commission and left Bandar for Multan.

In February 1849, news came of the death of Maria's nephew Thomas Lovell Beddoes, the son of her sister Anna and Dr Beddoes, the physician. Thomas, who is today hailed as an Elizabethan poet, committed suicide at the age of 45 after a stormy sentimental affair with a young baker named Degen at the Cigogne hotel in Basel. [123]

Maria Edgeworth's Death

At the beginning of the spring of 1849, Maria spent several days with the Butlers in Trim. Lucy Jane and Dr Robinson were also there.[124] It was the last time that she slept outside Edgeworthstown. Her last known letter was written on 2 May and was addressed to her friend Dr Crampton.[125]

The best description of Maria Edgeworth's last hours has not been used by her numerous biographers. It is revealed in a letter from Rosa Edgeworth to Dr Crampton written on 22 May, the day after her death. Rosa explained that, two days before, Maria was going with her to Trim to visit her cousin Margaret Ruxton and the Butlers, but as she did not feel well she asked Rosa to go back to Edgeworthstown. There, Rosa enlisted the aid of Frances Anne:

> We gave her brandy which she swallowed with avidity – after that great shuffle she was able to walk up to her little room, took one of her accustomed pills and appeared quite quiet. She did not pass a good night. I was with her at 6 o'clock yesterday. She appeared haggard, but I thought I had often seen her worse. She told me of various things she wished me to do. I went to my room to write, my mother [Frances Anne Edgeworth] came to my room saying 'She is much better. She is sitting in her bed and has begged me to go down and make breakfast.' No sooner had my mother disappeared when Maria's maid rushed into my room exclaiming 'Mrs Francis!'. I hurried to Maria's. She fixed her eyes on me and in an instant after she was no more![126]

Despite her old age, Maria was to her dying day the leading character in Edgeworthstown. All decisions of any importance, especially financial ones, were referred to her. She was active even in the education of Rosa's children, as we shall see in the next section. After Maria, Edgeworthstown would come to be managed by Frances Anne and Rosa, with the acquiescence and advice of Sneyd and Pakenham from London and Multan, respectively. As to the main bequest that Maria left – in 1842 she reported having £9011 and ten shares in the North Staffordshire Railway Co. – the main beneficiary of her will seems to have been her stepmother Frances Anne Edgeworth, since some years later Pakenham would mention in a letter to his mother 'the deed of gift from Maria' that Frances Anne was keeping in her files.[127]

1.11 ROSA EDGEWORTH'S CHILDREN'S DEVELOPMENT

Maria Edgeworth had been active in the care and education of Rosa's children. While she was alive, she had joined Frances Anne and Rosa in choosing their governesses and nursery maids. A 'Family Report' was kept about them, in which in 1846 a certain Miss M. Farlane is reported to be succeeded 'by the harsh Margaret Nugent probably early in 1847', by 'Miss Jopping probably late in 49 or early in 50, a reign of great beneficence' that lasted till the autumn of 1851 and was helped by 'Shaw as nursery maid'.[128]

After Maria's death, Rosa wrote to Pakenham that 'Maria had a peculiar tenderness for David, being so very like Francis. Baby [Ysidro Francis] is remarkably intelligent. She has taught him to read beautifully.'[129] Thus, we learn that Francis had been taught to read by the age of four years. In his adult days, Francis remembered Maria as 'a nice little old lady with a funny face'.[130]

In fact, Rosa was proud of her children's development, although apparently William, the eldest, was becoming a problem for the family. But she was particularly exultant about Francis, as she demonstrated in a letter to Pakenham when he was six years old:

> Francis is like David – solid, good, common sense – with great capacity for taking an idea and he always knows when he understands a thing or not. He pursues his plans with that calm but indomitable force which is the most striking peculiarity of his character. His affections are deep, his character commanding from the ascendancy he acquires over those he comes in contact with. He is guided by a sense of duty peculiar in so young a boy.[131]

Two years later, in 1853, Rosa described her youngest boys to Pakenham: 'David is most affectionate and kind and so honest and true . . . his fort is Science as that of Eroles is Classics. Francis is very clever both in languages and in Euclid. Richard has the most winning manners of them all.'[132]

After Rosa suffered from a serious illness during the spring of 1854, she reported to Pakenham in two letters dated June and July:

> So, David is in the naval fever, it is a usual stage for naval boys to go through. . . . Mr. Plowsman [the tutor] thinks Francis very clever and indeed he is. He makes with greatest ease verses in Latin on any subject and he reads Virgil with the greatest pleasure. He is equally at home in Greek and French. Richard is not equal to Francis in ability. David's tastes seem more mechanical. Mr. Plowsman is a good mathematician, which is a great advantage.[133]

Thus, the four younger Edgeworth boys and Mary were all educated by tutors at home. The names of their tutors are listed in the Family Report: Mr Jenkins, Mr Burke, Mr Plowsman and Mr Potterton taught them successively from 1851 to 1855.[134]

In November 1854, Rosa added in a letter to Pakenham:

> I send David to my friends at Kingstown, the LeClercs in order that he may thoroughly know whether he likes the sea as his future profession. Eroles is growing very, much . . . he is very fond of his books. Francis [nine years old] persists that he will enter the church and Richard in being a surgeon.[135]

Those were the days when Ysidro Francis Edgeworth learned Greek, Latin and English classical poetry by heart and 'complete books of Homer, Virgil, Milton and Pope would readily come to his memory'. He would remember them for the rest of his life. In fact, those were the days when he started to become a legend: 'It is narrated that in his boyhood at Edgeworthstown he [Francis] would read Homer seated aloft in a heron's nest. So, as it were, he dwelt always, not too much concerned with the earth.'[136]

Another characteristic shown by Francis from his early childhood was his complete attachment to his brother Richard. We have several testimonies to this attachment. First, we find it in the aforementioned Family Report, which recounts that in 1851, while their mother was away, 'Dicky and Francis ran away' in the hopes of overtaking her. And they 'ran away for the second time in the winter of 1852–3.' Secondly, in a letter from Rosa to Pakenham in July 1854, she informed him that Richard and Francis 'are very fond of each other.'[137] And third, a note written by Harriet Jessie Butler in 1926 stated:

> Dick and Frank were entirely devoted to each other, and by no means followed their elder brothers. They had great ploys of their own as growing boys. Once they ran away, when they were quite small, being caught, I think, entrenched somewhere in the lawn, where they defended themselves (or Frank did) with a carving knife. They were quite peaceable later on, but full of queer devices.[138]

1.12 THE INDIAN CONNECTION: WILLIAM EDGEWORTH'S TRIUMPH AND FALL

Harriet Jessie Butler, the writer of this note, was the second daughter of Pakenham and Christina Edgeworth and was born in Multan (West Punjab) on 9 October 1851, exactly fifty years after the birth of Pakenham's favourite sister, Harriet. Her complete name was Harriet Jessie Edgeworth, but her relatives familiarly called her Harrie. Just like her aunt Harriet, she became a Butler by marriage. She spent the first four years of her life in British India, which for her father Pakenham meant a lot of British socialising and some diplomatic and administrative work, at a time (1851–1855) when the East India Company ruled India and its employees were considered British civil servants.[139] Pakenham described his day-to-day life in his detailed *Journals*.

Rosa stepped up her correspondence with Pakenham in 1853 and tried to enlist his help to provide a new career for her eldest son, William, who had just entered the Army and was going to be sent to India. In a plan put forward by Frances Anne and Rosa Edgeworth with the support of Mariquita Tennant, they asked Pakenham to look after him in Multan once he had been trained as a civil engineer in London.[140]

Rosa and Frances Anne had kept in contact with the military committee that 'agreed to grant the permission', since, given the unstable political situation in India, it would be advantageous for the British Army to have him closer to where he might be needed. Pakenham agreed to the plan, albeit not very enthusiastically. William Edgeworth left for India in November 1854 and arrived in Multan in April 1855. His uncle Pakenham did not like his demeanour from the beginning. Despite that, he tried unsuccessfully to secure him a job under his friend Colonel Napier.

At that time Pakenham was busy with his job at the Punjab Commission and travelling around Multan during this particularly dangerous period. Then, in early July, he suffered from a 'sunstroke' that paralysed his right side. The stroke was not fatal and he eventually recovered the use of most of his limbs. At the end of December, he set sail with Christina and Harrie to spend his recovery leave in Great Britain.[141]

Pakenham and his family lived for some time in MacPherson's quarters in Scotland during 1856, and in September Pakenham learned from doctors 'that a return to India would be very dangerous'.[142] Before the end of that year, he had secured an early retirement, and he and his family settled in southeast London, first in Norwood and, after 1860, in Anerly. When visiting London, Rosa's children would quite often be their guests, especially Mary and Francis.

The Education of Eroles, Richard and Francis Edgeworth

In November 1855, Antonio Eroles Edgeworth entered Trinity College, Dublin. He was the first of three Edgeworths – Eroles, Richard and Francis – to study there, and he would become quite popular among their classmates. He was assigned Mr Salmon as his official tutor.[143] Eroles started the Edgeworth college tradition of preparing for the subjects without attending most of the lectures. They got excellent marks but sometimes had trouble with their tutors.

Meanwhile, as early as 1856, there had been a change of tutors for Richard and Francis back in Edgeworthstown. According to the Family Report, 'early in 1856 Mr Potterton was succeeded by Mr Floyd. Mamma 57 – Floyd and R[ichard] and F[rancis were] left at Edgeworthstown'.[144]

After having to stay at Edgeworthstown with Mr Floyd during all of

1856 and the first half of 1857, 'Richard and Frank went to Trim in the autumn of 57 and again in the Summer of 58'. In Trim, at the Vicarage, Dr Butler and Aunt Harriet saw to their education.[145] Richard Butler, the Vicar of the Anglican church of Trim and Dean of Clonmacnoise, was an excellent preacher and teacher at the Diocesan school where Hamilton had studied. With his wife Harriet, they made an extremely enlightened couple and kept an excellent library of classical authors. There is no doubt that the atmosphere in Trim, and especially Aunt Harriet, had a remarkable influence on Francis's intellectual development. He remained close to her all her life, while she in turn strove to keep track of all the family contributions to science, literature and poetry.

News from the Aunts and Uncles

After returning from Fiesole in 1843, Aunt Mariquita settled in Clewer, near Windsor, Berkshire, where she was known as 'Mrs. Tennant, the Spanish widow of a clergyman'. There, starting in June 1849, she became involved in sheltering rejected women in her home, 'to rescue fallen women from evil'. When this activity surpassed her own means, Mariquita asked the rector of St Andrew's Church, the controversial Thomas Carter, to help her. He contributed decisively by founding the House of Mercy. Mariquita Tennant was the first keeper of the house, and she befriended the man who was later to be British Prime Minister, William Ewart Gladstone, who was also interested in the fate of these outcast women. However, in February of 1851 she quit out of exhaustion and settled in Windsor.[146] Rosa went with Eroles and Mary to stay with Mariquita in the spring of 1857 because their sister and aunt was in critical condition. Despite the doctor's prognostications – 'the doctor has no hopes of Mrs. Tennant's recovery' – Mariquita recovered in less than a month.

In the meantime, in 1857 the health of Honora and her husband Rear Admiral Beaufort was deteriorating, and in October the latter died. Pakenham Edgeworth wrote to his sister Harriet that the day before he had received 'a most excellent letter from him so kind & affectionate & considerate & so exquisitely written'.[147] Sir Francis Beaufort wrote about two thousand letters as well as several journals. Portions of them were written in cipher, and when the code was cracked for biographical purposes,[148] they revealed an incestuous relationship lasting over three years – from 1835 to 1837 – with one of his sisters who, after he became a widower, kept his house until his second marriage. The journals were not decoded until 1973; therefore Francis Ysidro Edgeworth would never know that his much-admired uncle, a shining model for all scientists, had behaved so improperly.

The death of her husband magnified Honora's dislike of most of her relatives who tried to console her, especially Lestock Wilson and Rosa Edgeworth, and she became rather aggressive to them. After four months of family uneasiness, Honora died rather suddenly in February 1858.[149]

Another passing that concerned Rosa and her children was the death in June 1858 of her friend the general surgeon Sir Philip Crampton at the age of eighty-one. Crampton's death was a blow for Rosa, since her only friends were Crampton and Hamilton. On the other hand, she was constantly worried by her children's health and she was terribly anxious for news from her son William.

Anxiety about the Boys

When Pakenham and Christina Edgeworth left India at the end of 1855, William Edgeworth remained there on his own. He finally returned to the Army in 1856 and had a comfortable position as second lieutenant of the Fifth Lancers.[150] The revolt against the British started in 1857. William was posted to Lahore, and his mother and grandmother grew anxious about him because William was not writing. Rosa's anxiety was alleviated when she finally received a 'most satisfactory' letter in early October.[151] But then came the seizure of Delhi, and two of Pakenham Edgeworth's friends died. Fortunately, William escaped the slaughter.

Later in November, news came about another 'frightful carnage' in Delhi. This time, William Edgeworth was 'seriously wounded'.[152] The news about William's recovery from the wounds arrived slowly during December. Rosa received a letter from him 'written in his own right hand'.[153]

Moreover, Eroles, who was Rosa's main support, had been ill in November and Richard had had an accident. When Pakenham, Christina and Harrie went to Edgeworthstown to spend the 1857 Christmas holidays, they found Rosa 'decidedly ill' and Richard 'pulled down by his accident'. The good news was that 'Frank is much inspired and strengthened up' and 'David had got fees for his work on the rail'.[154] David had entered the Royal Navy as a midshipman in the summer of 1855 when he was just thirteen years old, having got a first on the entrance examination. After one and a half years partly spent aboard HMS *Firebrand* behind the front lines of the Crimean war, he left the Navy 'disliking the monotony of that life'. Then, at the end of 1857, he became a pupil of Mr G.W. Hemans (Vice-President of the Institution of Civil Engineers) in Dublin and trained as an engineer on several railway lines. When this tutelage finished in 1860, he was employed by Mr Hemans.[155]

William Edgeworth's Triumph

Fortunately, during the first half of 1858 everybody recovered their health and William came back triumphantly from India on leave. In July, the tenants of Edgeworthstown had the idea of paying homage to their hero of the Delhi battle, Lieutenant William Edgeworth, who was the first in line to become their landlord after Charles Sneyd Edgeworth. A big party was held in Edgeworthstown on 23 July 1858, in which the tenants offered William a sword. The party was hosted by Eroles, David, Richard and Francis Edgeworth, helped by Mr Floyd. Frances Anne, Rosa, Mary and Charles Sneyd Edgeworth, Mariquita Tennant, Rev. Butler and his wife Harriet, Sir William Rowan Hamilton[156] and his son William Edwin, Lord and Lady Essex Edgeworth (cousins from the Firmont branch of the Abbé Edgeworth) and a host of tenants, were present at the celebration. Pakenham and Christina Edgeworth were unable to attend the party, because they were on Eigg, an island located off western Scotland owned by the MacPhersons.[157]

Through a letter from Harriet Butler we know that at the party:

> everybody as they came in [was] received by my mother [Frances Anne]. Rosa did not at all put herself forward . . . the only person who showed any agitation was Frank who stood beside the greenhouse door corner, the whole time in the greatest anxiety. Eroles was excessively civil. Frank and Dick were carvers and David and Mr. Floyd roused out the beasts – saddles of mutton of portentous size.

We have no clue about the specific cause of this agitation of the 13-year-old Francis observed by his aunt Harriet.[158]

The second half of 1858 was the last placid period of Rosa's life. Her children were near her and everybody was in good health, even William, who was recovering from his injuries. After the anxieties of the past years, she could finally enjoy her life.

Then, in January 1859 William Edgeworth was sent back to his regiment – the Fifth Lancers – in India as first lieutenant. During this time, he wrote 'regularly and pleasantly from Kolapore' [Kolhapur, about 200 miles southeast of Bombay] and he only complains about 'the difficulties in sporting nowadays made by the regimental regulation'.[159]

Boys' Health Problems and Educational Track

In the spring of 1859, David contracted the measles and was recovering at Edgeworthstown.[160] There, in early 1859, Richard and Francis were given a new tutor, Mr Elms.[161] The former tutor, Mr Floyd, was kept in the Edgeworths' service as a right-hand man for Sneyd and Rosa. The arrival of Mr Elms could not have been less fortunate, since Francis fell ill, most likely with the measles as well, though it is only reported that 'he was very unwell and had difficulty of speech and swallow'.[162]

During the 1859 Christmas holidays, Richard became gravely ill, and he had a painfully slow recovery during the spring of 1860. Since Richard's illness might have been infectious, Rosa and Frances Anne decided that Eroles, David and Francis should stay in Trim at the Butlers' home.[163] So Butler and his wife Harriet once again acted as their tutors. Harriet had had some contact with Hamilton, also a good friend and former pupil of her husband, through which she obtained copies of all the poems that the astronomer had received from Francis Beaufort Edgeworth.

Around this time, on 21 February 1860, Aunt Mariquita died unexpectedly in Windsor while her sister Rosa was at Edgeworthstown. Mariquita Tennant was buried in the graveyard of St Andrew's Church, in Clewer, Berkshire.[164]

In the spring of 1860, the boys went back to Edgeworthstown. Richard and Francis had a new tutor, Mr Hobart, while Eroles obtained a Classics scholarship at Trinity College, Dublin.[165] Eroles was doing acceptably at Trinity, in spite of bouts of absences. The registers kept at Trinity College, Dublin, show that Antonio Eroles Edgeworth's performance as an undergraduate, from the day he entered in November 1855 until his final exam to earn his BA at the end of 1860, was quite irregular.

In August of 1860, Pakenham, Christina and Harrie Edgeworth went back to Edgeworthstown. There, they found that 'Dick looks better than he did when we left but does not of course feel robust. Eroles has turned extremely handsome. Frank has outgrown them all.'[166] Those were the times when, according to their cousin Harriet Jessie Butler, Dick and Frank:

> were full of queer devices, one being the culture of wasps. They dug up their nests (which abounded) and planted two or three of them together in the front lawn quite near the library window . . . Thence the wasps naturally spread into the house and frequently stung people, while they kept me in perpetual terror, which the boys naturally liked to work upon.[167]

Francis would remain interested in wasps and bees all his life.

William Edgeworth's Controversial Conduct

A few months after William had returned to India, his Army officers accused him of participating in a heavy gambling affair. Though he denied all the charges, William Edgeworth was sent back from Kolhapur to Berkshire, England, in May 1859, and an official investigation was opened. His mother visited him there in early June.[168] Sneyd had a conversation with him and he promised never to gamble again. It was agreed that Mr Floyd would accompany him whenever he was off duty.[169]

During the summer of 1860, William testified before the appointed Commission. He again denied all the charges and Charles Sneyd Edgeworth believed him.[170] Then, at the end of November 1860, the Army's judicial process reached the conclusion that, since there was not enough proof, William was acquitted with all his honours restored, and consequently had to be readmitted to his regiment. Everybody was relieved. However, the situation changed radically when Pakenham received a letter written by a certain Mr Robinson from India, painting 'the most unfavourable view of that youth character'.[171]

At that time, early January 1861, Rosa was worried about Francis's health, though she was soon informed that the 'worst is over'.[172] In February, after Rosa was apprised of the contents of Mr Robinson's letter, she was in such downcast spirits that she became 'greatly alarmed about herself'. However, she soon bounced back, although deep wounds remained.[173]

Meanwhile, in March 1861 William Edgeworth was officially readmitted into his regiment. However, this good news could not stop the deterioration of Rosa's health. She was reported by Mr Floyd to have 'violent pain in her back and throat and is very low'.[174]

Rosa's only consolation during the month of May 1861 was that her friend Sir William Hamilton and Dr Robinson, the husband of her sister-in-law Lucy Jane, were named Doctor Honoris Causa of Cambridge University in a joint session shared with several other remarkable men. The ceremony took place on 21 May, and Mary Edgeworth, who was spending a few days with Christina and Pakenham in Anerly, was present at the session, as noted by Hamilton's biographer.[175]

In the early summer of 1861, Pakenham and Christina travelled to Edgeworthstown. To their consternation, they realised that Eroles was backing William. So, they extended their resentment to Eroles as well and, to a lesser extent, to David. The only boys saved were Richard and Francis, who were 'exceedingly kind to and happy with the children [Harriet and her friend Annie] and have nothing other to distract their minds and doat upon croquet playing with them'.[176] Later on, Harriet Jessie Butler wrote about those summer days at Edgeworthstown that 'the young men teased me dreadfully' but 'Dick and Frank were always kind to us'.[177]

In September, William applied for a vacant position as captain in another regiment. The advancement in rank, promoting William Edgeworth to captain of The Third Dragoon Guards, was reported in the Gazette dated 9 October 1861. Nonetheless, the next week Sneyd had a heated argument with William and 'lost all command'[178] and Pakenham fervently hoped that he would be 'soon off to Bombay'.[179] However, he would remain in England for several months yet. In May 1862, Sneyd and Pakenham discovered through a report by Mr Floyd that William was drinking heavily, and they banned him from going home; he was not allowed to mix with his brothers and sister.

After they found out that William was also a heavy drinker, Pakenham and Charles Sneyd Edgeworth devised a plan to limit the effects of a second Lovell as Edgeworthstown's Esquire. The plan was hatched to warrant money to Eroles, David, Richard and Francis, just like those provided by Sneyd thus far because they suspected that William would be incapable of providing the necessary revenues for them. Since Rosa had no money whatsoever, the solution was to convince Frances Anne, who had a rather large fortune inherited mainly from Maria Edgeworth, to prepare a will that would ensure a decent living for her grandchildren. The agreement took place after 'chasing away all misunderstanding' with Sneyd and Pakenham.[180] Finally, at the close of March 1863, Frances Anne wrote to her son Pakenham that the deed had been signed on 17 March.[181]

However, now it is time for us to go back to the summer of 1861 and focus on Richard and Francis.

1.13 1861–65: GLOOMY TRINITY TIMES

In July 1861, Richard Lestock Edgeworth and Ysidro Francis Edgeworth entered Trinity College Dublin as pensioners. On 1 July, both brothers passed the entrance examinations. Francis's results were spectacular, as can be seen below in Appendix B.[182] Besides being first-ranked, Francis won the composition prizes in Greek verse and Greek prose. He was assigned Dr John Kells Ingram as his official tutor. Richard's results were not as spectacular as his younger brother's. He earned the 20th place, and he was assigned Dr George Salmon as his tutor, just as Eroles had been six years before. Both Richard and Francis were registered as having been educated by 'private tutors'. In fact, since the tutors in Edgeworthstown changed quite regularly, we should give the Butlers the most credit for Francis's and Richard's entrance results since both brothers had been at the Trim vicarage in the autumn term of 1857 and the spring term of 1858. Additionally, Francis was there once again with David and Eroles in the spring of 1860 while Richard was ill. When he and Francis entered Trinity, the era of the Trim vicarage was ending: on 17 July 1862, Richard Butler died, whereupon Aunt Harriet went to live at Edgeworthstown to assist her mother, Frances Anne, who was 93 years old.

After entering Trinity College in July 1861, Ysidro Francis Edgeworth is not mentioned on any list and did not pass any examination during the Michaelmas Term of 1861. As we have seen above, he had been gravely ill in January 1861, and the Family Report also mentions that 'Francis stayed part of next spring at Edgeworthstown ill with jaundice (1862)'. It is likely that he contracted jaundice some months earlier since sometimes the report makes mistakes in the seasons. Thus, we may infer that he was in poor health and could not attend the examinations at the end of autumn 1861. Meanwhile, Richard passed the Michaelmas exams with a 'second'.

During next term, the Hilary Term of 1862, Ysidro Francis Edgeworth registered as a Junior Freshman Pensioner and took the exams in the Trinity Term, which he passed with a 'first', and the right to sit for the exams for honours in Mathematics and Classics.

On the other hand, in 1862 Eroles managed to win a scholarship as a graduate student at Trinity College. He went to America that summer and came back at the end of the year to attend to his academic duties as a scholar.[183] Eroles's scholarship led to a reunion of the three brothers at Trinity.

In the Michaelmas Term of 1862, Francis, who was checked by the physician in early October and 'pronounced . . . quite well',[184] got third place in the exams for honours in both Maths and Classics, and passed with a 'first' the exams of his Junior Freshman class. In the meantime, Richard passed his Senior Freshman exams with a 'second'.

In the Hilary Term of 1863, Ysidro Francis Edgeworth booked as a Senior Freshman Pensioner and through the lecture attendance lists we find that he did not attend many of them. Of the Greek lectures of Professor J. Stack he attended just two classes out of 13; his level in Greek probably far exceeded the level of the lectures. In the Senior Freshman class exams, he got honours in Greek, Latin and Composition. In the next term, Francis, as a Junior Sophister Pensioner, won a scholarship in Classics. Richard Lestock Edgeworth is also mentioned in the registers of both terms as Junior Sophister Pensioner in the medical school, meaning that Richard chose Medicine as his subject for his years as Sophister.

William Edgeworth's Death

The growing concern of the three Edgeworth brothers about their careers was suddenly interrupted at the beginning of the long vacation of 1863 when news arrived from Bombay of the death of their brother William. After his departure to India – we only know that he left before Christmas of 1862 – his name was rarely mentioned by Sneyd, Pakenham, Christina or Harriet in their correspondence, and we have unearthed the cause of his death as cholera by looking at the *Register of the Burials at Back Bay from Saint Thomas Cathedral in Bombay in the year 1863*:

> [Died] 1863, June 29 . . . Cholera . . . William Edgeworth . . . 26 years
> Captain of H.M. 3rd Dragoon Guards . . . Buried 1863, June 30
> W.K.Fletcher, Jr., Chaplain.[185]

In spite of the gloomy atmosphere at Edgeworthstown during the summer, when Francis went back to Trinity for the Michaelmas Term of 1863 as a Junior Sophister Scholar, he earned his best results. In the examination for honours in Classics he got first place. In the Hilary Term of 1864, Francis attended most of the lectures. Apparently, as a scholar he felt more obliged to attend them. In the examination for honours, he again got first place in Classics. In the Junior Sophisters' class exam, Francis got lower results, with only honours in Greek and Latin. This tendency continued in the next term when he got just third place in the honours exam for Classics.

Death of Uncle Sneyd and Mother Rosa

During the Easter vacation, Eroles had to leave Ireland to attend the funeral of his uncle Sneyd, who had died on 31 March 1864 in Pakenham's house, where he had moved some days before.[186] Antonio Eroles Edgeworth was now the new landlord of Edgeworthstown, and Pakenham could not contain his suspicions about him:

> [Sneyd] died most peacefully after hours of suffering at about 7 and a half . . . Eroles made his appearance suddenly. I did not know him. He showed no good feeling, but nothing purely [apparent?]. I spoke to him to give him a [. . .hand?]. He said he would pay his brothers.[187]

Some days before Charles Sneyd Edgeworth's death, his stepmother Frances Anne had suffered symptoms of 'a failure in the circulation', but she recovered and, with the aid of her daughter Harriet, was able to visit Rosa in her room. Rosa Edgeworth was racked by 'dreadful sufferings' only soothed by opiates and other drugs.[188] In spite of their mother's terminal condition, Richard and Francis returned to Trinity College, where Francis took the examination for honours in Classics and got third place.

On 18 June 1864, nearly a year after William's death, Rosa Edgeworth died aged 49 years. After her death, Richard, who had inherited his father's poetic bent, wrote this poem to her:

> Oh mother, mother – like the softened swell
> of music in the melancholy night

Thy form arises that I loved so well
Thy mellow form of summer & delight
Now robed in robes of amaranthine white
Thou seem'd to smile over me thy sweet alarms
And oft thou did it when I, in childish fright,
Lay folded in thine everlasting arms
And cradled in the trembling rapture of thy charms.

Memories from Trinity

Despite his extreme sensibility, Richard Lestock Edgeworth became popular in Trinity College Dublin for his eccentricities. Alfred Percival Graves, who entered Trinity in 1863, described two of them. The first time that Richard 'obtained notoriety in the Medical School' was when he made 'a bet that he would carry a coffin round the College Park'. He won because 'he was not interrupted by any of the College officials'. Richard's second remarkable eccentricity was an impersonation. 'A great college function was due, to grant honorary degrees, and only a limited number of tickets were issued' since they were reserved for the Scholars. Both Eroles and Francis had the right to claim a ticket, but Eroles was absent:

> So Richard disguised himself with an artificial beard as Antonio [Eroles], and called upon Mr. Miller, the Registrar . . . and chatting with apparent familiarity over his experiences in the American Civil War in the course of which Antonio had acted as aide-de-camp to General Lee at the battle of Chancellorsville, Miller listened with great interest and handed him the ticket without demur.

Graves also describes an anecdote about Francis in Trinity:

> The only actual Town and Gown row that I recall was precipitated by the vagaries of a mad barrister of the name of Barnes, a great humorist with an astonishing gift of the gab, who was professing to stand for a Dublin constituency. He invaded Trinity College, and spoke in one of the quadrangles to a crowd of students, who in sheer fun cheered him to the echo and finally hoisted him on to the pedestal in front of the college gate shortly to be occupied by the statue of Burke. We then took him down and carried him outside, upon which the mob foul of us, and we of the mob; Francis Edgeworth, with whom I was walking arm-in-arm, was badly mauled in one of the charges against us.[189]

Last Contacts with Hamilton and Death of Grandmother Frances Anne

Eroles and David became involved in and worried about the execution of Charles Sneyd Edgeworth's will. Pakenham mistrusted both of them. David was working as a civil engineer on railways and earning a wage. Nevertheless, as we saw, Pakenham considered him 'a difficult one'.[190]

David was trying to become a sound professional and was keeping up with the technical advances in his field. In this sense, there is an interesting letter from January 1862 addressed to Sir William Rowan Hamilton, in which David asked him about a new engine invented in France. Hamilton replied to him a month later with a long letter setting out his latest theory on mechanics in 23 points that made use of differential equations and polar coordinates, which were then viewed as quite sophisticated techniques. These mathematical techniques would be used 15 years later by Francis.[191]

After Rosa's death, Hamilton wrote letters of condolence and later sent some of his recent work – which would turn out to be unexpectedly useful to Francis in the 1870s – to Frances Anne Edgeworth and Harriet Butler; she politely replied:

> I shall very gratefully accept your offer of letting Francis repossess whatever papers you may find, on the condition of copying them for you. . . . I remember the lines you sent to Rosa some years ago which she was much touched by. . . . I am extremely obliged to you for thinking of me and sending to mother [F.A.E.] rich Bill of Fare of your work . . . including a copy of the original *Essay on Quaternions.*[192]

So Francis, not yet 20 years old, was interested in reconstructing the documents containing the philosophical discussions between Hamilton and his father. This interest explains Francis's visit to Hamilton at the Dunsink Observatory in November 1864. As Hamilton reported to his colleague Dr Robinson, 'we all enjoyed much a recent visit from Francis Edgeworth – who permits me in conversation to call him "Frank" – for me there can be no second "Francis"'.[193]

Through a letter sent by Francis to Hamilton, we learn that he and Richard lived for a while in room 38 at the college,[194] at least before Eroles came back to Trinity as a graduate scholar. This confirms the impression given by Alfred Percival Graves's account that the Edgeworths lived fully the Trinity College atmosphere whenever their health and family problems permitted.

Now, let us take a brief look back at the data in the records. In the Michaelmas Term of 1864 Ysidro Francis Edgeworth ranked second in the honours' exam in Classics and also earned honours in Classics in the Junior Sophister Class exams; however, in the other subjects he got very low qualifications. The same happened to Richard, who did not pass his Bachelor's Degree exam.

In December of 1864, Antonio Eroles Edgeworth obtained his Magister degree at Trinity. Thus, he could add the Master of Arts initials after the Esq. as Esquire of Edgeworthstown. He was just in time because on 10 February 1865, his grandmother Frances Anne died, which left him deeply involved in much less academic affairs. She died peacefully after a fit of apoplexy on the first day of February in which she lost the power of speech, a fit that could have been related to the recent death of her dear sister, Harriet Beaufort.

In her funeral, a brief description of her life was distributed that is reproduced below as Appendix C. It is a rather interesting document which confirms most of the information about the Beaufort family.[195]

During the Hilary Term of 1865, Francis had not yet recovered his customary standards at Trinity, where he appears in the registers as a Senior Sophister Scholar. He was fourth in the honours' exam in Ethics and came within an inch of failing his class examination. He got honours only in Greek. In the next term, Richard Lestock Edgeworth appears as a Candidate Medical Bachelor and obtained his degree on 24 April.

The next blow for Francis was the death of Sir William Rowan Hamilton at just 60 years of age on 2 September 1865. In spite of this, Ysidro Francis Edgeworth, as a Candidate Bachelor Scholar, summoned all his powers and in the Michaelmas Term passed the final exams, in which he got his Bachelor of Arts degree as Number One, with honours in Physics, Greek, Composition and Latin.[196] These excellent results led him to be offered a Hibernian scholarship to study at the University of Oxford.

1.14 INFLUENCES ON EDGEWORTH IN TRINITY

In the years to come, Francis Edgeworth would refer to his Trinity and Oxford studies in applications for academic positions and detailed his achievements according to the specific field of each application. As a result, in these documents his Trinity studies appear less prominent than those at Oxford. It may be that this is why most of Edgeworth's biographers have understated his Trinity education.[197] What do we know about the professors and tutors at Trinity who most influenced Francis? By examining the testimonials that Francis Edgeworth asked for when he was candidate for a Greek Professorship in 1875, we infer that he had a good relationship with J.K. Ingram, his official tutor, with the young Fellow J.P. Mahaffy, his personal tutor during several terms, and with A. Palmer and R.Y. Tyrrell, who were just one or two courses ahead in Classics and became fellows at Trinity.[198]

John Kells Ingram

Ingram's influence on Edgeworth may have proved an unexpected challenge to him: to show that Ingram was wrong when he condemned the use of mathematics in the social sciences. Although Ingram was highly trained in mathematics, he wrote in the 'Political Economy' entry of the *Encyclopædia Britannica* (1880): 'The great objection to the use of mathematics in economic reasoning is that it is necessarily sterile . . . There is no future for this kind of study, and it is only a waste of intellectual power to pursue it . . . such researches . . . will in fact never be anything more than academic playthings.'[199]

The testimonial that Ingram wrote about Edgeworth at his request is expressed in a conventional style:

> I was Mr. Edgeworth's tutor during his highly distinguished courses in Trinity College Dublin, some years ago, and can therefore speak with confidence of his qualifications. He is a man of remarkable ability and when he was my pupil, was already an Excellent Classical Scholar. Since then he has pursued his studies.[200]

John Pentland Mahaffy

In contrast, Mahaffy's testimonial looks much more personal:

> He was very intimately known to me, as his College Tutor for several years in Dublin. During that time he always maintained the very highest character for ability, for learning and for high pure morals. He was considered the very ablest man in his class, and made himself deservedly popular by his genius as well as his kindliness.[201]

Mahaffy (1839–1919) had a bright career that culminated as Provost of Trinity Collage Dublin and he produced a remarkable collection of books such as *Twelve Lectures on Primitive Civilisation* (1868), *Prolegomena to Ancient History* (1871), *Kant's Critical Philosophy for English Readers* (1871), *Social Life of Greece from Homer to Menander* (1874), *Rambles and Studies in Greece* (1876), *Descartes* (1880) and *The Art of Conversation* (1887). Yet he is perhaps better known as Oscar Wilde's tutor. We may have a better glimpse of Mahaffy through Oscar Wilde himself. He wrote to his friend Frank Harris:

> I got my love of the Greek idea and my intimate knowledge of the language at Trinity from Mahaffy and Tyrrell; they were Trinity to me. Mahaffy was especially valuable to me at that time (1872–73). Though he was not as good a scholar as Tyrrell, he had been to Greece, had lived there and saturated himself with Greek thought and Greek feeling. . . . He was a delightful talker, too, a really great talker in a certain way – an artist in vivid words and eloquent pauses.[202]

Francis was possibly influenced by Mahaffy's witty and ironic language, though not to the extreme of Oscar Wilde, as we shall see when we come to his reviews. Some of Edgeworth's sentences are comparable to Mahaffy's most mordant dictums. Two of Mahaffy's most famous ones are: 'Joyce is the living embodiment of my conviction that the native Irish are unfitted for education' and 'I always said we should never have given a university to the aboriginals of this island'.

Robert Yelverton Tyrrell

Tyrrell's testimonial for Edgeworth is also unconventional:

> He obtained all the highest honours and prizes . . . In Trinity College, there are two verdicts pronounced on every student of eminence; one that of the Lecturers and Examiners, the other that of the élite of students. The latter verdict was pronounced in favour of Mr. Edgeworth with an enthusiasm to which I can remember no parallel.[203]

Tyrrell (1844–1914), a remarkable scholar, would become professor of Latin at Trinity College Dublin in 1871, Greek in 1880 and Ancient History in 1900. In conjunction with Purser, he edited *Correspondence of Cicero*.

Arthur Palmer

Last but not least, Palmer's testimonial is interesting in that it emphasises the combination in which Edgeworth excelled after Trinity and Oxford:

> I have known Mr. F.Y. Edgeworth for more than twelve years. . . . rare abilities, great industry . . . He highly distinguished himself both in Classics and Mathematics in the University of Dublin. . . . He is indeed a very rare example of a combination of the highest mathematical and literary attainment.[204]

Therefore, we see that Edgeworth's personality ripened considerably at Trinity. We may even glimpse some of Francis's traits in the description that Mahaffy makes of the Trinity College prototype:

> Trinity College men . . . may not be as a rule, as polished as an Oxford or Cambridge man, but there is a rough strength about them that atones for other deficiencies. As Irishmen, they are fluent talkers, and as Trinity College men they are independent talkers, free to utter their opinions, not guided by precedent, differing readily, even from their teachers. A man is judged by his conversation, by his ability to take in new ideas, by a thousand things which cannot be enumerated, but which are taken as evidence against all artificial tests.[205]

1.15 FAMILY DISPUTES: GRANDMOTHER BEAUFORT'S INHERITANCE

Harriet Jessie Butler wrote the following recollections many years after the fact, probably around 1938:

> My grandmother died in February 1865, and then Eroles was the heir to Edgeworthstown and my father found him impossible to deal with. My father had with needless generosity paid him a large legacy from Uncle Sneyd's estate which ought not to have been paid, and this led to legal proceedings and many disputes, of which I only know that, after two or three years of legal controversies, the final verdict went against my father, and Eroles and David remained triumphant and entirely unfriendly.[206]

We can perceive the influence exerted by her parents Pakenham and Christina upon her. Once back from India, Pakenham had to fight against his natural propensity to feel jealous of the future heir of Edgeworthstown. He tried generously to repress his jealousy as best as he could, but he was not always successful. As we have seen, he took a belligerent attitude towards William Edgeworth, which he then extended to Eroles and David even before William's death. And, as we have seen above, he had convinced his mother, Frances Anne, to sign a deed to exclude the heir from any benefit in her will.

In September 1864, Frances Anne Edgeworth had amended her will after the deaths of William, Sneyd and Rosa. After her own death in February 1865, the will was made public. The main heirs were her three younger grandsons, David, Richard and Francis. Her granddaughter Mary got some stock and her children Harriet, Lucy Jane and Pakenham some furniture. Pakenham also inherited the silver. The will is reproduced below as Appendix D.[207]

This document would become essential for Francis's financial independence in the years to come. The three executors were confirmed, but according to Christina, David contested Pakenham's appointment as executor.[208] Moreover, Mary, who of all of Rosa's children was the closest to Pakenham, quarrelled with her brothers Eroles and David.[209]

Meanwhile, David used part of his inheritance to establish an engineering firm, and both Eroles and Francis helped him financially. In January 1866 he wrote to his Aunt Lucy Jane from 46 Upper Sackville Street, Dublin, offering the services of the engineering firm Edgeworth and Sandford.[210]

After being occupied with the legal battle over the personal estates of Frances Anne and Charles Sneyd Edgeworth, with David living in

Dublin and Francis having the chance to go to Oxford as a Hibernian Scholar, Eroles decided to close the Edgeworthstown Manor House. He went to live in France as an absentee landlord, probably around 1866. Aunt Harriet found accommodation at her sister's home at the Armagh Observatory, where Dr Robinson was still living.

Throughout this period, Pakenham Edgeworth maintained his resentment of Eroles and David.[211] This resentment was conveyed to his daughter Harrie, who wrote in her 1938 report about Eroles Edgeworth:

> Eroles had lost too much money to remain at Edgeworthstown, if he had wished to do so, and he lived in Paris and elsewhere, gambling continually. We saw him in Paris in 1866–67, when my father tried to be friendly. Eroles looked thoroughly dissipated, and we naturally saw no more of him. He was in Paris during the Commune [1871], and was made by them to carry ammunition or something in the streets. Then he lived on in different places gambling. Maxwell Fox [Sophy's eldest son] took his sick wife [Florence Jane *née* Buchanan] to Pau, and Eroles was there gambling heavily.[212]

Thus, in the family tradition, Antonio Eroles Edgeworth was momentarily at the very centre of history, living in the streets of Paris in 1871, the days of the Commune. They were the same streets that had seen the Abbé Edgeworth accompanying Louis XVI on his way to the scaffold and his uncle Lovell Edgeworth come back from Verdun as a prisoner of war.

1.16 1867–69: FRANCIS EDGEWORTH AT OXFORD

Francis went to Oxford as Hibernian Scholar at the beginning of 1867. Let us look at his record in the Balliol College Register book (1833/1933):

> EDGEWORTH, Prof. Francis Ysidro, born Feb. 8, 1845, 5th son [*sic*, he was the 6th son] of Francis Beaufort Edgeworth of Edgeworthstown, Ireland. Education: Private; Trinity College Dublin (Scholar), B.A., first class honours; Exeter, 1867; Magdalen Hall (Scholar), 1867; Balliol Hib., 1868–69; BA (1873); MA, 1877; Barrister Inner Temple, 1877.

From the record we can infer that he had some trouble getting a place that suited him, since he spent part of 1867 in Exeter College but moved within that year to Magdalen Hall, only to change again in 1868 to Balliol, where he remained until his final exam in Literæ Humaniores during the

Michaelmas Term of 1869. In fact, he did not impress his tutors in the way he had done at Trinity College Dublin.[213]

Benjamin Jowett

It seems that Francis Edgeworth's most important influence in Oxford by far was his tutor at Balliol College, Benjamin Jowett (1817–1893). After becoming a tutor at Balliol and a clergyman in 1842, Jowett was appointed to the Greek Professorship in 1855. Thus, by the time Francis was at Balliol in 1868, Jowett had had extensive experience as a tutor, a job which suited him. Most of his pupils became his friends, and Edgeworth was no exception as we can see from their correspondence, which Francis kept his entire life. There are three letters in Nuffield College's files showing this friendship. In the first, dated 22 April (but no year, though it was most likely 1868), Jowett thanked him for the gift of some 'charming volumes' and for giving him the opportunity to meet his family, which was 'related to Maria Edgeworth herself'. In the second letter, dated 15 August (again no year, most likely also 1868), Jowett invited Edgeworth to go to his place, the Palleyard Cottage.[214] We shall refer to the third letter, the most interesting one of all, in the next section.

After being named Master of Balliol College in 1870, Jowett became famous for his proverbs. There are two among them that fit perfectly the attitude that Francis displayed after his Balliol experience: 'Never retreat, never explain. Get it done and let them howl' and 'The way to get things done is not to mind who gets the credit for doing them.'

Though Jowett advised Francis Edgeworth on many subjects besides Greek, among them ethics, philosophy and political economy, he was rather conventional in his testimonials as on 13 November 1875, when Edgeworth was applying for a Professorship of Greek at Bedford College:

> I have known Mr. Edgeworth for several years. He is a nephew of the celebrated Miss Edgeworth. He is a good scholar and obtained a First Class. I believe him also to be a good teacher and think him well fitted to be a Classical Professor.

In March 1887, Edgeworth applied for an Examinership of Political Economy at London University:

> I know Mr. F.Y. Edgeworth to be a man of ability and of considerable originality. He gained a first class in Literæ Humaniores in the year 1869. Since that

> time he has devoted himself to Mathematics and Logic and is the author of several curious investigations in both those branches of knowledge. I believe him to be quite competent to examine in Political Economy.[215]

We know for sure that Edgeworth's 'curious investigations' in mathematics and logic were not prompted by Jowett, nor were they – as we have seen – encouraged by Ingram. Edgeworth himself wrote much later on, in 1925, about Jowett:

> Jowett, for instance, as I can testify, much as he liked Marshall, disliked his mathematical apparatus. The authority of Jowett on the question of method was indeed not particularly great, for he had not realized that the use of curves and symbols does not imply the use of exact calculation. But Jowett was representative of cultivated opinion.[216]

From the testimonials that Francis requested afterwards we can infer that, besides his tutor Jowett, he also was on friendly terms with T.H. Green (1836–1882) and T. Fowler (1832–1904).

Thomas Hill Green

Green, who had obtained First Class Honours in Classics at Balliol in 1859 and was appointed Fellow there in 1860, became Balliol's first lay tutor in 1866. He held these positions when Francis was there; later, in 1877, he was named Whyte's Professor of Moral Philosophy. There are two testimonials signed by Green. On 18 February 1874, Edgeworth was applying for a Professorship of Greek at Bedford College:

> A diligent and successful student [at Balliol]. He [. . .mastered] a very wide range of information, wider than most men who, like him, obtain the highest classical honours at Oxford, and is at the same time a good writer.

On 7 October 1881, Edgeworth was applying for a Professorship of Logic, Mental and Moral Philosophy and Political Economy at the newly founded University College, Liverpool:

> In respect both of knowledge and talent, he is thoroughly competent to undertake the instruction of young men in Philosophy, and owing to a certain ingenuity and versatility and brightness of mind by which he is characterised in a high degree, I should expect him to prove a very attractive and effective teacher.[217]

Thomas Fowler

In Francis Edgeworth's times, Fowler was tutor at Lincoln College and examiner in the Literæ Humaniores Schools in Oxford. He later became Professor of Logic in 1873 and President of Corpus Christi in 1881. Here is his testimonial from 7 October 1881, when Edgeworth was applying for a Professorship of Logic, Mental and Moral Philosophy and Political Economy at University College, Liverpool:

> The very favourable impression I formed of your interest in, and knowledge of philosophy, so long ago when I examined you in the Literæ Humaniores Schools in Oxford, has been sustained whenever I have had the opportunity of talking with you on these subjects since that time.[218]

It may seem odd that Edgeworth trusted one of his examiners more than most of his tutors at the Oxford Colleges he attended. However, this may be explained by the fact that Francis's best performance in Oxford was his final exam in Literæ Humaniores, taken in the autumn of 1869. In 1926, Harriet Jessie Butler told John Maynard Keynes about it:

> There is a tradition in Oxford concerning his [Edgeworth's] *'Viva'* in the Final Schools. It is said that, being asked some abstruse question, he enquired, 'Shall I answer briefly, or at length?', and then spoke for half an hour in a manner which converted what was to be a Second Class into a First.[219]

What Keynes did not explain, because he did not know, was that after May 1869, Francis was like a broken toy. As had happened to him at Trinity before, he had to overcome his own dejected spirits to get such an excellent academic result. What was the cause of this dejection?

1.17 THE SHELLEYAN STORY OF RICHARD AND FRANCIS EDGEWORTH

In 1867, when Francis Edgeworth went to Oxford, his brother Richard decided to travel to the Continent. It was going to be the first long period of their life that they would be apart. In those voyaging years, Richard sent poems about love, friendship and life to Francis.[220] Let us have a look at some of their contents:

> 'Song' (excerpts)
>
> Oh me! I had a friend I loved so tenderly
> That in his face I scarce could bear to look

And when he talked to me – soft and slenderly
His fingers wound to mine – and like a brook
His delicious voice along my spirit shook
I scarce could speak – but arm in arm we wandered
Or lying in some ferny forest nook
We mingled there our souls while crimson onward
Per Elysian plains our thoughts flew as we fondled.

And he was beautiful beyond description –
with soft brown eyes – and luxury of hair –
And lips on which was writ the sweet inscription
Of purity of love – and his whole air
was something sunny wonderful and rare.
. . .
And he was like the echo of my soul
And he was like my life – but only dearer
and every thought I had, without control,
To him was known, his eyes reflecting clearer
than the words the linked charm that drew us nearer
And all our time we spent in dreams together
Each of the other's heart – the heart revered
Forever vying who from thought's far tether
The last and loveliest wreath of worded light could gather.

Oh more – oh more than brother!
My heart – Oh my own life's Herat –. This song
is silent – and no more to one another
Oh, never, never more our love can we discover.

Other parts of this long poem are apparently much lower in spirit:

Alas! those lips of life are silent now,
Those liquid eyes are sealed in silver sleep,
The kiss of death is on that pale white brow
The darkness of the night doth round him creep,
That beauty – on which women hung with deep
And awful love – is motionless and cold.
But still for thee, my bosom-friend I weep
Still lingering by the moss-grown stones which hold
Thy form I loved so well – I ne'er shall more behold.

And thou art gone – but I remain –. Relief
I may not find – and though I may not smother
My sweet youth pines away – but not the grief
I feel for one I may no more recover –
Ah me! no more to clasp my heart's own brother
No more with thee to walk through life along
And never more to thee my love discover
And yet I live – and getting days prolong
Though silent is your love – and silent sleeps thy song.

Richard sent the poem to his brother possibly on the occasion of Francis's 23rd birthday on 8 February 1868, along with a bouquet of violets, since in F.Y. Edgeworth's personal files there is a copy of the following ironic poem written by Francis at Oxford on that date and signed Y.F.E.:

'Roses for a matron's head
Pansies for a brided bead
Violets for a maiden dead.'
Violets dost thou send to me?
Sacred to a maiden's fate
Symbol of the maiden state,
Violets inviolate,
Dost thou violets send to me?

Thoughtest those the Old Year fled
Was to me a maiden dead
When in some far flower-bed
Violets there culled'st for me?
Barren of achievement high
Dead bride without a sigh,
Cold and buried, let her lie
In oblivion's apathy.

Or have I thy thought misread,
And in Fancy's flower bed
Some coy violet lie, hid,
Which thou bidd'st me see?
Howsoever Shelley sings,
Violets tell me happier things,
The returning violet brings
Spring and thy return to me.

And, in wandering among
Flowers of that familiar song,
Which imbued our spirits young
With the *deus* of poesy.
(Shelley's gleam, and Byron's glow
Kindling us or wild Rousseau,
he, whose solemn sentence slow
Builds an Imperial History.

Thus Ideal Beauty wooing,
Charms of Classical pace pursuing,
or the chains of creeds undoing,
Still unwearied wandered we.)
Thoughts of thee my breast came over,
And sweet hopes that evermore
Still together we might soar
O'er life's troubled sea.[221]

In fact, Francis was parodying 'Remembrance', a poem by Percy Bysshe Shelley in which violets are associated with a dead maiden in the last stanza:

Lilies for a bridal bed
Roses for a matron's head
Violets for a maiden dead
Pansies let my flowers be.

In July of 1868, from Styria, Richard Lestock Edgeworth sent his brother Francis a long, apparently more impersonal poem called 'Onora'.[222]

Excerpt from 'Onora'

The hope of an eternal song, forever
As with her speechless charm the evening star
My soul inspires – and tho' I know that never
I may approach that heaven of thought afar
Mere crowned with vital bay the Deathless are.
Yet I must sing for ceaselessly I hear
In wind and wave – in life's discordant jar
And in the dirge of each departing year
A god-voice youngly whispering still to persevere.
. . .
Oh weep – oh weep my heart – big bloody tears
Tears for the happiness forever flown
Tears for the love that withered with the years
Tears for the friend too fondly, fondly known
Oh weep – and could thy bitter sweat atone
For deeds once done – for words unkindly said
when crimson bannered youth was all thine own
Thou still went happy – but the past is dead
And on thy core the rime of frozen hope is spread.

In the first excerpt, Richard ironically expressed his regret at Francis's decision to study at Oxford, since 'I know that never I may approach that heaven of thought afar; mere crowned with vital bay the Deathless are.' But it looks to him that by persevering he hoped to redress the situation. In the second excerpt, the poet quite desperately showed his remorse at 'words unkindly said'.

In his brotherly relationship with Richard, Francis had apparently adopted an attitude of submission; from childhood he had always followed Richard's lead, and they had even run away from home twice before Francis was eight. In both cases they left home to pursue their mother, who had left Edgeworthstown for a brief period. Both Francis and Richard adored their mother, and Francis seems to have inherited his predisposition to submission from her. Rosa Florentina Edgeworth was always the patient wife who was wholly devoted to her husband and children, always concerned with their health, education and professional careers, and always placing duty before the promptings of her heart. Francis, too, had an uncompromising sense of duty, to such an extent that sometimes he subdued his naturally dominant temper to enjoy the advantages of submission to affective authority. Let us recall that when he was just six, his mother, Rosa, wrote about him that 'his affections are deep, his character commanding from the ascendancy he acquires over those he comes in contact with. He is guided by a sense of duty peculiar in so young a boy'.[223]

Let us now go back to the story of the two brothers during the summer of 1868. From Styria, Richard went to Naples and continued writing poems during his travels through Italy. He was accompanied by a friend named Constance, and judging from his poems she seems to have exerted some influence on their contents:

Oh, summer! amorous – drunken summer, sighing
Lust sighing from the youth of green heart
with sounds which music like some strange replying
To thoughts that from the stillest heart-wells start.
Come – breathe on me the soul-enchanting art
And voice my lip with thy mysterious dole
That I may be to her of thee a part –
That from my eyes to hers thy power may roll
Until I grow like thee into her ebriate soul.

In Christmas of 1868, Francis composed some verses in the guise of a Latin inscription, nearly an epitaph, for Richard:

Richardo

Shelleium de violis videre Iubenti
Venerunt violæ tuæ
Vidi apud Shelleium laud' tuas
credo violas:
Vici infaustum venere
dulci et spe et memoria tui.

Die Natali. MDCCCLXVIII.
Ball. Coll. [Balliol College] [224]

As we can see, by mentioning Shelley and the violets and explicitly dedicating the verses to Richard, Francis confirmed that his English poem from February was indeed addressed to Richard.

At this stage, early 1869, Richard described his ideal way of life, which was not so different from the mythicised family life of his parents with Aunt Mariquita and David Reid in Florence, which had been so cruelly broken:

With music, painting, poesy and art
And with one heart-beloved bosom-friend
And one soft woman loving me apart
From fortune or from other worldly end
But loving me because mine eyes might lend
Her gentle joy me near to be.
With such companions were it sweet to bend
With life might be but love – and earth a memory.

Curiously enough for his time, and totally unrelated to John Stuart Mill's feminist manifesto published in the same year, Richard Lestock Edgeworth also composed a feminist poem.[225]

After he arrived in Naples with Constance, Richard worked on new poems. One of them is dedicated to this city, just as Shelley had composed his famous 'Stanzas written in Dejection, near Naples' in the same city half a century earlier. But Richard does not seem to have been at all dejected when he proceeded to compose a joyful, rhythmic song that could easily serve as the hymn of a rowing team. Nonetheless, the last poem in Richard's notebooks is rather melancholy, as we can observe in these excerpts:

A boy sat by a spring
and looked at a wreath of flowers
and he saw it torn away
Swimming in the watery dance
'Sad so flee my days
Onward, as the restless spring
Sad so pales my youth
Quickly as the wreath fades'.

Ask not why I am sad
In the happy time of youth
Others enjoy themselves and hope
for the time when Spring returns
But these thousand voices
Of awakening nature
Wake in the depths of my bosom
Only heavy sorrow.[226]

And, then, unexpectedly, on 8 May 1869, Richard and Constance were 'torn away' themselves, like that 'wreath of flowers'. While 'swimming in the watery dance', they both drowned in the bay of Naples. Their deaths appear to have been accidental, the result of a sudden storm, just like Percy Bysshe Shelley's death in the bay of La Spezia in 1822. At least this is what Harriet Butler believed when she wrote elliptically to her brother Pakenham of 'the fatal, fatal day', mentioning that such a 'dark cloud on the bright summer of Rick and Constance is a most unexpected calamity'.[227]

News about Richard's death quickly reached Oxford. Francis was on the verge of desperation. His future with Richard, which he had so hopefully imagined at the end of his 1868 poem – in fact the only poem by Ysidro Francis Edgeworth found so far – had vanished in the sea he mentioned:

> Thoughts of thee my breast came over,
> And sweet hopes that evermore
> Still together we might soar
> O'er life's troubled sea.

Moreover, he had been playing a Shelleyan game with Richard that ended with death, the same kind of death as Shelley's. He wanted to know the details of the misfortune and travel to Naples immediately, but his Balliol tutor Benjamin Jowett stopped him:

> My dear Edgeworth,
> I was very sorry to hear of the calamity which has befallen you.
> I am very doubtful about your getting off if you can sum strength and courage to go on. I do not expect that the Proctor will give you leave or can and if this is the case I would strongly urge you to delay your visit to Naples for a few weeks.
> These sort of trials are painful. I suppose that they should make us feel both respecting ourselves and those who are gone that we are in the hands of God. Believe me. Ever yours, B. Jowett.[228]

There is no proof that Francis Edgeworth ever went to Naples during his lifetime.

1.18 1870–72: FRANCIS EDGEWORTH'S MELANCHOLY INTERMEZZO

We have found only scattered information about Francis Edgeworth in the years immediately after his Oxford studies in Literæ Humaniores.

What we know, as he later declared, is that he 'studied Mathematics for some years'[229] but on an informal basis. He also noted that he had 'studied Law' at the Inner Temple of London, just like his grandfather, and had been called to 'the English Bar' in 1877.[230] This same year he also acquired a Master of Arts degree at Oxford University.

But the period from 1870 to 1877 was excessively long by all academic standards. What happened to him to explain this delay?

Francis probably spent 1870 at Oxford starting his studies in law. We do not know which college he attended or even if he lived in one. In 1869 Francis had asked Robert Williams, a friend of his sister Mary, about the best place in Oxford to study law. Williams told Francis that he should go to Corpus Christi College, where Francis's grandfather had studied.[231] There is also a letter from James Bryce, Regius Professor of Civil Law at Oxford from 1870 to 1893 and Irish Secretary in 1905, who wrote to him in June 1870 from the Temple to congratulate him for his 'first class' and to express his hope that 'we may soon have to congratulate you on a fellowship also'.[232] Therefore, he might also have been preparing himself for a Balliol Fellowship that never materialised.

On the other hand, through Christina Edgeworth we learn that around the summer of 1870 Mary 'looks very ill' and 'she reports badly of poor Francis always of late years a very delicate lad, they have no health among them'.[233] And, inasmuch as Edgeworthstown was closed during 1870 and 1871, we suppose that Francis spent part of his long vacation in Ireland with his brother David at Dublin.[234] He also most likely visited his aunts Harriet Butler and Lucy Jane Robinson in Armagh.

Two Weddings

The happiest family event of those years was his sister's wedding. On 11 April 1871, Mary married John Sanderson, rector of Winchfield, an English village on the road from London to Salisbury. This marriage came as a surprise to Pakenham Edgeworth and family, as his daughter Harriet Jessie Butler reports:

> At one time we saw something of Mary, and were surprised at her marriage to the really respectable and good Jack Sanderson. Why she did it I can't think, but it was the saving of her, and she turned into an excellent mother, and devoted herself to her four girls [Mary actually had five girls, but one of them died in infancy].[235]

Nine days after this wedding, William Waller Fox, one of Aunt Sophy's children, married his cousin Emma Louisa Fox of Fox Hall. His older brother Maxwell had married Florence Buchanan in 1865, and after her death in 1882 he married another cousin, Edith Susan Marian, the eldest daughter of Rev. Essex Edgeworth of Kilshrewly, from the Abbé Edgeworth's branch of the family. Emma and Waller lived at Fox Hall and Maxwell and Edith lived at Kilshrewly. They were both important contacts in Ireland not only for Eroles, but also for Harrie and Francis all of their lives.

David Edgeworth's Death

But 1871 did not come to an end without another major blow: on 14 October, David Reid Edgeworth, one of Francis's two remaining brothers, died in Ireland 'from bronchial inflammation' as reported in the 'Memory of David Reid Edgeworth' published in the *Annual Report of the Institution of Civil Engineers*, 17 December 1872 with a summary of his educational background and professional activities, which is reproduced below as Appendix E.

By the end of 1871, Francis's family circle had been drastically reduced, as in less than eight years his grandmother, mother, three brothers – William, David and Richard – and two uncles – Uncle Sneyd Edgeworth and Uncle Lestock Wilson – had died. The last had died on 17 July 1869, a mere two months after his nephew Richard Lestock Edgeworth. Francis's only remaining close relatives were his old aunts Harriet Butler and Lucy Jane Robinson, who lived in Armagh, his uncle Pakenham Edgeworth and family living in Anerly in the south of London, his Fox cousins at Kilshrewly and Fox Hall, near Edgeworthstown, his sister Mary Sanderson living at the old rectory in Winchfield and his brother Antonio Eroles Edgeworth, who at this time was purported to be in France.

But the deaths of Richard and David Edgeworth, though extremely painful for Francis, were at the same time the solution to his economic problems. By the end of 1871, he was the only survivor of the three heirs to the bulk of his grandmother's fortune. After the executor simultaneously issued the statements on David's and Richard's estates, their only debts were the ones with their brothers Eroles and Francis, who had helped David to establish himself as a civil engineer. There were also lands through which Francis could get his money back, in case of need. These two documents are reproduced below in Appendix F.

With the new influx of income from his grandmother's principal he could afford to live in London closer to the Temple and could try to find some academic work there. He could even think of ending his bachelorhood by getting married. Thus, in around 1872 he proposed marriage to his cousin Harriet Jessie Edgeworth, then 21 years old. We are aware of this marriage proposal through her own recollections, written when she was in her eighties:

> Francis used to visit us at Anerly and always kept up a friendly appearance. I do not think he joined his brothers in their bad ways, though he condoned them. . . . Francis also settled down, and, unluckily, took it into his head to propose to me, or rather for me; and my mother wrote that I could not think of such a step; after which he became, or rather remained, my very good friend.[236]

Harriet's mother had some clear reasons for advising her daughter to refuse him, including Francis's poor health and his propensity to back Eroles and the late David against Pakenham. But the main reason for Christina and Pakenham's refusal was that they had no faith in Francis's future. In November 1871, Christina wrote to her sister-in-law, Harriet Butler:

> He [Francis] is to be one of the 'thinkers' for the rest of the world . . . unless the thoughts are made accessible I am at a loss to see their extreme usefulness . . . Francis said to me he did *think* of teaching as a profession but he is too old to commence a work requiring such flexibility and adaptability of mind to others – qualities wholly absent or underdeveloped as yet in him.[237]

It seems that Francis, at the age of 26, had given up his early intention, expressed back when he was nine, of entering the church, and instead had a clear design on becoming a professor. He even had in mind a theory on which to base his future research. But of course, Uncle Pakenham and Aunt Christina did not have any faith in his scheme nor in any process leading young Ysidro Francis Edgeworth up to the making of the academic Francis Ysidro Edgeworth.

NOTES

1. Keynes is mistaken when he writes in 1926 that 'F.Y. Edgeworth himself was the fifth son of a sixth son' because he was the sixth son of a ninth son. See Keynes (1926) in vol. X of Keynes (1971–1989), p. 251.
2. See Butler and Butler (1927).
3. R.L. Edgeworth's multifaceted work has been described and analysed by Paterson (1914), Clarke (1965) and Lyons (2003).
4. See Schofield (1963), Uglow (2002).
5. Edgeworth (1820), Ch. XIII, p. 207.
6. Edgeworth (1820), Ch. XIV.
7. Macdonald (1970).
8. Butler (1972).

9. Butler, M. (1972), p. 428.
10. Letter from Maria Edgeworth (M.E.) quoted in Edgeworth, F.A. (1867).
11. M.E. to Frances Anne Edgeworth (F.A.E.): February 1822 (EP BODLEIAN UO, Ms. Eng. lett. c698).
12. Letter from Harriet Butler (H.B.) to Francis Ysidro Edgeworth (F.Y.E.), 12 June 1888 (EP NUFFIELD C UO College. Box D 1).
13. Letter from M.E. to Mrs Ruxton, 9 March 1822. Included in Sraffa and Dobb (1965), X, p. 179.
14. Letter from Fanny Wilson (F.W.) to M. Pakenham Edgeworth (M.P.E.) (EP BODLEIAN UO, Ms. Eng. lett. c725). There is another version of this anecdote by M.E. that was published by F.Y.E. in the *Economic Journal*, 1907 and appears in Sraffa and Dobb (1965), X, pp. 32–3.
15. Letter from David Ricardo to M.E., 26 May 1823, included in Sraffa and Dobb (1962), IX, pp. 295–6.
16. Graves (1882), I.
17. Letter from W.R. Hamilton to Eliza, 23 September 1822. Quoted in Graves (1882), I, p. 114.
18. See Section 1.3 above, *Aunts*.
19. W.R. Hamilton to his sister Eliza, from Edgeworthstown, August 1824 (Graves, 1882, I, p. 164).
20. Graves (1889), III, pp. 179–81.
21. Letters between W.R. Hamilton and M.E., April and May 1828. EP NL IRELAND, MS. 11132(6).
22. Hankins (1980), p. 37.
23. Quoted in Graves (1882).
24. Mozley (1882).
25. Report by Harriet Jessie Butler (H.J.B.) sent to J.M. Keynes on the occasion of the death of F.Y. Edgeworth in 1926 (KP KING'S C CAMBRIDGE MA/*EJ*/6/6).
26. Mozley (1882).
27. Letter from Francis Beaufort Edgeworth (F.B.E.) to Lucy Edgeworth (EP BODLEIAN UO, Ms. Eng. lett. c746).
28. At Cambridge, Francis Beaufort Edgeworth also probably met Robert John Tennant, who was at Trinity from 1827 until 1835, when he got his M.A. (*Alumni Cantabrigiensis, Vol. 1752–1900*). In 1838, Tennant would marry Mariquita Eroles, Francis's sister-in-law.
29. Letter from H.E. to Charles Sneyd Edgeworth (C.S.E.), 26 May 1829 (EP BODLEIAN UO, Ms. Eng. lett. c739).
30. Graves (1882).
31. Letter from M.E. to Mrs Ruxton, 27 September 1829 (EP BODLEIAN UO, Ms. Eng. lett. c718).
32. Graves (1882).
33. Letter from F.B.E. to W.R. Hamilton, August 1829 (EP BODLEIAN UO, Ms. Eng. lett. c746).
34. Hankins (1980), p. 101.
35. Mozley (1882).
36. Graves (1882) and Hankins (1980).
37. Hankins (1980).
38. Butler and Butler (1927), p. 248.
39. Letter from F.B.E. to M.P.E., 21 December 1831 (EP BODLEIAN UO, Ms. Eng. lett. c729).
40. Letter from F.B.E. to W.R. Hamilton, December 1831 (EP BODLEIAN UO, Ms. Eng. lett. c746).
41. Letter from H.E. to C.S.E., 5 December 1831 (EP BODLEIAN UO, Ms. Eng. lett. c739).

42. Letter from M.E. to F.W., 6 December 1831 (EP BODLEIAN UO, Ms. Eng. lett. c707).
43. Letter from C.S.E. to H.E., 15 December 1831 (EP BODLEIAN UO, Ms. Eng. lett. c739).
44. Letter from Lord Holland to C.S.E., December 1831 (EP BODLEIAN UO, Ms. Eng. lett. c741). Partly quoted in Hicks (1984b).
45. Joaquín de Ibáñez-Cuevas, Baron of Eroles (1784–1825), son of the military governor of Talarn (Lleida, Catalonia) volunteered in 1808 against the French troops. After being taken prisoner in Girona, he escaped from France, was appointed Commander of the Army of L'Empordà, raised to General and named Captain General of Catalonia in 1813. He kept this position until 1814. In 1822 the Baron, became the leader of the absolutist revolt against the constitutional government, took part in the Regency of Urgell and, after being defeated, fled to France. There, he was appointed chief of the Spanish vanguard of the *cent mil fils de Sant Louis*, commanded by the Duke of Angoulème, which in 1823 restored the absolute power of Ferdinand VII, who instituted the terror against the Constitutionalists. After once again serving as the Captain General of Catalonia for one year, Ibáñez-Cuevas was appointed to rebuild the army in Madrid and died mad in Daimiel (Ciudad Real) in 1825. See Camp (1945).
46. Lord Holland's report was also accepted by Sir John R. Hicks who, on the advice of Raymond Carr, erroneously identified General Eroles as a brother of the Baron of Eroles fighting on the absolutist side. See Hicks (1984b).
47. Francisco Espoz y Ilundain, widely known as General Espoz y Mina using his father's names, or simply as Mina, hero of the 1808–1814 war against Napoleon, leader of a liberal conspiracy in Pamplona that failed in 1814, exiled until 1820 and 'Capitán General de Cataluña' from September 1822 to November 1823 under the liberal government. After *les cent mil fils de Saint Louis* invaded Spain commanded by the Duke of Angoulème and with the Baron of Eroles leading the Spanish absolutist troops, Mina went into exile again, this time to London.
48. Letter from C.S.E. to H.E., 18 December 1831 (EP BODLEIAN UO, Ms. Eng. lett. c739).
49. Rosa Florentina Edgeworth (née Eroles) has been ignored by Maria Edgeworth's biographers. She is mentioned once in all the biographies of M.E. (Kavanagh (1862), Edgeworth (1867), Casey (1877), Oliver (1882), Hare (1894), Lawless (1904), Hamilton (1904), Inglis-Jones (1959), Butler (1972) and so on) and it is on the occasion of her marriage to F.B.E. The rest of her life, even her years in Edgeworthstown with M.E. and F.A.E. from 1842 to 1849, is never mentioned. Most of M.E.'s biographies are incomplete after 1833, when her last novel *Helen* was published.
50. London Metropolitan Archives, Reel X030/031, No. 851.
51. F.B.E. to F.A.E., December 1831 (EP BODLEIAN UO, Ms. Eng. lett. c737).
52. 'Francis Ysidro Edgeworth', in Murphy, ed. (1984). The other biographical references to Edgeworth – Bonar (1926), Price (1926), Bowley (1934), Hildreth (1968), Stigler (1978), Creedy (1986), Newman (1987) and Mirowski (1994) – do not delve further than Keynes and Hicks concerning his grandfather Antonio Eroles.
53. Translated from Spanish, Espoz y Mina (1852), I, pp. 387–388.
54. See letter from F.A.E. to Louisa C. Beaufort, 16 December 1831 and F.W. to F.A.E., December 1831 (EP BODLEIAN UO, Ms. Eng. lett. c738 & 735).
55. Organyà, a village in the Alt Urgell next to the Segre river, at an altitude of 558 m and 20 km south of La Seu. It had 535 inhabitants in 1787 and 1195 in 1860 (*Gran Enciclopèdia Catalana*).
56. These results have been obtained by the economist and local historian Amadeu Rocamora from the *Arxiu Diocesà* of La Seu d'Urgell.
57. El Pla de Sant Tirs, a village in the Alt Urgell near the Segre river, at an altitude of 702 m, 8 km south of La Seu and 12 km north of Organyà. It had 426 inhabitants in 1787 and 473 in 1860 (*Gran Enciclopèdia Catalana*).

58. Antonio's half brother Tirs (1751–74) died before he was born; his half sister Anastasia (born around 1753) married and left the house probably before Antonio's birth; his half brother Marc (1760–1824) inherited his father's house in 1796, worked as farmer and stonecutter and remained in Pla de Sant Tirs, just like Antonio's sister Rosa (1776–1843), who married Isidro Serra, a gelder (godfather of Antonio's first son). Sister Margarita (1781–?) married Miquel Niell, a farmer from Arfa, a small village nearby; and sister Teresa died in infancy (1784–85) (Arxiu Diocesà de La Seu d'Urgell, Parròquia de Sant Tirs, I).
59. F.Y. Edgeworth's Catalan grandmother Rosa Eroles had the same maiden name – Eroles – as her husband's name.
60. Arxiu Diocesà de la Seu d'Urgell, Pla de Sant Tirs's Register of Christenings. Was this Antonio, Rosa's deaf-mute brother mentioned by H.J. Butler (in Section 1.6) and by F.B. Edgeworth (see below, in this Section)? Very likely, as there is no document found where both 'Antonio' and 'the deaf and dumb brother' are mentioned as separate people, and the inscription in the Register would explain the deaf-muteness.
61. Galli (1831), p. 117. This source, however, was not mentioned by F.B. Edgeworth, who copied the paragraph in French. The translation into English is ours (EP NUFFIELD C UO, Box E 3/7).
62. Mariano Zorraquín, appointed 'jefe de la plana mayor' by general Mina: Espoz y Mina (1852).
63. EP NUFFIELD C UO, Box E 3/7.
64. Her niece Harriet Jessie Butler in a report written around 1938 described her as 'though a very admirable woman in many ways, she had not the English standard of truth' (EP NUFFIELD C UO, Box E 3/10/2).
65. See Sánchez de la Campa (1851) and López, Peruga and Tudel (1988).
66. Most smugglers – *contrabandistes* – were from Andorra. The people from Alt Urgell were mostly poorly paid conveyors – *paquetaires*. At any rate, this activity would explain how Eroles might have been acquainted with smugglers or conveyors such as Blasi. Therefore, General Zorraquín's decision to choose Eroles to seize Blasi can be viewed as concealed blackmail.
67. Espoz y Mina (1852), I, pp. 387–8.
68. Information supplied by Amadeu Rocamora.
69. Edgeworth (1867).
70. Butler (1972).
71. F.A.E. to Louisa C. Beaufort, 24 January 1832 (EP BODLEIAN UO, Ms. Eng. lett. c738).
72. M.E. to M.P.E., March 1832 (EP BODLEIAN UO, Ms. Eng. lett. c715).
73. F.B.E. to F.A.E.,1832 (EP BODLEIAN UO, Ms. Eng. lett. c737).
74. F.B.E. to F.A.E., December 1832 (EP BODLEIAN UO, Ms. Eng. lett. c737).
75. F.A.E. to M.P.E., January 1834 (EP BODLEIAN UO, Ms. Eng. lett. c723).
76. H.B. to M.P.E., 28 January 1834 (EP BODLEIAN UO, Ms. Eng. lett. c726).
77. M.E. to F.W., 1 January 1834 (EP BODLEIAN UO, Ms. Eng. lett. c708).
78. H.B. to M.P.E., 28 January 1834 (EP BODLEIAN UO, Ms. Eng. lett. c726). Website of the English Cemetery of Florence www. florin.ms/cemetery.html, accessed 24 July 2009. Mary Reid, died on 29 December.
79. H.B. to M.P.E., 17 March 1834 (EP BODLEIAN UO, Ms. Eng. lett. c726).
80. M.E. to M.P.E., 9 April 1834 (EP NL IRELAND, Ms. 8145).
81. Graves (1885), II.
82. F.A.E. to M.P.E., August 1834 (EP BODLEIAN UO, Ms. Eng. lett. c723).
83. F.B.E. to F.A.E., 14 December 1834 from London (EP BODLEIAN UO, Ms. Eng. lett. c737).
84. F.B.E. to W.R.H., 3 September 1837 (EP BODLEIAN UO, Ms. Eng. lett. c746).
85. Part II, Ch. IV. Partly quoted by Keynes (1926).
86. Mozley (1882), I, p. 52. Partly quoted by Keynes (1926).

87. M.E. to F.A.E., 2 April 1837 (EP BODLEIAN UO, Ms. Eng. lett. c701).
88. M.E. to Sophy Ruxton, 19 July 1837: Edgeworth (1867).
89. EP BODLEIAN UO, Ms. Eng. misc. c899.
90. Clarke (1949), p. 168. See also Friendly (1973).
91. H.E. to C.S.E., 23 September 1838 (EP BODLEIAN UO, Ms. Eng. lett. c739).
92. Butler (1972).
93. F.B.E. to Rosa Florentina Edgeworth (R.F.E.), 8 and 11 January, 10 and 28 February, 12 March 1841, from Edgeworthstown (EP BODLEIAN UO, Ms. Eng. lett. c746).
94. Balaguer (1863).
95. University of Barcelona's Historical Archives. 'Expediente de D. Isidro Eroles y Eroles', 1842.
96. F.B.E. to R.F.E., 10 April and 8 May 1841, M.E. to F.W., 15 June 1841 (EP BODLEIAN UO, Ms. Eng. lett. c746 and c710).
97. Quoted in Edgeworth (1867).
98. Edgeworth (1867).
99. M.E. to F.W., May 1844 (EP BODLEIAN UO, Ms. Eng. lett. c710).
100. M.E. to Honora Beaufort, May 1844 (EP BODLEIAN UO, Ms. Eng. lett. c705).
101. M.E. to F.W., 15 February 1845 (EP BODLEIAN UO, Ms. Eng. lett. c711).
102. Dante Alighieri, *La Divina Commedia*, Paradiso, Canto, XV, vv. 47–48.
103. F.B.E. to Lucy Robinson, 2 April 1846 (EP BODLEIAN UO, Ms. Eng. lett. c746).
104. M.E. to H.B., 30 May 1846 (EP BODLEIAN UO, Ms. Eng. lett. c714).
105. M.E. to F.B.E. and R.F.E., 28 July; 13, 16, 21, 24, 28 and 30 August; 6, 13 and 15 September 1846 (EP BODLEIAN UO, Ms. Eng. lett. c714).
106. EP BODLEIAN UO, Ms. Eng. misc. b432.
107. EP BODLEIAN UO, Ms. Eng. lett. c731.
108. M.P.E. to F.A.E., 26 September 1846. (EP BODLEIAN UO, Ms. Eng. lett. c724).
109. M.E. to C.S.E., December 1846 (EP BODLEIAN UO, Ms. Eng. lett. c703).
110. F.A.E. to H.B., 4 May 184? (EP BODLEIAN UO, Ms. Eng. lett. c736).
111. EP NL IRELAND, Ms. 989.
112. M.E.'s correspondence with the Central Relief Committee, 1, 2, 5, 15 and 18 February; 16 April and 5 and 8 May 1847 (EP NL IRELAND, Ms. 989).
113. M.E. to the Central Relief Committee and back, 22 October, 20, 22 and 29 December 1874; 14 March and 16 May 1848 (EP NL IRELAND, Ms. 989).
114. M.E. to Richard Jones, 18 March 1846 and 14 June 1847 (EP NL IRELAND, Ms. 22822).
115. M.E. to Richard Jones, 15 August 1847, April, 1849 (EP NL IRELAND, Ms. 22822).
116. Graves, (1885), Vol. II.
117. M.P.E. to F.A.E., November 1846 (EP BODLEIAN UO, Ms. Eng. lett. c 724).
118. R.F.E. to M.P.E., 1847 (EP BODLEIAN UO, Ms. Eng. lett. c 729).
119. See R.F.E. to William Rowan Hamilton, 18 February 1847 (EP NL IRELAND, Ms. 11132, no 65).
120. Graves, (1885), II.
121. H. B. to Elizabeth MacPherson (EP BODLEIAN UO, Ms. Eng. lett. d 532).
122. M.P.E to FA.E., 10 June 1848 (EP BODLEIAN UO, Ms. Eng. lett. c 724).
123. T.L. Beddoes (1803–1849): after graduating from Oxford, he studied medicine in Göttingen during 1825 to 1829, after which he unsuccessfully tried to convince Francis to study philosophy in Germany. Beddoes had discovered that he enjoyed studying medicine; nevertheless, he started writing versified plays in the Elizabethan style and became a poet. After 1830, he spent most of his life in Bavaria and Switzerland. Beddoes's plays – *The Brides' Tragedy*, *The Second Brother*, *The Old Ghost*, *The Oviparous Tailor*, *Death's Jest Book* – were quite gothic. A biographic

sketch of Thomas Lovell Beddoes is contained in Lytton Strachey's article, 'The Last Elizabethan'. See Strachey (1922).
124. Edgeworth (1867).
125. M.E. to Sir Philip Crampton, 2 May 1849 (CP TRINITY CLD, Ms. 4178).
126. R.F.E. to P. Crampton, 22 May 1849 (CP TRINITY CLD, Ms. 4178).
127. M.P.E. to F.A.E., 21 March 1863 (EP BODLEIAN UO, Ms. Eng. lett. c 724).
128. The report is attributed to Lucy, but it looks as if she copied it or received the information from one of Rosa's children (EP NUFFIELD C UO, Box E3/7).
129. R.F.E. to M.P.E., 3 June 1849 (EP BODLEIAN UO, Ms. Eng. lett. c 729).
130. Butler and Butler (1927), p. 244. Quoted by Keynes (1926), fn. 6.
131. R.F.E. to M.P.E., August 1851 (EP BODLEIAN UO, Ms. Eng. lett. c 729).
132. R.F.E. to M.P.E., 21 September 1853, from London (EP BODLEIAN UO, Ms. Eng. lett. c 729).
133. R.F.E. to MPE, 10 June and July 1854 (EP BODLEIAN UO, Ms. Eng. lett. c 729).
134. EP NUFFIELD C UO, Box E3/7.
135. R.F.E. to M.P.E., November 1854. (EP BODLEIAN UO, Ms. Eng. lett. c 729).
136. Keynes (1926), in Vol. X of Keynes (1971–1989), pp. 254 and 266. Reported by Harriet Jessie Butler.
137. R.F.E. to M.P.E., July 1854 (EP BODLEIAN UO, Ms. Eng. lett. c 729).
138. Notes by Harriet Jessie Butler written around 1938 (EP NUFFIELD C UO College Box E3/9).
139. The East India Co. was nationalised by the British Government in 1858.
140. R.F.E. to M.P.E., 21 September 1853, from London. (EP BODLEIAN UO, Ms. Eng. lett. c 729).
141. M.P.E. to H.B. (copy by CE), Jullundhur, 19 September 1855 (EP BODLEIAN UO, Ms. Eng. lett. c 728).
142. M.P.E. to F.A.E., 7 September 1856 (EP BODLEIAN UO, Ms. Eng. lett. c 724).
143. George Salmon, mathematician and theological author, who became Regius Professor of Divinity in 1866 and Provost of Trinity College Dublin in 1888.
144. Family Report (EP NUFFIELD C UO, Box E 3/7).
145. Family Report (EP NUFFIELD C UO, Box E 3/7).
146. Carter (1884), Ch. IV.
147. M.P.E. to H.B., 17 October 1857 and December 1857 (EP BODLEIAN UO, Ms. Eng. lett. c 728).
148. By Alfred Friendly when preparing *Beaufort of the Admiralty: The Life of Sir Francis Beaufort*, Random House, 1973.
149. In re of Honora Beaufort, 1858 (EP BODLEIAN UO, Ms. Eng. lett. c 730).
150. C.E. to H.B., 31 January 1857 (EP BODLEIAN UO, Ms. Eng. lett. c 731).
151. M.P.E. to F.A.E., 5 October 1857 (EP BODLEIAN UO, Ms. Eng. lett. c 724).
152. C.E. to H.B., 16 November 1857 (EP BODLEIAN UO, Ms. Eng. lett. c 731).
153. M.P.E. to F.A.E., 12 December 1857 (EP BODLEIAN UO, Ms. Eng. lett. c 724).
154. M.P.E. to H.B., December 1857 (EP BODLEIAN UO, Ms. Eng. lett. c 728).
155. Memory of David Reid Edgeworth, published in the *Annual Report of the Institution of Civil Engineers*, 17 December 1872 (EP BODLEIAN UO, Ms. Eng. lett. c 900).
156. Graves, Hamilton's biographer, includes a description of the party which is rather interesting because it shows the solid bonds between the Hamiltons and the Edgeworths (Graves, 1889, Vol. III, pp. 102–103).
157. In 1827, Dr Hugh MacPherson, father of Christina, bought the whole island, where his ancestors came from, for £15,000 and he used it as a summer home. He had died on 20 March 1854, but his thirteen children still spent most of their summers on Eigg, which was owned by the eldest brother, Norman, Professor of Scots Law at Glasgow University.
158. H.B. to M.P.E., 23 July 1858 (EP BODLEIAN UO, Ms. Eng. lett. d 529).

159. Lucy Robinson to M.P.E, 24 January 1859 (EP BODLEIAN UO, Ms. Eng. lett. c 729).
160. R.F.E. to William R. Hamilton (EP NL IRELAND, Ms. 11132, nr. 49).
161. Family Report (EP NUFFIELD C UO, Box E 3/7).
162. H.B. to M.P.E, 1859 (EP BODLEIAN UO, Ms. Eng. lett. d 529).
163. Family Report (EP NUFFIELD C UO, Box E 3/7).
164. Carter (1884), Chap. IV. Mariquita Tennant's charitable work has recently been earning public recognition: in November 2005, the Mayor of the Royal Borough of Windsor and Maidenhead erected a blue plaque in her honour at The Limes, the house where she had lived in Clewer.
165. Family Report (EP NUFFIELD C UO, Box E 3/7).
166. C.E. to H.B., no date, [by content and filing place, summer 1860] from Edgeworthstown (EP BODLEIAN UO, Ms. Eng. lett. c 732).
167. Notes by Harriet Jessie Butler written in 1926, after F.Y.E.'s death (EP NUFFIELD C UO, Box E3/9).
168. M.P.E. to H.B., 4 June 1859 (EP BODLEIAN UO, Ms. Eng. lett. d 529).
169. M.P.E. to H.B., 15 Wed. (February 1860?) (EP BODLEIAN UO, Ms. Eng. lett. d 529).
170. M.P.E. to H.B., September 1860 (EP BODLEIAN UO, Ms. Eng. lett. d 529).
171. M.P.E. to H.B., 3 January 1861 (EP BODLEIAN UO, Ms. Eng. lett. d 529).
172. M.P.E. to H. B., 18 January 1861 (EP BODLEIAN UO, Ms. Eng. lett. d 529).
173. C.E. to H.B., 19 February 1861? (EP BODLEIAN UO, Ms. Eng. lett. c 732).
174. C.E. to H.B., 19 February 1861? (EP BODLEIAN UO, Ms. Eng. lett. c 732).
175. Graves (1889), Vol. III, p. 132.
176. C.E. to H.B., June 1861, from Edgeworthstown (EP BODLEIAN UO, Ms. Eng. lett. c 732).
177. Harriet Jessie Butler's 1938 report (EP NUFFIELD C UO, Box E 3/9/1).
178. M.P.E. to H.B., 18 October and 2 November 1861 (EP BODLEIAN UO, Ms. Eng. lett. d 529).
179. M.P.E. to H.B., 9 October 1861 (EP BODLEIAN UO, Ms. Eng. lett. d 529).
180. M.P.E. to F.A.E., 1861 (EP BODLEIAN UO, Ms. Eng. lett. c 724).
181. F.A.E. to M.P.E., 24 March 1863 (EP BODLEIAN UO, Ms. Eng. lett. c 724).
182. All the academic results mentioned in this section or the corresponding appendix about Eroles, Richard or Francis Edgeworth at Trinity College Dublin (TCD) are taken from the TCD register books V/23/6, V/24/4, V/30/21 to 24, V/31/3.
183. Family Report (EP NUFFIELD C UO, Box E 3/7).
184. M.P.E. to H.B., Oct. 6, 1862 (EP BODLEIAN UO, Ms. Eng. lett. d 529).
185. Oriental and India Office Collections, British Library, London (N/3/37/Folio 121).
186. H.J.B.'s report around 1938 (EP NUFFIELD C UO, Box 3/9).
187. M.P.E. to H.B., 31 March and 4 April 1864 (EP BODLEIAN UO, Ms. Eng. lett. d 529).
188. C.E. and M.P.E. to H.B., March 1864 (EP BODLEIAN UO, Ms. Eng. lett. c 733).
189. Graves (1930), pp. 128 and 130.
190. M.P.E. to H.B., 18 February 1863 (EP BODLEIAN UO, Ms. Eng. lett. d 530).
191. EP NL IRELAND, Ms. 11132.
192. HB. to W.R. Hamilton, 9 July and 16 October 1864 (EPNLI, Ms. 11132 (6)).
193. Graves (1889), III, p. 168.
194. Y.F. Edgeworth to W.R. Hamilton, 20 November, 1862?, from 38 Trinity College (EP NL IRELAND, Ms. 11132(6))
195. Memory of Frances Anne Edgeworth (EP BODLEIAN UO, Ms. Eng. misc. c 899).
196. TCD registers V/23/6, V/24/4, V/30/21 to 24, V/31/3.
197. Keynes wrote in 1926 that 'he went to Trinity College, Dublin, at the age of seventeen', when, in fact, he entered at sixteen, and the erroneous year of 1862 instead of 1861 for his entrance is the one quoted by Bowley (1934), Kendall (1968), Stigler (1978), Hicks

(1984b), Creedy (1986), Newman (1987) and Mirowski (1994). It may be because, after getting his entrance and the Composition Prizes in Greek, he could not attend the lectures in the Michaelmas Term of 1861. On the other hand, none of the aforementioned authors have found his final exam as a Candidate Bachelor in the Michaelmas Term of 1865, when he graduated with the top qualification of his class. So, they are not in a position to tell how many years or terms Edgeworth studied at Trinity College, Dublin and if they try to do so, they reduce the real time.

198. EP NUFFIELD C UO, Box D 5.
199. According to Stigler, 'Ingram's views were common at that time.' See Stigler (1978), p. 320.
200. John K. Ingram, 17 November 1875 (EP NUFFIELD C UO, Box D 5).
201. John Pentland Mahaffy, 22 December 1873 (EP NUFFIELD C UO, Box D 5/2).
202. Quoted in Merlin Holland, *Wilde Album*, 1997, London, 4th Estate, p. 27.
203. Robert Yelverton Tyrrell, 12 November 1875 (EP NUFFIELD C UO, Box D 5).
204. 24 March 1874 (EP NUFFIELD C UO, Box D 5/4).
205. Quoted by Graves (1930), p. 122.
206. Notes by H.J. Butler around 1938 (EP NUFFIELD C UO, Box E 3/9).
207. EP BODLEIAN UO, Ms. Eng. misc. c 900.
208. C.E. to H.B., 21 March 1865 (EP BODLEIAN UO, Ms. Eng. lett. c 733).
209. C.E. to M.P.E. (EP BODLEIAN UO, Ms. Eng. lett. c 729).
210. EP BODLEIAN UO, Ms. Eng. lett. c 747.
211. M.P.E. to H.B., 13 July 1866 (EP BODLEIAN UO, Ms. Eng. lett. d 530).
212. Notes by H.J. Butler around 1938 (EP NUFFIELD C UO, Box E 3/9).
213. See Stigler (1978), p. 289.
214. B. Jowett to F.Y.E. in Balliol, 22 April and 15 August, no year (1868?) (EP NUFFIELD C UO, Box D 2 / 1/37 & 38).
215. EP NUFFIELD C UO, Box D 5 & D 8. For more on Edgeworth's research into mathematics and logic, see below.
216. Included in Pigou (1925), p. 66.
217. EP NUFFIELD C UO, Box D 5 / 3 & D7/14.
218. EP NUFFIELD C UO, Box D 7/14.
219. Keynes (1926 and 1983), in Vol. X of Keynes (1971–1989), p. 254. *Viva*, for *viva voce*, the popular name for the oral final exam.
220. These poems are kept at the Bodleian Library, Oxford (EP BODLEIAN UO, Ms. Eng. poet. e 187). Thus we infer that Richard sent some or all of them to Francis, because if he had sent them to their Aunt Harriet they would now be housed in the National Library of Ireland. Moreover, see n. 222.
221. EP NUFFIELD C UO, Box E 1/3.
222. A copy of 'Onora' housed at the Bodleian Library of Oxford has the following heading: 'Copy of "Onora" written out in Styria and sent to England, July 1868.' (EP BODLEIAN UO, Ms. Eng. poet. e 187).
223. R.F.E. to M.P.E., August 1851 (EP BODLEIAN UO, Ms. Eng. lett. c 729).
224. It can be translated as: 'To Richard / who wants me to look at Shelley about violets./ Your violets arrived / I saw that Shelley's work praise / – I think – your violets /I vanquished misfortune / with sweet love, hopes and memories of you' (EP NUFFIELD C UO, Box E 1). Look at the structure 'Veni, vidi, vici' (Julius Cæsar) of the original poem. I thank Joan Carbonell for the translation from Latin and the comment on Cæsar's sentence.
225. EP BODLEIAN UO, Ms. Eng. poet. e 187.
226. Pages 162 and 176 of the notebooks. EP BODLEIAN UO, ms. Engl. Poet. e529.
227. H.B. to M.P.E. from Armagh, May 1869. EP BODLEIAN UO, ms. Engl. Lett. d187.
228. B. Jowett to F.Y.E, 14 May 1869, from Oxford to Balliol College (EP NUFFIELD C UO, Box D 2 / 1 /37).
229. Report to the Electors for the Drummond Chair of Oxford submitted at the end of 1890. See below, Section 3.5 and Appendix H.

230. Report to the Council of King's College submitted in June of 1888. See below, Section 2.18.
231. 1869 (date of the postage stamp): Robert Williams to F.Y.E. in Balliol (EP NUFFIELD C UO, Box D 2 /1).
232. J. Bryce from Temple to F.Y.E., 9 June (EP NUFFIELD C UO, Box D 2 / 1).
233. C.E. to H.B., no date, 1870? (EP BODLEIAN UO, Ms. Eng. lett. c 733).
234. C.E. to H.B., 13 August 1870? (EP BODLEIAN UO, Ms. Eng. lett. c 733).
235. H.J.B.'s report around 1938 (EP NUFFIELD C UO, Box 3/9).
236. H.J.B.'s report around 1938 (EP NUFFIELD C UO, Box 3/9).
237. C.E. to H.B., November 1871 (EP BODLEIAN UO, Ms. Eng. lett. c 733). Quoted by Creedy (1986), p. 12.

Francis Ysidro Edgeworth in about 1880

Reproduced with permission from Dr David Edgeworth Butler

2. The making of Francis Ysidro Edgeworth

As one,
Who versed in geometric lore, would fain
Measure the circle; and, though pondering long
And deeply, that beginning, which he needs,
Finds not: e'en such was I, intent to scan
The novel wonder, and trace out the form,
How the circle fitted, and therein
How placed: but the flight was not for my wing;
Had not a flash darted athwart my mind
And, in the spleen, unfolded what it sought.
(Paradise, Canto XXXIII, vv. 133–141).
Dante Alighieri, *The Divine Comedy*.

2.1 1872–76: THE SAVILE AND THE ATHENÆUM

After the death of his brother David in October 1871, Francis Edgeworth settled in London. He joined two clubs, the Savile and the Athenæum, and he took rooms in Hampstead. These three places and the Library of the British Museum were the venues where he was to spend most of his time in London over the coming years.

Edgeworth joined the Savile Club in 1872 or perhaps in 1871 at the earliest. This club had been founded as the Eclectic Club in 1868. It then changed to the New Club, and when it moved to number 15 Savile Row in 1871 it was renamed the Savile Club. In 1880, the club moved yet again, this time to number 10 Piccadilly, but it kept its name. Although the purpose of the club was simply good fellowship, from the start there were quite a few members interested in literary and scientific pursuits. One of the founders was Edgeworth's classmate at Trinity College, Alfred Perceval Graves, who was probably the one who introduced Francis to the club. In 1872, Edgeworth seconded the nomination of his brother Eroles as a Savile member.[1]

Thus, Eroles Edgeworth had returned to London after his adventurous stint in France. Two years later, he went back there to marry a wealthy French widow, as Harriet Jessie Butler reports:

> But there [in Pau] he met Mme . . . [Françoise Delcher, the daughter of a French Colonel], a rich widow who, strange to say, married him [in 1874] and kept him in order, forcing him to return to Edgeworthstown in 18 . . ., [1881, see below] and refurnishing the house with her money, made him live there quite respectably.[2]

But let us go back to the Savile Club, which became an important place for Edgeworth. He used this club as his personal mailing address[3] and dispatched his correspondence from there as well. The club was a fine place for meeting people, most of them young, who were interested in literature, science and the arts, like Arthur Balfour, the future prime minister, Thomas Hardy, Charles Darwin and Robert Louis Stevenson. Stevenson described the atmosphere at the Savile, where 'gather daily those young eaglets, the swordsmen of the pen, who are the pride and wonder of the world, and of the effete pensionnaires of the Athenæum. They are all young . . . and they are all rising.'[4]

The Athenæum had much more clearly scientific and artistic leanings, and it was much older than the Savile. It had been established in 1823 by the famous bibliomaniac Richard Heber, whose personal library had nearly 150 000 volumes, as well as by the Irish writer and Secretary of the Admiralty John Wilson Croker, the Member of Parliament Joseph Jekyll, the sculptor Sir Francis Legatt Chantrey, the painter Sir Thomas Lawrence and others such as the physicist Michael Faraday, who was appointed Secretary. The Athenæum's library has always been the best club library in Great Britain. In 1829, the club moved to its present headquarters at Pall Mall in a new building designed by Decimus Burton.

According to Charles Dickens Jr in his *Dickens's Dictionary of London*, in 1879 the entry fee at the Athenæum Club was 30 guineas, and the yearly membership fee was eight guineas. The Savile Club was more modest, as its entry fee was a mere ten guineas and the yearly membership fee was only four guineas.

With regard to his settlement in London, we know neither the precise date when Edgeworth chose to establish residence in Hampstead, nor why he chose to do so. Perhaps he intended to find the healthiest spot near London, a place from whence he could hike through the woods and fortify his own health, which had declined so abruptly in 1870 and 1871. The two rooms that he rented at number 5 Mount Vernon, at the top of Hampstead Heath, were just perfect for him. He kept them as his living quarters in London for the rest of his life.

Another reason could have been that Hampstead was becoming a home to many scientists and artists; some of his new friends at the clubs lived there and encouraged him to do so. In fact, another member of the Savile Club lived at number 7 Mount Vernon, next door to Francis, for a short period, namely the aforementioned writer R.L. Stevenson. However, there is no evidence that they ever became close friends.

As to the year when Edgeworth settled in Hampstead, it was most likely before 1874, since there is a letter from Christina Edgeworth to Harriet Butler in November 1874 saying that he had visited them in Anerly and 'looked very well', and that 'it is quite remarkable to see how he improves as he takes to practical work'.[5]

What is Christina referring to by this 'practical work'? Does she refer to the fact that Francis took exercise around Hampstead? Yet at that time, physical exercise was seldom referred to as 'practical work'. Therefore, she is probably alluding to Edgeworth's teaching jobs. According to the testimonials submitted at the end of 1875 when he applied unsuccessfully for a professorship of Greek at Bedford College, London, he was teaching classics at Mr William Baptiste Scoones's private academy, where the prospective Indian Civil Service candidates were trained. 'Mr Edgeworth's lectures in Greek were very highly appreciated by a proverbially hypercritical class of students',[6] wrote Mr Scoones in his testimonial. From another testimonial issued in 1881, we learn that Edgeworth had also been teaching the same type of student in Mr Walter Wren's private academy for two years. However, at Wren's he was teaching metaphysics and the moral sciences, namely 'Logic, Mental and Moral Science',[7] a subject in which he became increasingly involved. Mr Wren reported that 'Mr Edgeworth's punctuality, zeal, and earnestness with his classes matched well with his extensive learning'. Francis may well have found these first teaching jobs through his Uncle Pakenham, the former top in his class of the East India College and very well known in British-Indian circles.

At the same time, Edgeworth was studying law at the Inner Temple. Based on his earliest publications, it is clear that in this period he was also delving intensely into ethics and mathematics.

As mentioned above, in 1875 he applied for the professorship of Greek at Bedford College, London, and he submitted testimonials from B. Jowett and T.H. Green (Balliol), John Mahaffy, A. Palmer, R.N. Tyrrell, J.K. Ingram (all four from Trinity College, Dublin) and the aforesaid Mr Scoones and Mr Wren.[8]

2.2 HAMPSTEAD LIFE WITH JAMES SULLY

At the Savile Club, Francis met James Sully, a psychologist under the academic protection of Alexander Bain, then Professor of Logic at Aberdeen University.[9] Sully would become famous around 1875, the year he published *Sensation and Intuition*. In *My Life and Friends*, Sully later wrote: 'the Savile Club helped to enlarge my literary and scientific circle. It was here where I was introduced to S. Hodgson, W.K. Clifford and F.Y. Edgeworth, whose many-sidedness of mind refuses to be contained even under two categories.'[10]

Though this is the only explicit mention of Francis in Sully's book, he is later referred to as the author's beloved 'bachelor chum'.[11] Why James Sully hid Francis Edgeworth's identity when he referred to this close friendship we shall never know. In fact, Sully became Edgeworth's best friend and was one of his main academic influences during the period from 1875 to 1890, when they both lived permanently in Hampstead.

Sully, three years older than Edgeworth, earned a Master of Arts in philosophy and studied in Göttingen and Berlin. After being married in 1868, he lived in Italy and then in the London Borough of Brent, specifically in the district of Harlesden. Around 1875 to 1876 he settled in Hampstead, because of 'my bachelor chum, who attracted me to Hampstead and I selected a cottage only a few steps from his abode'.[12] Sully then described their daily life:

> We met almost daily, now for a short after-breakfast walk, now for a longer tramp,. . . and now on a skating excursion to Hendon and Elstree. Later on the bicycle came upon the scene. In another bodily exercise I was quite unable to join him – the early-morning dip in the bathing-pond – even in winter. Our lines of study overlapped somewhat and this, together with the multiplicity of his interests, made conversation fluent.

And Sully paid homage to their friendship:

> Although my junior, he secured to take special care of me. This was due in part to the sweet courtesy of his nature, for he came from the island of hospitality. I spent a week with him in his ancestral home a summer week made the more fragrant by his delicate attentions. Propinquity no longer makes it easy for me to get possession of his genial and sustaining comradeship.[13]

Sully also described other Hampstead friends from his first years there, who were probably also acquaintances of Francis, such as:

> Charles Lewes[14] and family; Miss Coats and Miss James (Upper Terrace); Canon Ainger, who lived two doors from Miss James and his beloved Charles Lamb, Du Maurier and wife and children and St. Bernard dog 'Chang' (who lived a few steps from the Flagstaff). These acquaintances ripened into long friendships.

Sully was also a good friend of Robert Louis Stevenson. Once when Stevenson was sick at his country house at Skerryvore, Bournemouth, Sully visited him several times.

This network of friendships, in which Sully was a central figure, made Hampstead a very agreeable place to live. And it became even more exciting after 1878, when Edgeworth and Sully met William Stanley Jevons. But before we delve into this important encounter we first have to describe the events of 1877, a glorious year for both Sully and Edgeworth.

2.3 1877: *NEW AND OLD METHODS OF ETHICS*

Sully and Edgeworth's intellectual interaction becomes quite clear if we list the references used by Francis in his early works, which include a brief note and a small book. The brief note, just two paragraphs long, published in *Mind*, the journal that had just been founded by Alexander Bain in 1876, is entitled 'Mr Matthew Arnold on Bishop Butler's Doctrine of Self-Love'.[15] The short book, entitled *New and Old Methods of Ethics*, was published in 1877 by James Parker, Oxford and London, at Francis's expense.[16] The most significant references include: Bain, *The Emotions and the Will*; Barratt, *Physical Ethics*; Butler, *Sermons on Compassion*; Fechner, *Element der Psychophysik*; Helmholtz, *Popular Lectures on Scientific Subjects*; Hume, *Treatise on Human Nature*; James Mill, *Analysis of the Phenomena of the Human Mind*; John Stuart Mill, *A System of Logic* and *Utilitarianism*; Sidgwick, *Methods of Ethics*; Spencer, *Principles of Psychology*; Sully, *Sensation and Intuition* and *Pessimism*; Tyndall, *Scientific Materialism*; Wundt, *Principles of Physiological Psychology*; and last but not least, Todhunter, *Researches in the Calculus of Variations*.

With the exception of this last mathematical text, it is obvious that the remainder of Edgeworth's readings were very familiar to Sully, and he may well even have recommended some of them directly to Francis, and later discussed them.Therefore, it is not surprising that on the first page Edgeworth acknowledged that his 'friend Mr James Sully' had 'revised and corrected the following pages' and thanked him 'for many suggestions'.

Moreover, Edgeworth added in a footnote on page 14 that some peculiar relations concerning non-hedonistic actions 'were pointed out to the present writer by Mr Sully'.[17]

Despite the fact that *New and Old Methods of Ethics* was practically his first work, all the elements of his unique style are present. Greek and Latin quotations are blended together with English poetry, of which the following verses from Alexander Pope's *An Essay on Man* are among his favourites:

> Trace science then, with modesty thy guide;
> First strip off all her equipage of pride.

Deduct what is but vanity or dress,
Or learning's luxury, or idleness:
Mere tricks to shew the stretch of human brain.[18]

The most novel aspect of his style was the reference to advanced mathematics, not for the purpose of obtaining exact results but mainly as a precise language for describing certain types of abstractions or representing dynamic processes. Edgeworth was not addressing himself to the common reader; rather he took for granted that his readers were as educated as he was in ancient and modern languages, as well as mathematics. This is why John Maynard Keynes ironically remarked: 'quotations from the Greek tread on the heels of the differential calculus, and the philistine reader can scarcely tell whether it is a line of Homer or a mathematical abstraction which is in course of integration'.[19]

New and Old Methods of Ethics, a 92-page long monograph, is divided into two parts. In the first one, Edgeworth set out, criticised and compared Alfred Barratt and Henry Sidgwick's then-recent theories on ethics which were contained respectively in *Physical Ethics* and *Methods of Ethics*.[20] The main question at hand was whether all ethics can be built out of the physical argument developed by Barratt:

> Moral science is a section of that division of Physics which treats of animate nature, and its special subject is the relation which exists between the active and passive elements of that nature. The fundamental principle, therefore, from which it starts, is the ultimate correlation of the two primary qualities of organized matter, irritability and contractility. . . . The law of action. . . comes to be this, that a pleasurable state produces a reflex motion of acceptance, and a painful state, one of repulsion and avoidance; all motion being comprehended under these two classes. Hence the physical correlation of contraction and irritation, when expressed in the inner language, comes to mean the aiming at pleasure and the avoidance of pain.[21]

According to Barratt, Edgeworth argued, 'Moral Science is a section of Physics', but 'upon the whole, Mr Barratt's arguments . . . have not much controversial value as against the moralist, who finding in his consciousness intuitions of beauty and truth and duty and the good of others, is determined to pursue those objects rather than the subjective feeling of pleasure'. Edgeworth realised that 'the dawn of Physical Ethics is not yet at hand' and hence, 'morality might be no more injured by physical science than music by acoustics'. Therefore, 'we ought not in the meanwhile to neglect the domestic light of introspection' as Professor Sidgwick does in the *Methods of Ethics*, 'one of the most brilliant sources of that light'.[22]

In the second part, Edgeworth used the high mathematics of calculus of variations for the first time in this field in order to develop the various specific mathematical meanings of what he called 'exact utilitarianism, as distinguished from Hume's non quantitative principle of utility, and the not very explicit greatest-happiness principle of Bentham and his followers, including J.S. Mill'. He ascribed 'exact utilitarianism' to Fechner and Sidgwick. According to Edgeworth, 'the doctrine of Fechner and Sidgwick [. . .is about] the greatest happiness of sentients, exclusive of number and distribution – an end to which number and distribution are but means'. Following the laws stated by Weber, Fechner, Delbœuf, Hemholtz and Wundt, Edgeworth conceived individual happiness as a dependent variable that increases at a decreasing rate in response to each stimulus, and therefore as a function with a maximum for empirically relevant values. In fact, as we shall see in the next section, without being cognizant of it, he was dealing with that sentient characteristic which political economists referred to as 'utility'.[23]

Moreover, the most amazing thing was the formal way Edgeworth dealt with the functional forms concerning the individual happiness U and V of two different persons when approaching the question 'to find the greatest quantity of one thing *(U)* of, or in relation to, the greatest quantity of another thing *(V)*'. Instead of using a specific function depicting the Fechnerian law, which states that the happiness of each sentient individual is a function of the logarithms of the stimulus, he generalised to any function in which 'the first differential is positive, the second differential negative for values of the variable, or at least all with which we are concerned'. Therefore, the curve representing the function should be concave to the abscissa. Nevertheless, he admitted that 'the form of the pleasure-curve as delineated by Wundt[24]. . . is not continously concave to the abscissa, as our condition demands; the lower part is convex, the upper concave'. However, Edgeworth thought that 'there is reason for supposing that the *upper* part of the curve is alone that capable of being employed in cases of maximum pleasure, and therefore alone concerning us here'. Thus, through Wundt, Edgeworth described a function of individual happiness that greatly resembles the total utility function displayed in modern-day textbooks.[25]

After posing the question about jointly maximising the individual happiness U and V, he then pondered the possible dependence between them and sketched this dependence through a function $W(U,V)$.[26] In another of the possible cases of dependence, he considered the problem of the maximisation of U subject to given values of V and for the first time in a text on social science introduced the Lagrangian function with its corresponding Lagrangian multiplier.[27]

The final and main part of the book, 'Proof of Exact Utilitarianism', poses the utilitarian question in a form rather familiar to present economists:

> Problem I. Given a certain quantity of stimulus to be distributed among a given set of sentients (with the condition that every element is to have *some* stimulus) to find the law of distribution productive of the greatest quantity of pleasure.

Edgeworth solved this problem mathematically in several specific cases, such as:

> Case I. Where all the elements are equal as touching sensibility and capacity for pleasure. . . . the law of distribution is equality.
>
> Case II. Where the sensibility only varies. . . . The same conclusion is deduced.
>
> Case III. Where the capacity-for-pleasure only varies. . . . Unto him that hath greater capacity for pleasure, shall be added more of the means of pleasure.
>
> Case IV. Where both sensibility and capacity vary. . . . The same conclusion, as in the preceding case, is deduced.

Then Edgeworth dealt with Problem II, in which he dropped 'the condition that every element is to have *some* stimulus' and considered each sentient as 'infinitesimal with regard to the whole' since, 'strictly speaking, each pleasure element consists of the indefinitely small element of the whole sensory tract to which stimulus is applied.'[28]

By introducing the 'whole sensory tract', his intention was not purely formal since he added that 'the whole sensory tract is generally considered as made up of tracts belonging to different sentient individuals, e.g., different animals'. By doing so, Edgeworth tried to include exact utilitarianism into the theory of evolution since capacity for pleasure is supposed to vary among creatures according to their position in the evolutionary order.[29] Edgeworth's solution to Problem II in the cases 'where the capacity-for-pleasure only varies' may shock modern readers not immersed in the vogue of evolutionism of those times:

> Unto him that hath higher development shall be added more of this world's goods. This deduction agrees with common sense, as exhibited in the approved dealings of men with animals, of civilized with savage races, in the privileges of aristocracy approved in ages when aristocracies really represent a higher order of evolution.[30]

After the publication of the book, Edgeworth sent it to the most renowned authorities on the subject. We know that one of them was Professor Bain, whose work he knew and admired through Sully. Bain encouraged him by replying: 'it is not often that Ethical problems are so well conceived or handled with so much verve. I duly appreciate the boldness of your attempt at a Hedonistic Calculus, and wonder that it has not been tried former[ly]'.[31]

2.4 'THE HEDONICAL CALCULUS'

In this same year, 1877, James Sully published *Pessimism: A History and a Criticism*.[32] In the preface he acknowledged 'the assistance lent by my friend Mr F.Y. Edgeworth, M.A. of Balliol Coll., to whose careful perusal of the proof sheets I am indebted for numerous improvements both in the argument and in the style of the work'.

So, as good friends who were proud of each other, they shared the task of revising the proofs of their respective books. In fact, according to Keynes, this exchange was both a practical and a wise decision, at least on Edgeworth's behalf, since 'he was quite incapable of detecting misprints in what he wrote himself, but he had an exceptionally sharp eye for other people's'.[33]

Master of Arts and Barrister-at-Law

The year 1877 was a productive year for Edgeworth in earning titles. He acquired the degree of Master of Arts from Oxford and was called to the Bar by the Inner Temple. This means that he had successfully completed his studies in law at the professional level. However, by then he had decided not to pursue a legal career. He probably studied law influenced by the tradition of the Anglo-Irish landowners' families that his grandfather had followed and Maria Edgeworth had emphasised in her books.[34] Once he had begun these studies, Francis had to finish them since with a professional title under his belt he would feel less anxious about playing academic roulette. Moreover, the most original subject in his research so far had been what he had called 'Exact Utilitarianism', which was close to theoretical research in natural law.

Harriet Jessie Edgeworth's Marriage

The last news that we have from 1877 is the marriage of Edgeworth's cousin Harrie, then nearly 26 years old. Harriet Jessie Edgeworth married the Rev. Arthur Gray Butler and went to live in Oxford. So, following in the footsteps of her aunt Harriet, she became the second Harriet Butler. The bridegroom, 20 years Harrie's senior, was the third son of the Rev. George Butler. After a brilliant career at Oxford, A.G. Butler was elected Fellow of Oriel College in 1856, and he served as assistant master under Frederick Temple at Rugby. He was very successful afterwards as the first Headmaster of Haileybury and was ordained a priest in 1862. Five years later, he resigned his Haileybury position due to a nervous breakdown and pursued a new career as a tutor in Oriel from 1875 to 1895. Arthur Gray Butler's family had a tradition of being headmasters. His father, Rev. George Butler, had been Headmaster of Harrow and Dean of Peterborough, his brother George was Headmaster of Liverpool College, his brother Henry Montagu was Headmaster of Harrow, and another brother, Spencer Percival, was a barrister.[35] With regard to Francis Edgeworth, it is interesting to note that Arthur Butler's sister Louisa had married Francis Galton, grandson of Erasmus Darwin, who became a famous scientist and statistician and would come to be one of Edgeworth's main supporters in academic circles.

'The Hedonical Calculus'

In July 1879, Edgeworth published in *Mind*, 'The Hedonical Calculus', in which he followed his 1877 investigations into 'Exact Utilitarianism' in *New and Old Methods of Ethics* by posing a broader mathematical problem:

> PROBLEM. To find (α) the distribution of means and (β) of labour, the (γ) quality and (δ) number of population, so that there may be the greatest possible happiness.[36]

To approach it, Edgeworth stated a set of 'appropriate postulates', of which the first is: 'The rate of increase of pleasure decreases as its means increase'. And he remarks:

> The proposition thus stated is evidenced by every-day experience; experience well focused by Buffon in his 'Moral Arithmetic', Laplace in his 'Essay on Probabilities', William Thompson in his 'Inquiry into the Distribution of Wealth' and Mr Sidgwick in the 'Methods of Ethics'.
>
> This empirical generalisation may be confirmed by 'ratiocination' from simpler inductions, partly common to the followers of Fechner, and partly peculiar to Professor Delbœuf. . . . The very parameter in virtue of which such functional variation occurs is exhibited by Professor Delbœuf in the case of eye-sensations; that a similar variation holds good of pleasures in general is Bain's Law of Accommodation. Increase of means, then, affording proportionately increased repetition of the conditions of pleasure, does not afford proportionately increased pleasure.

Thus, Edgeworth was dealing with the law that some authors interested in political economy had named the 'law of decreasing final utility'. Indeed he was working in the same field of research as Jevons, Marshall or Walras, yet he was unaware of this fact since, despite his early readings of the classical political economists, he thought that he was on a more general level that would encompass the different social sciences.[37]

The complete text of the article was reproduced by Edgeworth in his second and most famous book, published two years later, entitled *Mathematical Psychics*. The only changes were the title, 'Utilitarian Calculus' instead of 'Hedonical Calculus', and two additional footnotes in which he mentioned Jevons and Marshall.[38] In another footnote on page 34 of the new section on 'Economical Calculus', Edgeworth pointed out that some of the laws there stated had already been stated in the Utilitarian Calculus. And he added that 'the proofs were offered in *Mind*, without acknowledgement, because without knowledge, of the cumulative proofs already adduced by Prof. Jevons'.

The 'cumulative proofs adduced by Prof. Jevons' – which we interpret as meaning those published in his 1871 book – were quotations from Jeremy Bentham, Alexander Bain, Nassau Senior, Richard Jennings and T.E. Banfield. Edgeworth's independent references and arguments were interesting enough to be taken into consideration, but in this case, as he would always do in the future, he declined to take credit for any kind of intellectual contribution and preferred to rely upon 'authority', even upon 'the undemonstrated dicta and opinions of the wise, who have a power of mental vision acquired by experience'.[39]

These new footnotes added in 1881 therefore prove that Edgeworth had not met either Jevons or Marshall before writing the 1879 article.

2.5 SULLY, JEVONS AND EDGEWORTH

William Stanley Jevons, ten years older than Edgeworth, had moved from Manchester to London because he wanted to deliver fewer lectures and have more time to pursue his scientific research. Thus, he resigned his post at Owens College to establish himself in London. However, while he was in the process of moving, he was offered a professorship in Political Economy at University College, London, with rather light lecturing duties. He accepted, yet after only four years he resigned because he again felt too constrained in his research due to his feeble health. In order to live in London, he found a house in Hampstead where he started living in October 1876. The house was named The Chestnuts and was located in Branch Hill, East Heath Drive, less than one mile from Edgeworth's home.[40] He had already published most of his works, namely his texts on the value of money (the 1863 opuscule, *A Serious Fall in the Value of Gold Ascertained, and its Social Effects Set Forth*, and the 1865 article, 'On the

Variation of Prices, and the Value of the Currency since 1782'), on the theory of exchange (contained in his famous book from 1871, *The Theory of Political Economy*) and on logic and method (the 1874 book, *Principles of Science: A Treatise on Logic and Scientific Method*), all of which would influence Edgeworth's work.[41]

Around the year 1879, Jevons met Sully, who later introduced him to Edgeworth. Let us eavesdrop on Sully's impressions:

> I found Stanley Jevons a particularly helpful colleague and, as he lived near us, acquaintance soon ripened into friendship between the families. He was a valuable companion, full of all sorts of knowledge, gained not only from books, but from shrewd personal observation. He was given to rummaging in bookshops searching for out-of-the-way pamphlets, sometimes in most unpromising-looking suburban quarters. He collected a large number of such papers, which he carefully pigeon-holed. A common interest drew him and my chum together, and so we made a trio in many a pleasant walk and skating excursion.[42]

Jevons may have met Edgeworth either in Hampstead or at the Savile or the Athenæum, since all three men, Sully, Jevons and Edgeworth, were members of both clubs. We have already seen that Sully and Edgeworth were members, and Jevons's membership is evinced by a letter from November 1878, in which he boasted to his brother Tom: 'I have just had a pleasant lunch at my little club in Savile Row with Harry Roscoe [a cousin] and Huggins the astronomer.'[43] Jevons and Edgeworth first met before the end of 1879, since there is a letter by Aunt Harriet Butler, from Armagh, dated 'Dec. 27, 1879' in which she comments that 'you were fortunate in finding a Hampstead neighbour in Mr Jevons able to appreciate your theories and an agreeable companion besides. . . . Dr. Robinson is prostrate and suffering under an anthrax on his back which had to be lanced.'[44]

Leaving aside the ailments of Dr Robinson, then 87 years old, we realise that Francis had not informed Aunt Harriet that Jevons was the grandson of William Roscoe. As mentioned above, Jevons's grandfather had been a great friend of Francis's grandfather, Richard Lovell Edgeworth, who used to visit him in Liverpool. There is no comment about this friendship in any letter from Aunt Harriet, so perhaps Jevons and Edgeworth were never aware of the closeness of their grandfathers.

With regard to the date of their acquaintance, they may have already met by April 1879, since Jevons wrote a letter to Harold Rylett in which he stated:

> My interest in Ireland is rapidly increasing, and when I had a run of a week through some parts [he refers to his August 1875 trip to Dublin and Tipperary] I resolved to come again. . . My impression is increasing to the effect that

> landlordism is a terrible burden on the country, and that the just laws of England are rather a myth.[45]

Those were the days when the Irish members of the Parliament, Biggar and Parnell, were starting a political movement that centred British attention on Ireland by obstructing the work of the House of Commons. And shortly thereafter, in October 1879, Davitt and Parnell founded the Irish National Land League in Dublin. The main objectives of the league were to ensure that tenants paid a fair rent and had a fixed tenure. The long-term aim was for farmers to own the land, thus becoming 'peasant proprietors'. The Land League rapidly became an extremely popular movement that taught the Irish farmers to stand up for their rights.

In two passages from Jevons's writings we also find some proof that around 1880 they were well acquainted. The two passages refer to Edgeworth and his Irish grandfather:

> Richard Lovel [*sic*] Edgeworth's wind chariots are as likely to supersede railway trains as the flying machines of more recent inventors.[46]

> Cantillon winds up the subject prophetically by suggesting that it is a question whether it is better for a kingdom to be filled with a multitude of very poor inhabitants, or with a less considerable number of better maintained persons. Here is a forecast of the most recent hedonic speculations of Mr F.Y. Edgeworth.[47]

Edgeworth's friendship with Sully was more intense. Sully's description of his own life in Hampstead helps to cover the gaps arising from Francis's total lack of concern for his future biographers:

> Walking was my *pièce de résistance*. There were solitary walks, walks with my chum or other friends, as well as with a smaller or larger company (Leslie Stephen, Colonel Osborn). There were skating parties in the winter to Hendon or Elstree. Later there came the mixed cycling party and tennis parties on the Library Court.[48]

Sully went to concerts at St James's Hall or in Hampstead, to the theatre to see Sarah Bernhardt and Coquelin père, and preferred Wagner's musical dramas over Italian opera. Sully also read novels in French, Spanish and English – as did Edgeworth – and even in Norse. He visited Norway during holidays, and Francis might have accompanied him on some of these holidays. In May 1880, Aunt Harriet wrote back to him that she hoped that Francis would 'be able to tramp in Norway'.[49] We have no confirmation of this journey from other sources. Jevons also went there in July 1879, 1880 and 1881, not with them, but with close relatives.[50]

It was reported by Noel Annan that both Sully and Edgeworth had joined the 'Sunday Tramps' founded by Leslie Stephen after his second marriage. 'Stephen, the lawyer Frederick Pollock, and Croom Robertson, the first editor of the philosophical journal *Mind*, were in the habit of taking long Sunday walks and began to invite others'. In addition to Edgeworth and Sully, other members were 'the Positivist James Cotter Morison; Scrutton and Romer represent[ing] the law, W.P. Ker criticism, Maitland history and Robert Bridges poetry. . . . Twenty miles was an average stroll and the rule of the order was high thinking and plain living.'[51]

Leslie Stephen, the founder of the Sunday Tramps, had married his second wife, Julia Duckworth, in March 1878 and was to become known, despite himself, as the father of Vanessa Bell and Virginia Woolf, his two daughters from this second marriage. At that time, he was writing a book called *The Science of Ethics* that would be published in 1882; its title suggests that at the tramps meetings he might have been a fine conversational partner for both Sully and Edgeworth. Stephen was also a proselytiser for The Alpine Club, whose biannual bulletin, *The Alpine Journal*, he edited during the years 1869–1872. Francis became a member of the club, but he never contributed to the journal in spite of his mountaineering trips to Switzerland and Italy.[52]

2.6 ENTER MARSHALL AND FOXWELL

For Edgeworth, Alfred Marshall (1842–1924) was first a name uttered by Jevons, as he reported after Marshall's death:

> Alfred Marshall first became for me a notable name when Jevons (in 1879 or 1880) conversing about mathematical economics, recommended as the latest contribution to that subject the now celebrated papers on the *Pure Theory of Foreign Trade and Domestic Values*. At the same time Jevons highly praised the then recently published *Economics of Industry*. Eagerly studying these writings, I discerned a new power of mathematical reasoning, not only in the papers bristling with curves and symbols, but also in certain portions of the seemingly simple textbook.[53]

The Pure Theory of Foreign Trade and Domestic Values was originally a manuscript that Marshall used to lecture on foreign trade. Henry Sidgwick somehow got a copy and at the end of 1878 requested permission to print a few more for private circulation among friends, Jevons among them. *The Economics of Industry*, published in October 1879 by Alfred Marshall and his wife Mary Paley Marshall, is an introductory text in political economy which sold quite well.[54]

In spite of their mutual respect, Jevons's relationship with Marshall was never intense. During their life, they exchanged only three letters each, two in 1875 and the last in 1879. Jevons wrote to Marshall for the first time in January 1875, while Marshall was a fellow at Cambridge. Jevons, then a professor at Owens College, Manchester, had been visiting Cambridge as one of the examiners for the Moral Science Tripos, which had begun as always on the last Monday in November:

> Ever since our most agreeable visit to Cambridge I have been intending to write to you and say how much I was interested in the [Marshall's students'] answers in Political Economy. While at Cambridge indeed, I called at your rooms in hope of seeing you but found you had gone down.[55]

Edgeworth admired those early writings by Marshall and started corresponding with him around 1880, when Marshall and his wife Mary Paley were living in Bristol. Francis sent him his *Hedonical Calculus*,[56] to which Marshall replied:

> I had heard of your paper in *Mind* and had intended to read it, but I had forgotten your name. I have now nearly read all the book you sent me and am extremely delighted by many things in it. . . . As to the interpretation of the Utilitarian dogma, I think you have made a great advance: but I still have a hankering after a mode of exposition in which the dynamical character of the problem is made more obvious; which may in fact represent the cultural notion of happiness as a process rather than a statical condition.[57]

One of the subjects that Edgeworth raised in his correspondence with Marshall was the use of mathematics in political economy. Marshall's point of view, which Edgeworth neither directly criticised nor thoroughly subscribed to, was explained in a letter to Francis in 1881:

> I may say generally that in writing the *Economics* [*of Industry*] my chief difficulty was to be mathematically accurate without introducing apparently pedantic limitations and indeed without suggesting difficulties that would only perplex the nonmathematical. . . . [For pedagogical purposes] I intend never to use [mathematical] analysis when I can use geometry. . . . My experience of the exact treatment of supply and demand in inference has been disappointing.[58]

Marshall's mathematical approach was to be described afterward by Edgeworth as 'bearing under the garb of literature the armour of mathematics'.[59]

A copy of the article *Hedonical Calculus* was also sent by Edgeworth to Foxwell, Professor of Political Economy at University College, London. Foxwell was four years younger than Edgeworth and a friend of Jevons and Marshall, who had lectured him in economics in 1870. In a testimonial written in October 1881, Foxwell affirms:

> [Around 1880] I had the pleasure of making the acquaintance of Mr Edgeworth through our common friend Professor Jevons, at whose house and elsewhere we have had many discussions, chiefly upon economics, but frequently also upon ethical and philosophical subjects. The impression made upon me by Mr Edgeworth's writings was strengthened by these conversations, in which he showed great speculative ability, singularly wide and various acquirements, and marked originality and vigour of expression.[60]

Through Jevons and Marshall, Edgeworth immersed himself in the study of the works by the new political economists, including Walras, and their predecessors like Cournot. Through Foxwell, he realised that the classical authors who were already familiar to him, including Ricardo, Senior and Mill, should be kept in mind and used for his lectures on political economy. In 1880 he started to lecture in the Ladies' Class of King's College London at Kensington. At the same time, he was also teaching logic, which encompassed mental and moral science, in the evening boys' classes at the same college. This is corroborated by the Principal of King's, Alfred Barry, in a testimonial, where he adds:

> I need not, of course, speak of Mr Edgeworth's ability and knowledge of the subjects with which he deals. But I may perhaps be allowed to say how thoroughly well he proves himself able to impart knowledge, and to control and interest a class. He is himself an enthusiast for knowledge and for teaching, and this spirit he always infuses into his students.[61]

Edgeworth's connection with King's College London would span eleven years, from 1880 to 1890.[62] These King's College lectures were poorly paid, and although his inherited private income allowed him to survive comfortably, he repeatedly tried to secure a better academic position. However, in order to succeed, he first needed to bolster his curriculum vitae and build a reputation in academic circles through his publications.

2.7 *MATHEMATICAL PSYCHICS*

Few books have been so widely ignored when they appeared and so highly valued a century later as Edgeworth's *Mathematical Psychics*, a small book that was privately published by the author in early 1881.

The first page sets forth the research programme that Edgeworth was planning to undertake throughout his life. Thus, in spite of Edgeworth's apparent diversity of concerns, *Mathematical Psychics* offers the key that gives unity to the ambitious and apparently heterogeneous programme designed to apply mathematics to the moral sciences, that is, 'the application of mathematics to *Belief*, the calculus of Probabilities [. . . and] the Calculus of *Feeling,* of Pleasure and Pain'.[63]

In this work, he concentrates on the second part of the programme, the calculus of feeling or the 'Calculus of Pleasure', which may be subdivided into two fields, 'namely, Economics and Utilitarian Ethics. The Economical Calculus investigates the equilibrium of a system of hedonic forces each tending to maximum individual utility; the Utilitarian Calculus, the equilibrium of a system in which each and all tend to maximum universal utility.'[64]

Preliminary Considerations

The first point that Edgeworth elaborates is an apologetic defence of the use of mathematics in social sciences, based on the argument that, though in social sciences numerical values may be unappraisable, we do not need to attach numerical values to reason in terms of general functional relationships. In a particularly eloquent statement, Edgeworth affirms:

> The science of quantity is not alien to the study of man, it will be generally admitted, in so far as actions and effective desires can be *numerically* measured by way of statistics – that is, very far, as Professor Jevons anticipates. . . . Where there are data which, though not *numerical* are *quantitative* – for example, that a quantity is *greater* or *less* than another, *increases* or *decreases,* is *positive* or *negative,* a *maximum* or *minimum,* there mathematical reasoning is possible and may be indispensable.[65]

Though Edgeworth refers to Jevons's *Theory of Political Economy*, it is only to add that 'the view adopted in these pages is expressed by Cournot, *Recherches*'. Edgeworth had probably arrived at Cournot through the preface to the second edition (1879) of Jevons's book.

Then Edgeworth argues that the application of mathematics to the social sciences entails the choice of a 'particular salient feature' that will give body to this application. The 'salient feature' which will 'countenance the application of Mathematics to the world of soul' will be 'Pleasure', through 'the hypothesis that Pleasure is the concomitant of Energy. *Energy* may be regarded as the central idea of Mathematical Physics; *maximum energy* the object of the principal investigations in that science.'[66] And Edgeworth, by analogy with the application of the Lagrangian method to dynamics developed by his admired friend William Rowan Hamilton, sketches the 'central conception' of his 'Economical and Utilitarian Calculus' in the following way:

> [The] accumulation (or time-integral) of energy which thus becomes the principal object of the physical investigation is analogous to that accumulation of pleasure which is constituted by bringing together in prospect the pleasure existing at each instant of time, the end of rational action, whether self-interested or benevolent. The central conception of Dynamics and . . . in general of Mathematical Physics is *other-sidedly identical* with the central conception of Ethics; and a solution practical and philosophical, although not numerical and precise, as it exists for the problem of interaction of bodies, so is possible for the problem of the interaction of souls.[67]

But Edgeworth admits that 'in Physical Calculus there is always a potentiality, an expectation of measurement; while Psychics want the first condition of calculation, *a unit*'. Then, without any previous definition, he enters into the quantitative analysis of 'Pleasure', by introducing a 'Utility' based upon Jevons:

> Utility, as Professor Jevons says – in reference to Economics – has two dimensions, *intensity* and *time.* The unit in each dimension is the just perceivable increment. The implied equation to each other of each *minimum sensibile* is a first principle incapable of proof. It resembles the equation to each other of undistinguishable events or cases, which constitutes the first principle of the mathematical calculus of *belief*. . . . Such is the unit of *economical* calculus.[68]

At this point Edgeworth mentions in a footnote Laplace's *Essai philosophique sur les probabilités*. We find, therefore, from the very beginning of Edgeworth's main work, one of his favourite topics, namely, the intertwining nature of utilities and probabilities. And he adds a new requirement when we arrive at the utilitarian calculus:

> For moral calculus a further dimension is required; to compare the happiness of one person with the happiness of another, and generally the happiness of groups of different members and different average happiness. Such comparison can no longer be shirked, if there is to be any systematic morality at all. It is postulated by distributive justice. . . . In virtue of what *unit* is such comparison possible? It is here submitted: Any individual experiencing a unit of pleasure-intensity during a unit of time is to 'count for one'.

And, suddenly, he displays, under the influence of Æschylus and Tennyson, his poetical inspiration:

> *Atoms of pleasure* are not easy to distinguish and discern; more continuous than sand, more discrete than liquid; as it were nuclei of the just-perceivable, embedded in circumambient semi-consciousness. We cannot *count* the golden sands of life; we cannot *number* the 'innumerable smile' of seas of love; but we seem to be capable of observing that there is here a *greater*, there is a *less*, multitude of pleasure-units, mass of happiness; and that is enough.[69]

So, despite the cardinal nature of 'Utility' and the problems associated with its quantification, Edgeworth is optimistic since the use of mathematics will lead to finding 'a solution practical and philosophical, although not numerical and precise, as it exists for the problem of the interaction of bodies,. . . for the problem of interaction of souls.' And this applies, not only 'to the utilitarian problem of which the object is the greatest possible sum total of universal happiness', but also to the economic problem of 'contracting agents actuating uniquely by self-interest'.[70]

Economical Calculus

Edgeworth starts the *Economical Calculus* by outlining a set of definitions:

> The first principle of Economics is that every agent is actuated only by self-interest. The workings of this principle may be viewed under two aspects, according as the agent acts *without,* or *with,* the consent of others affected by his actions. In wide senses, the first species of action may be called war; the second, *contract*.[71]

To complete the concept of 'contract', Edgeworth specifies that the content of the '*articles* of contract' is the 'amount' of each agent's contribution and 'the principle of distribution'. He then goes on to characterise 'economic *competition*' through an analogy with a poem by Tennyson:

> 'Is it peace or war?' asks the lover of Maud, of economic *competition*, and answers hastily: It is both, *pax* or *pact* between contractors during contract, *war*, when some of the contractors *without the consent of others recontract*. . . . The *field of competition* with reference to a contract, or contracts, under consideration consists of all the individuals who are willing and able to recontract about the articles under consideration. . . There is free communication throughout a *normal* competitive field. A *perfect* field of competition professes in addition certain properties. . . Any individual is free to contract . . . and to recontract with any out of an indefinite number, [. . . even]*without the consent* of any third party.

Therefore, a perfect competition allows no pre-contract among any group of agents limiting these properties. To complete the definitions, Edgeworth outlines the difference between two types of contracts: 'A *settlement* is a contract which cannot be varied with the consent of all the parties to it. A *final settlement* is a contract which cannot be varied by recontract within the field of competition. Contract is *indeterminate* when there are an indefinite number of *final settlements*.'[72]

After this first instrumental stage, Edgeworth poses the problem of 'how far contract is indeterminate'. And 'the general answer' is: '(α) Contract without competition is indeterminate, (β) Contract with *perfect* competition is perfectly determinate, (γ) Contract with more or less perfect competition is less or more indeterminate.'

Barter

To prove these assertions, Edgeworth considers first the problem of the bilateral exchange of two goods by two self-interested agents, a problem treated by Jevons and Walras, as Edgeworth duly acknowledges.[73] Jevons and Walras had dealt with price-taking agents by introducing the 'law of indifference of the open market' – in Jevons's terminology – or the 'uniformity of price' – in Walras's terminology.[74] This law simply adds the requirement that the ratio between the quantities exchanged, x and y, should be the same for all units, from infinitesimal sizes, dx, dy to real sizes, x, y. This means that $dx/dy = x/y$, or in other words, the ratio of proportionality of exchanges that expresses their relative price should be uniform with quantities exchanged. This also means that the two agents are price-takers and have no power to bargain. In this situation Jevons and Walras prove that exchange is perfectly determinate.[75]

Edgeworth takes a longer route to prove his three assertions, starting with the analysis of pure barter. To do so, he develops in a few pages some of the most used and useful devices in current economics: the set of indifference curves, the contract curve and a graphical representation of them that led to the Edgeworth box. And, moreover, in the process he pioneers the use of Lagrangian multipliers and determinants.[76]

To begin with, he introduces an important instrumental improvement, consisting of the fact that instead of using four functions of additive (fully cardinal) utilities, each of them depending on a single commodity as Jevons and Walras did, he uses two functions of utilities, one for each agent, which are the result of the two commodities taken jointly. This permits him to define the 'line of indifference' among bunches of commodities that result in the same utility level and the 'line of preference', perpendicular to the line of indifference. Once he has introduced these lines, he reasons that the agents:

> will consent to move *together* in any direction between their respective lines of indifference, in a direction *positive* as it may be called for *both*. At what point then will they refuse to move at all? When their *lines of indifference* are coincident (and *lines of preference* not only coincident but in opposite directions).

Edgeworth calls the locus of points at which the respective utilities of the two agents cannot increase together – and therefore refuse to move jointly – the 'contract-curve'. Each point on the contract-curve is an equilibrium point that 'may be described as a *relative-maximum*, the point at which the utility of one party being constant, the utility of the other party is a maximum' or vice versa. Therefore, 'equilibrium is attained when the *total pleasure-energy of the contractors is a maximum relative*', and consequently, 'the *total utility of the system is a relative maximum at any point on the pure contract-curve*.'[77]

We find, then, that in barter between rational, self-interested agents, there is not a single solution to the problem of determining the quantities exchanged, but a range of 'final settlements' located along the 'contract-curve'. And the precise amount of goods exchanged is a result of the bargaining power exerted by the exchangers.

Edgeworth represents these concepts graphically in a diagram where the axes express the amounts x and y of the goods exchanged. In this diagram he draws some curves of indifference and the contract-curve of the barterers. This graph already contains all the relevant ingredients of the geometrical device known as the Edgeworth box that has become a popular pedagogical tool.[78]

Edgeworth also shows that the price-taking solution of Jevons and Walras is one of the 'final settlements' located on the 'contract-curve'.[79]

The Role of the Number of Traders

After the analysis of this case of pure barter, Edgeworth goes on to demonstrate that '(γ) Contract with more or less perfect competition is more or less determinate', and '(β) Contract with *perfect* competition is perfectly determinate'.

To do so, Edgeworth proceeds by increasing pair by pair the exchangers of two given commodities and by making use of the possibility of 'recontracting', that is, declaring null a previous agreement by assenting to a new contract. This possibility implies that any exchanger who may be better off by leaving an old trade and contracting with new partners who offer better conditions, may do so. This will assume the forming of a coalition which entails a fairer distribution of the gains from trade among their members and the exclusion of the trader who does not accept this redistribution of gains. Each time we consider a new couple entering into the exchanging set, the extreme settlements at the contract curve forced by a greedy exchanger will have to be abandoned because of the risk of a coalition being formed against him.

And Edgeworth shows that in order to avoid coalitions among three, five, seven or more exchangers through recontracting, as the number of exchanging couples grows to infinity, the contract curve reduces to a single point. As he indicates, 'proceeding by degrees from the case of two isolated bargainers to the limiting case of a perfect market, we see how *contract is more or less indeterminate*'. Therefore, by gradually augmenting the number of similar exchanging couples, the rate of exchange between the two commodities tends towards a determinate value and, consequently, their relative market price is formed. So, Edgeworth shows that by adding similar traders, barter leads to replication of the price-taking solution of Jevons and Walras.[80]

Utilitarianism as a Principle of Arbitration

After obtaining this result of determinateness of relative quantities exchanged under perfect competition, Edgeworth closes the section 'Economic Calculus' with a call to correct the imperfections in the field of competition through arbitration:

> If competition is found wanting . . . economics would be indeed a 'dismal science' and the reverence for competition would be no more. There would arise a general demand for a *principle of arbitration*. . . . Throughout the whole region of in a wide sense *contract*, in the general absence of a mechanism like perfect competition, the same essential indeterminateness prevails; in international, in domestic politics; between nations, classes, sexes. The whole creation groans and yearns, desiderating a principle of arbitration, an end of strifes.[81]

Edgeworth wonders then about where to seek a principle of arbitration to resolve the prevailing indeterminateness. 'In *justice*, replies the moralist. [. . . But] the star of justice affords no certain guidance . . . unless it reflects the rays of a superior luminary – utilitarianism'. And he proves mathematically that, in the two exchangers' case, the application of the principle of utilitarianism – 'the greatest possible total utility of the contractors' – leads to a settlement, that is, a point in the contract curve. Edgeworth also proves that if two contractors actuate 'by a sympathy with each other's interests' then each of them tends to maximise, not his own utility, but the weighted sum of both utilities according to individual coefficients of effective sympathy – these coefficients reflecting how much they care about each other's utilities. Through this maximisation we get a contract-curve that is the old one 'between narrower limits'. And, 'as the coefficients of sympathy increase and utilitarianism becomes more *pure,* the *contract-curve narrows down to the utilitarian point*'. Edgeworth was very pleased with these results since they pointed out one singular peculiarity of the principle of utilitarianism as a principle of arbitration:

> In this direction, it may be conjectured, is to be sought the required principle. For the required basis of arbitration between economical contractors is evidently *some* settlement; and the utilitarian settlement may be selected, in the absence of any other principle of selection, in virtue of its moral peculiarities:

its satisfying the sympathy (such as it is) of each with all, the sense of justice and utilitarian equity.[82]

We find, therefore, that 'competition requires to be supplemented by arbitration, and the basis of arbitration between self-interested contractors is the greatest possible sum-total utility'. Consequently, through this principle of arbitration 'the *economical* leads up to the *utilitarian calculus*'.[83]

The section 'Utilitarian Calculus' is a transcription of the 1879 article 'Hedonical Calculus', with the addition – as mentioned above – of two footnotes referring to Jevons and Marshall. As we saw in Section 2.4, the mathematical problem is similar to the problems posed in *New and Old Methods of Ethics*, but here it is generalised 'to find the distribution of means and of labour, the quality and number of population, so that there may be the greatest possible happiness'.

Edgeworth also makes a significant point that is worth stressing here: the competitive solution and the utilitarian solution may not be necessarily equal. In 1889 he insisted on this point: 'The utilitarian determination is clearly discerned to be by no means necessarily coincident with the settlement towards which competition tends. . .theoretically it is tenable that there is an adjustment of contracts more beneficent than that which the mechanical play of competition tends to establish.'[84]

Moreover, Edgeworth reasons that in the case of two parties confronted with a particular contract-curve, where their final settlements consequently 'lie in a *reverse order of desirability* for each party'; and 'in the absence of any definite principle of selection', each party 'has about as good a chance of one of the arrangements as another'. In this case if parties are risk-averse, Edgeworth asserts: 'rather than resort to some process which may virtually amount to tossing up, both parties may agree to commute their chance of any of the arrangements for the certainty of one of them, which has certain distinguishing features and peculiar attractions as above described – the utilitarian arrangement'.[85]

In fact, through this argumentation Edgeworth justifies utilitarianism by basing it on a 'social contract'. For him this was a very important result due to his deep conviction that utilitarianism was the 'sovereign principle'. And with this conclusion he completes 'the passage from the principle of self-interest to the principle, or at least the practice, of utilitarianism', a passage that 'neither Helvetius, nor Bentham, nor any deductive egoist has made clear'.[86]

Appendices

The book finishes with seven appendices conceived as 'illustrations' of 'the main body of the work'. The first two, entitled 'On Unnumerical

Mathematics' and 'On the Importance of Hedonical Calculus', are devoted to the defence of Edgeworth's position on the proper use of mathematics in the 'moral sciences', a subject that he would later elaborate on in his 1889 Presidential Address. In another appendix, 'On the errors of the ageometrists', Edgeworth criticises several classical writers – Bentham, J.S. Mill, Cairnes, Spencer, Sidgwick – for not using any mathematics, which could have been helpful in their work.

The most quoted and praised author throughout the appendices is Jevons. Edgeworth dedicates the third appendix, 'On Hedonimetry', to elaborating the concept of utility upon Jevons's cardinal utility. The fifth appendix is, 'On Professor Jevons's Formulæ of Exchange'; here as we have already noted, Edgeworth tries to convince the readers that Jevons's unique solution for bilateral exchange is because the exchangers are price-takers, since in Jevons's theory 'an individual dealer only is presented, but there is presupposed a class of competitors in the background' and therefore we are in the case of perfect competition. This defence is based in the fact that Jevons introduced the exchangers as 'trading bodies' and Edgeworth interprets them as 'representative particulars' in the sense of Berkeley.[87]

Applying the Contract Curve to the Irish Situation

Mathematical Psychics has a final appendix called 'On the present crisis in Ireland', in which Francis tries to reason about the 'political contract between two classes of society, the landlord and the tenant class' in terms of a contract-curve and a bargaining process. Edgeworth tries to show that an underlying principle of 'equality *per se*', as proposed by the defendants of 'peasant proprietorship', and 'expropriation of landlords' should be replaced by the 'Universalistic Hedonism or Principle of Utility', which leads to the perception that the contract between the two classes of society is a matter of not only mutual conflict but also mutual interest.[88] But Edgeworth's baroque style makes it difficult to follow the argument, immersed as it is in lateral reflections about the power of the mathematical method:

> Considerations so abstract it would of course be ridiculous to fling upon the flood-tide of practical points. But they are not perhaps out of place when we remount the little rills of sentiment and secret springs of motive where every course of action must be originated. It is a height of abstraction in the rarefied atmosphere of speculation that the secret springs of action take their rise, and a direction is imparted to the pure fountains of youthful enthusiasm whose influence will ultimately affect the broad current of events.

And, a few pages further on, he presses the point, by saying that 'Human affairs have now reached a state of regular complexity necessitating the aid of mathematical analysis; and the lights of unaided reason – though sparkling with eloquence and glowing with public spirit – are but a precarious guide unless a sterner science fortify the way.'[89]

In fact, though he also affirms that the 'mathematical method makes no ridiculous pretensions to authority in practical politics',[90] through Edgeworth's entire paraphernalia of words and formulæ he shows that his heart was on the side of the enlightened landlords and that he cared about the political movements that were striving to change Ireland.

2.8 THE RECEPTION OF *MATHEMATICAL PSYCHICS*

Edgeworth distributed copies of *Mathematical Psychics* among economists and social scientists such as Jevons, Marshall, Foxwell, Sidgwick and Sully. As a reply, Foxwell wrote:

> I do not feel quite able to express an opinion even on that part of the book which appeals to my 'ungeometrical mind'.
>
> I may say generally in the first place that I find your style in spite of its piquancy, very hard to follow. It seems to proceed by hints, excessively allusive, and roams about the field of human knowledge and fancy with an exuberance and discontinuity of metaphor which is quite Carlylese. I do not expect to understand their mathematical part, but I doubt whether even mathematicians will follow it easily.
>
> Generally, I should say I like your Utilitarianism better than your Economics. I thoroughly and heartily agree with your 'aristocratic' treatment or application of Greatest Happiness – though I consider myself a sound philosophical Radical.[91]

Jevons's and Marshall's Reviews

Jevons and Marshall offered their opinions in the reviews section of journals. Marshall published his review in *The Academy* and expressed mixed feelings:

> This book shows clear signs of genius, and is a promise of great things to come. [But] it will be interesting, in particular, to see how far he succeeds in preventing his mathematics from running away with him, and carrying him out of sight of the actual facts of economics.[92]

Edgeworth took this constructive criticism as a stimulating comment and tried to meet Marshall personally. In fact, they met for the first time just after the publication of this review, and from that moment on Edgeworth held Marshall in high regard and often expressed unlimited admiration for him in laudatory terms that were, however, not exempt from hints of irony.

Jevons's review, which appeared in *Mind*, was laudatory with regard to the conceptions and methods involved but criticised mildly Edgeworth's style:

> Whatever else readers of this book may think about it, they would probably agree that it is a very remarkable one. The fearless manner in which Mr Edgeworth applies the conceptions and methods of mathematical physics to illustrate, if not solve, the problems of the hedonic science is quite surprising. As the invisible energy of electricity is grasped by the marvellous methods of Lagrange, so may the invisible energy of pleasure admit of similar handling. . . . The book throughout proceeds upon the conception of Man as a *pleasure machine*. . . . There can be no doubt that in the style of his composition Mr Edgeworth does not do justice to his matter. His style, if not obscure, is *implicit*, so that the reader is left to puzzle out of every important sentence like an enigma.[93]

Galton's Encouragement

After reading Jevons's review, the famous geneticist, psychobiologist and statistician Francis Galton read the book, and when he finished it in October, he immediately wrote to Edgeworth the following:

> Permit me to express the very great interest with which I have been reading your powerful book on *Mathematical Psychics* and especially those parts of it that claim the right of Mathematics to deal even with the loosest quantitative data. I write more especially because I was led to a knowledge of your book by an article of Prof. Jevons in *Mind* in which he happens to speak of its being an unnecessarily difficult book to read. With that verdict I am totally at issue. It strikes me that you have handled topics very difficult in themselves, with great lucidity and vivacity, and I do sincerely hope that you will not bother yourself to be discouraged by that verdict.[94]

Though encouraging for Edgeworth, the letter does not speak very positively for Galton. The attack on Jevons is based on praising one of the most controversial peculiarities of Edgeworth: his style of composition. Edgeworth showed the letter to Jevons, who replied:

> I hope that Mr Galton is quite mistaken in thinking that my notice of your book disparaged it. While saying what I conceive to be the truth about the manner of composition, I thought that I implied a very high opinion of its merits otherwise. I have [asserted] your success quite as much as Mr Galton, but I believe it would be mistaken kindness to pass over every fault and praise only the excellences.
>
> I have no doubt that by plenty of care and practice you may attain lucidity and skill in writing and I am sure it is essential to success as an author.[95]

It is rather touching to realise that this discussion took place between Francis Ysidro Edgeworth, the grandson of Richard Lovell Edgeworth, and William Stanley Jevons and Francis Galton, the grandsons of two of Richard's best friends, William Roscoe and Erasmus Darwin, respectively.[96]

In any case, Edgeworth reacted to Galton's letter by keeping contact with him in the future. As we shall see below, in Section 3.8, this contact centred mostly on the field of statistical inference.

Academic Discussions with Jevons

Meanwhile, in their private conversations and correspondence, Jevons and Edgeworth discussed the concept of capital. Here is Jevons's answer to Edgeworth's comments:

> I have read your remarks on capital with care and interest; you will excuse my saying that you seem to be still deep in the fallacies of Mill. I fear you have not yet approached to a comprehension of my theory of capital as involving solely the element of time. I now see that the whole theory of the matter is implied in the expression for the rate of interest as given on p. 266 of my second edition [*Theory of Political Economy*, 1879]. Some of my other expressions may be misleading. Indeed, as long as you speak of 'capital' instead of 'capitalisation', I think you are pretty sure to go wrong. However, the matter is too difficult to discuss in a letter. [97]

Edgeworth did not drive his point of view home. Independently, Marshall would also object to Jevons's theory of capital by pointing out that Jevons's construction fixed only the expression for the rate of interest and did not help at all to its determination through the relevant factors. In other words, it was not a theory but a bare definition of the rate of interest.[98]

Some years later, when after Jevons's death Edgeworth reviewed his *Pure Logic and other Minor Works* in 1890, he reacted to Jevons's criticism of John Stuart Mill's system of logic:

> As we read the deeply interesting passages in which Jevons intimates his own belief about the duties and destinies of man, we feel how true of himself is what he said of Cairnes, that his own opinions were much more valuable than his criticism of other people's opinions.[99]

J.M. Keynes's Criticism of *Mathematical Psychics*

Not all opinions were as constructive as those of Jevons and Marshall. In Edgeworth's obituary, J.M. Keynes dogmatises about Mathematical

Psychics as a field of investigation and indirectly debases Edgeworth's book:

> Mathematical Psychics has not, as a science or study, fulfilled its early promise. In the 'seventies and 'eighties of the last century, it was reasonable, I think, to suppose that it held great prospects. When the young Edgeworth chose it, he may have looked to find secrets as wonderful as those which the physicists have found since those days. But . . . this has not happened, but quite the opposite. The atomic hypothesis which has worked so splendidly in physics breaks down in psychics. We are faced at every turn with the problem of organic unity, of discreteness, of discontinuity – the whole is not equal to the sum of the parts, comparisons of quantity fail us, small changes produce large effects, the assumptions of a uniform and homogeneous continuum are not satisfied.[100]

2.9 1881: THREE UNSUCCESSFUL ACADEMIC APPLICATIONS

Also in 1881, James Sully published another book, *Illusions*, which contains several references to books and articles by authors who were familiar to Edgeworth, such as Wundt, Lewes, Bain, Spencer, Galton, Venn; however, his 'chum' is not mentioned in the acknowledgements. Sully also worked on two of the *Encyclopædia Britannica*'s entries – 'Æsthetics' and 'Evolution' – and he also entered the academic market because his personal income had changed:

> The years 1879 and 1880 were anxious ones for me. My father wrote he was unable to continue my allowance, and shortly afterwards he lost the whole of his fortune. . . I had to trouble my friends Bain, Jevons and others with inquiries as to possible openings to work. New lecturing employment came opportunely. Croom Robertson helped to get me appointed for a year as Lecturer on the Theory of Education at the College of Preceptors, an engagement which was made permanent a year later. I also gave a series of lectures on Art and Vision at the Royal Institution.[101]

This pedagogical activity helped Sully to overcome his financial crisis and cemented the friendship between him and Edgeworth. Francis's stimulating and mature attitude in response to the challenges of life was an invaluable help for Sully, especially because the latter had lost several close friends, including George Henry Lewes (1878) and his partner Mary Ann Evans (the great novelist George Eliot, 1880) and the mathematician Clifford (1879).

Encouraged by Edgeworth's example, between 1879 and 1881 Sully applied for the Chair of Philosophy at Trinity College Dublin, the Chair of Logic at Aberdeen – vacant upon Alexander Bain's retirement – and the

Chair of Logic, Mental and Moral Philosophy and Political Economy at the new University College of Liverpool, created in 1881, a position that Edgeworth also solicited.

In 1881, Edgeworth also applied for three academic positions. The testimonials exhibited above and his own reports supply some valuable information – partially mentioned above – about Edgeworth's activities and the extent to which he was academically supported by his personal acquaintances. In one of the reports he submitted, we learn that Edgeworth had accepted additional teaching assignments outside King's College, as he declared:

> I lecture at King's College, London upon Logic; at Queen's College, London, upon Political Economy and Logic; and Political Economy at the Kensington Ladies Branch of the King's College. I have been appointed lecturer upon Philosophy and Political Economy by the London Society for the Extension of University Teaching and I am lecturing for that Society at Battersea upon Political Economy.[102]

He had already left Mr Wren's Academy for candidates to the Indian Civil Service, as Walter Wren reported in his testimonial.

The three applications were the following: in March 1881, Edgeworth applied unsuccessfully for a Professorship of Philosophy at King's College, London.[103] Also in March 1881, Edgeworth applied – again unsuccessfully – for a Professorship of Political Economy at University College, London. This time, Foxwell was the candidate chosen.[104] Finally, in October of the same year, Edgeworth applied for the aforementioned Professorship of Logic, Mental and Moral Philosophy and Political Economy at the newly founded University College of Liverpool, where Sully was also a candidate.[105] Neither of them was appointed.

Generally speaking, the testimonials were always more flattering than the reviews, since they were addressed to the candidates and often mentioned the positive aspects only, reporting directly – if required – a more complete opinion to the corresponding committee of electors. Marshall wrote in his second testimonial letter: 'You seem to have gone straight to the root of the fundamental difficulties of pure economic theory and have dealt with them with great originality and masterly power.'[106]

The third of Marshall's testimonial letters was sent from Paris, on his way to Palermo with his wife Mary Paley to spend the winter and, incidentally, discover the concept of elasticity.

> I have derived very great pleasure from reading your *Mathematical Psychics* and your *New and Old Methods of Ethics.* You seem to me to have shown great originality and power of the highest order in handling some of the fundamental

> difficulties of the Pure Theories of Ethics and Economics. Should you succeed in fully working out the lines of thought which you have suggested, you will, I think, exercise a lasting influence on the growth of these sciences.[107]

Jevons's testimonial letter stated:

> I have no hesitation in saying that his published works on *New and Old Methods of Ethics* and on *Mathematical Psychics*, are among the most remarkable and original contributions to the social sciences which the last few years have produced. I expect to see other works of equal ability and greater talent proceed from an investigator of such unquestionable power.[108]

Edgeworth appreciated Jevons's words, although Jevons explicitly showed his preferences in the letter attached to the testimonial by wishing Francis success 'especially if Sully declines to stand'. However, Edgeworth did not balk at Jevons's preference for Sully as he also wished the best for his friend. And Jevons finally reported to the University College London committee that 'his [Edgeworth's] studies have not long been devoted to political economy, and the evidence which he presents of teaching capacity is not conclusive'.[109]

2.10 PAKENHAM EDGEWORTH'S DEATH AND OTHER FAMILY STORIES

Yet another relevant event for Edgeworth, albeit of a very different nature, took place in this busy year of 1881, namely the death of his uncle Pakenham Edgeworth at 69 years of age on 30 July on MacPherson's estate on the Isle of Eigg. After Harriet's marriage, her parents had moved from Anerly to London, and Francis visited them often. But it was during their sojourn on Eigg during the summer holidays that Pakenham suddenly died, as reported in the biographical note transcribed as Appendix G.[110]

His death was especially felt by Francis and his sister Mary, now happily married to Jack Sanderson and living at the old Winchfield rectory. Both of them had been in regular contact with their uncle Pakenham and his wife Christina, their closest relatives in England. Pakenham's loss rekindled Francis's need to be in touch with his family. He visited Mary from time to time and was happy to spend time with the Sandersons' growing family. A flow of five Sanderson girls – Maria Edgeworth (Mia), Frances Harriet (Fay), Helena Laura, Felicia and Rosa Benedicto – arrived, and Francis was delighted to play the charming uncle with them.

Aunt Harriet Butler's Letters

Francis also increased his contact with Aunt Harriet Butler and Aunt Lucy Jane Robinson. Through Aunt Harriet's letters we can infer that during the summer of 1881 Eroles had once again opened Edgeworthstown, where he was living with his French wife. We can also surmise that Edgeworth had met Henry Fawcett, the Professor of Political Economy at the University of Cambridge since 1863, who had also been a liberal Member of Parliament since 1864 and Postmaster General since 1880. He was a good friend of Leslie Stephen and the author of a *Manual of Political Economy*. Having been blinded by a shooting accident when he was 25 years old,[111] he invited Edgeworth to a peripatetic conversation, arm in arm, along Trumpington Road. From Aunt Harriet's letters we also learn that Francis had travelled through Scotland, and she recommended that he visit the Giant's Ring next time. In the same letter, Harriet commented on Francis's plans to visit Italy, 'so grand in itself and personally interesting to you from your Father and Mother having lived so much there'. And she told him that 'to be beyond the Alps and not be at Rome and Naples and Pompeii would be a trial'.[112]

Edgeworth does not seem to have gone to northern Italy in September 1881, since he was with his sister's family in Filey, a seaside resort in Yorkshire.[113] Later, as we have already seen, he had a busy October preparing his candidacy for the new University College of Liverpool, and at Christmas he visited Edgeworthstown and Armagh. After he returned to Hampstead in January 1882, he received Aunt Harriet's latest letter written from Armagh, where we learn that Francis suffered much at sea, and that she was 'rejoiced that you crossed without a storm'.[114]

In February 1882, Thomas Romney Robinson died at Armagh. His widow, Aunt Lucy Jane, had to leave the Observatory, and moved with her sister Harriet Butler to Kingstown,[115] a fashionable seaside resort a few miles south of Dublin.

This was not the last family death in 1882. Pakenham Edgeworth's widow Christina survived him only one year. She also died in 1882 at the age of 63. With her death, we also lose a lot of potential information gleaned from her correspondence with her sister-in-law, Aunt Harriet Butler.

2.11 JEVONS'S DEATH

On 13 August 1882, while he was on holiday with his family at Galley Hill on the Sussex coast, Jevons went to take a swim and was drowned in the sea. 'There is no doubt that the shock of the cold water was too severe for his enfeebled health', and this shock 'rendered him, after the first plunge, quite unconscious and powerless to help himself', his widow Harriet A. Jevons reported in 1886.[116]

When Edgeworth found out about Jevons's death, he immediately wrote to her:

> I cannot refrain from expressing my deep regret at the loss of my venerated friend. I shall always remember with gratitude the kind encouragement and a peculiar intellectual sympathy which he extended to one whose studies were in the same direction however immeasurably behind his. It is difficult to realize that I shall never more meet Mr Jevons on the ice or heath, be fascinated by his philosophical smile and drink his words. I shall always regard it as one of the privileges of my life to have come under the influence of his serene and lofty intellect.[117]

Edgeworth also wrote an obituary note that was published before the end of August in *The Academy*, where he praised Jevons's work:

> Of his work may be said what was said by Herschel of Laplace – that, if it alone of modern writings should survive, it would 'convey to the latest posterity an impression of intellectual greatness' exceeding the furthest attainments of the ancients. In him an antique boldness of theory was complemented by the cautious spirit of Baconian investigation.[118]

In his memoirs written in 1918, Sully would express that he had 'lost almost an older brother, so unfailing kind, so wisely helpful had he been from the first'.[119]

What was Jevons's influence on Edgeworth? According to J.M. Keynes:

> Of the younger men with whom he [Jevons] was intimate, he fully converted Prof. Foxwell to his point of view, and it was a bond of sympathy. . . . Jevons may have played a significant part in drawing both Wicksteed and Edgeworth to economics. Both had been educated in classics, Edgeworth began his academic work by lecturing on English Language and Literature at Bedford College and on Logic at King's College and I have no evidence that his interest in economics antedated his contact with Jevons. Wicksteed, Edgeworth and Foxwell may be considered Jevons's offspring, but his contact with all three came some time after they had taken their degrees.[120]

Differing from Keynes's opinion, we have seen that Edgeworth's utilitarian calculus had been firmly entrenched prior to his 1879 encounter with Jevons. At the same time and in tune with Keynes, what we have learned from *Mathematical Psychics* confirms that Edgeworth's interest in the theory of exchange came from Jevons, that Edgeworth based his notion of utility on Jevons's cardinal utility and that, thanks to Jevons, Edgeworth discovered his affinity with Cournot's conception of mathematics as applied to social sciences.

On the other hand, it appears that Jevons had an important posthumous influence on Edgeworth's work on index numbers, as we shall see below. Another influence from Jevons came indirectly through Marshall, who wrote to Edgeworth in September 1882:

> I sometimes wish that I had published before Jevons's book [*Theory of Political Economy*] came out: as I did not, I determined to put off publishing till I could do so with satisfaction to myself; and as the cruel fates would have it, I did hardly any new work at economic curves between 1872 and 1881. Now I am going on again on a slow job trot: I hope to publish my justification in the course of two or three years.[121]

And Edgeworth was quite probably the admirer of Jevons of whom Wicksteed wrote:

> One who had the advantage of knowing him, and frequently experienced the help which his powerful mind brought to the solution of any economic problem, remarked of his conversation: 'It was more like talking with an early Greek philosopher, if one can realise what that would have been, than with one of our contemporaries'.[122]

2.12 TURNING TO PROBABILITY AND STATISTICS

The Sunday Tramps started to meet once again in the autumn of 1882. Leslie Stephen wrote to Edgeworth 'to ask whether the Hampstead contingent is ready',[123] and the athletic routine was revived. Francis spent that autumn and winter of 1882–83 going through the motions of performing his daily routine – delivering his lectures at King's, practising athletics, exchanging ironic thoughts with Sully, socialising at the clubs.

Earliest Articles on Probability and Statistics

In April 1883, Francis was elected fellow of the Royal Statistical Society, and the event was followed by a substantial flow of articles by him on probability and statistical inference: in the period from 1883 to 1885 he published eleven articles on these subjects.[124]

What was the traceable origin of this copious output? Who were the authors upon whom Edgeworth relied to base his work on probability and statistics? The author most oft-quoted by far is Laplace, followed by Venn, Boole, De Morgan, Donkin and Glaisher.[125] Other familiar names in the field crop up less often, such as Bayes, Quetelet, Poisson, the philosopher Hume, the astronomer Herschel and three political economists with specific works on logic and inference, namely Cournot, J.S. Mill and Jevons.[126]

What was the purpose of this output? First of all, we should recall that the 'calculus of belief' had already been announced in *Mathematical Psychics*. Edgeworth's immediate purpose was to explore the conceptual symmetry between the 'calculus of feeling' and the 'calculus of belief', between utility and probability. The calculus of belief would somehow reinforce the economic and the utilitarian calculuses. Indeed, when we read these articles dating from 1883 to 1885, we realise that Edgeworth persistently highlighted this underlying symmetry that gives unity to his work. For instance, in his article 'The Philosophy of Chance' we read: 'As the regime of contract becomes predominant. . . it may be found necessary often to fall back upon those analogies confirmed by generalized experience which constitute general probability.'[127]

To grasp this particular point, let us have a closer look at these first articles on probability and statistics from the period 1883–1885.

Probability

In 'The Philosophy of Chance', Edgeworth writes first that 'probability may be described as importing partial incomplete belief [. . . and] belief is of the nature of volition', in Bain's sense of 'preparedness to act'. Francis then goes on to assert:

> The Calculus of Probabilities is concerned with the estimation of degrees of probability; not every species of estimate, but that which is founded on a particular standard. That standard is the phenomenon of statistical uniformity. . . . Thus the object of calculus is probability as estimated by statistical uniformity: the partial belief about some unknown occurrence, as the throw of a die, *together with* the observed fact, or full belief, that any one face is thrown about as often as another.

Here we have the appearance of the concept of a priori probabilities, which Francis would define in 1884 as those probabilities which being 'prior to observation' are 'not determined by statistics'. Nevertheless, they 'underlie many important calculations both in Physics and Social Science'.[128] Edgeworth often assumed equal a priori probabilities, as we have already seen in *Mathematical Psychics* where, in dealing with his social contract approach to utilitarianism, he argued that 'in the absence of any definite principle of selection', each party 'has about as good a chance of one of the arrangements as another'.[129]

Then Edgeworth goes on with the idea of measuring 'gradations of belief', an idea that was generally not accepted by Venn. However, according to Edgeworth, 'gradations of belief can be discerned with some precision in simple cases', such as the case of an urn containing black and white balls, where we can increase the proportion of white to black balls. He then proceeds to more familiar fields:

> The felt probability, the quantity of belief, that a white ball will be obtained at a single drawing continually increases, other things being constant, as the proportion of white to black balls. This measurement of a subjective feeling is like the measurement of felt heat by the thermometer. It is very like the Fechnerian measurements of sensation. Like the Fechnerian measurements it is directly applicable only to simple cases.

Thus, Edgeworth finds out that both gradations founding utility and probability are based on similar laws derived from Fechner's law. In the same article, he alludes to 'problems in inverse probability', in which 'as Boole and Donkin point out, constants are generally introduced'. To solve this type of inverse problem, certain knowledge about each constant – in fact, the knowledge of a priori probabilities – is required. In those days, Bayes's method was widely criticised by Boole, Venn, Pearce and others, but Edgeworth was brave enough to maintain that 'the much decried method of Bayes may be employed to deduce from the frequently experienced occurrence of a phenomenon the large probability of its recurrence'. He also complained that 'the ridicule which has been heaped upon Bayes's theorem and the inverse method will be found only applicable to the pretence, here deprecated, of eliciting knowledge out of ignorance, something out of nothing'.[130]

Statistics

These first articles on probability were accompanied by some forays into the more matter-of-fact methods of statistics. In his 1883 article, 'The Method of Least Squares', Edgeworth tries to discern the distinctions between Gauss's law of error and Laplace's method of least squares. He remarks:

> [T]here is another distinction, more interesting to the philosopher and less familiar to the mathematician, namely, that in the Law of Error we are concerned only with the objective quantities about which mathematical reasoning is ordinarily exercised; whereas in the Method of Least Squares, as in the moral sciences, we are concerned with a psychical quantity – the *greatest possible quantity of advantage*.[131]

Laplace's distinction between the most probable and the most advantageous value, which somewhat reflects Daniel Bernoulli's distinction between *espérance mathematique* and *espérance morale*, became one of Edgeworth's preferred topics in the years to come.

In the article, he remarks that in order to get the measurement of the central value of a variable, that is, to estimate its mean value through a sample of observations, the established criteria of Laplace and Gauss were based on minimising the 'integrated disadvantage' produced by error, that is, the total disadvantage 'incurred in the long run by employing any particular system of values'. This integrated disadvantage was measured by either the most probable error, the mean error or the mean square of the error. Edgeworth then goes on to analyse the deviations in the results for asymmetrical distributions in which the most probable mean – the weighted arithmetical mean – does not coincide with the most advantageous mean – with weights based on the method of least squares starting from observations distributed along a previously defined 'facility-function'. In the course of this analysis, Francis alludes to an asymmetrical distribution of 'subexponential form' that depicts the subsequent distribution of the sample mean of a normal variable, obtained from a limited number of observations when the dispersion of the population is unknown. This type of asymmetrical distribution would be ignored until 1908, when W.S. Gosset, 'Student', independently rediscovered it and popularised its use in relation to small samples under the name of '*t*-distribution'.[132]

One of the most interesting early works on statistics by Edgeworth is his 1884 article 'Chance and Law' where he asserts that out of an unpredictable 'elementary' level shaped by chance, you may build an 'aggregate' level subject to certain regularities, ruled by law:

> Suppose that the aggregate of actions of a certain species obeys the Law of Error. Then, it is a fair presumption from the mathematical analysis of that law, that the aggregate quantity say E, may be regarded as the sum of a great many terms, such as $l_1 e_1 + l_2 e_2$ + &c.; where l_1, l_2, &c., are constants, and e_1, e_2, &c., are quantities ranging each under a definite, though not in general simple, law of facility. But the Calculus of Probabilities affords no proof that the elementary quantities, e_1, e_2, &c., are predictable with regard to each particular event as well as the long run.[133]

During these years, Edgeworth was also ripening a definite plan to accommodate the methods of statistics to the needs of social sciences, as he expressed in his review of Jevons's papers, which were posthumously published in 1884 as *Investigations in Currency and Finance*.[134] In 1885, Edgeworth wrote four papers on statistics that he read at different societies: 'Observations and Statistics; an Essay on the Theory of Errors of

Observation and the First Principles of Statistics', was read on 25 May to the Cambridge Philosophical Society; 'Methods of Statistics', on 23 June, to the international meeting to celebrate the Jubilee of the Statistical Society; 'On Methods of Ascertaining Variation in the Rate of Births, Deaths, and Marriages', and 'Progressive Means', were read at meetings of the British Association for the Advancement of Science on 12 September and 10 October 1885 respectively.[135]

'Methods of Statistics' is a survey that begins to indicate Edgeworth's main objectives in statistics, that is, 'the Science of Means, where the latter word includes measures of both location and dispersion'. He says that this science deals with two main problems: '(1) To find how far the difference between any proposed Means is accidental or indicative of a law. (2) To find what is the best kind of Mean to use.' Specifically, in (1) Edgeworth was considering the question of whether a difference in figures is 'significant', that is, it really comes from a difference in fact. In this article, Edgeworth uses the term 'significant' in a statistical sense three years before John Venn used it.[136]

As a curious illustration of his findings' wide range of applicability, in 'Methods of Statistics' Edgeworth applies his results to the frequency of dactyls in successive extracts from Virgil's hexameters of *Æneid.* In his 1909 *Treatise on Probability*, John Maynard Keynes would criticise:

> These authorities [Lexis and Edgeworth] are at fault in the principles, if not of probability, of poetry. The dactyls of the Virgilian hexameter are, in fact, a very good example of what has been termed *connexité*, leading to subnormal dispersion. The quantities of the successive feet are not independent, and the appearance of a dactyl in one foot diminishes the probability of another dactyl in that line.[137]

The third and the fouth papers of the series of lectures of 1885 are also path-breaking: the article on the rate of births, deaths and marriages includes an analysis for a two-way classification which anticipated the analysis of variance, and 'Progressive means' consists of an argumentation about the use of linear least squares for detrending time series.[138]

In these early articles we can see that Edgeworth was playing around his two favourite subjects in the field, which were enumerated by A.L. Bowley in 1934:

> First subject: Every judgment based on mathematical chance is related to a priori probability. . . . The second subject which Edgeworth developed from a specially distinctive view is the Law of Error. . . . It was a favorite theme for Edgeworth that the normal law of error, or its generalized expression, is prevalent in nature and in the subject matter of Economics, however much disguised; or, if not in the raw material, then in the play of averages.[139]

Because of the change from political economy to statistics, in these articles Edgeworth seldom quotes Jevons, except in the five-page note 'On the Method of Ascertaining a Change in the Value of Gold', where Edgeworth pays homage to and criticises several of Jevons's works in which he defends the use of the geometrical mean. Edgeworth examines the properties of different means and concludes that 'the (weighted) arithmetical mean is certainly the most probable and probably the most advantageous method of reducing observations' since it fulfils Laplace's criterion of least squares whenever the observations follow the normal distribution. Therefore, 'the arithmetic mean is to be preferred in the economic investigation upon the ground of convenience in the absence of any other ground of preference'.[140] In fact, this analysis was more than fifty years ahead of the econometric research on best estimators.

The Rationale of Exchange

During this period, Edgeworth's only contribution to economics was the short – two-page long – yet significant note entitled 'The Rationale of Exchange', in which he compares his limit theorem in a progressively more competitive setting of exchanges with the 'law of great numbers, or law of error' – best known nowadays as the 'law of large numbers':

> We do not here with Professor Jevons start from the fact of price and make a short step to the unity of price; but starting higher up from the abstract definition of the economic man we reason down to the fact as well as the unity of price.
>
> Now the *rationale* of this deduction, the reason why the complex play of competition tends to a simple uniform result – what is arbitrary and indeterminate in contract between individuals becoming extinct in the jostle of competition – is to be sought in a principle which pervades all mathematics, the principle of limit, or law of great numbers as it might perhaps be called. . . . [In] physics it continually occurs that from whatever initial circumstances we start, and it may be often added by whatever steps we move, we arrive ultimately at the same position. [. . .T]he law of great numbers, the law of error, is at least as applicable to social as physical phenomena.

Then Edgeworth cautions against the confusion between this 'law of great numbers' and the 'law of competition', which works in a different way:

> The field of competition may be compared to a fanciful system consisting of two groups of particles in a plane, each particle tending to its own position of maximum energy, and so peculiarly constrained, that the sums of the resolved paths of one set of particles, are respectively equal to the sums of the resolved paths of the other set; and that, while no particle of one set can move without one or more of the other set, no work can be done against any particle.

> Equilibrium, which is indeterminate in the case of a finite number of particles, becomes determinate in the limit.[141]

Besides Jevons, the note contains mentions of the economists Marshall, Sidgwick and Walras. And by publishing it in the *Journal of the Royal Statistical Society*, Edgeworth indicated that economics and statistics should go hand in hand.

2.13 1883–87: FAMILY AND FRIENDS

In early 1883 Edgeworth read the first volume of Sir William Rowan Hamilton's biography by R.P. Graves. His attention was especially drawn by a paragraph in which Graves recounts that in mid-December 1831, Hamilton had received a letter from his friend Francis Beaufort Edgeworth announcing his wedding. This letter also contained a poem of remarkable beauty, inspired by the young Spanish refugee who was going to be his wife.[142] Since this poem was unknown to Francis, he asked his Aunt Harriet Butler about it. She replied that she did not have his father's poem entitled 'To a Spanish Refugee' and had never set eyes on it.[143]

Around this time, Aunt Lucy Jane Robinson reported that in 1834 her brother Francis Beaufort Edgeworth had written a letter to his mother Frances Anne containing some information about his father-in-law, Antonio Eroles, which included the story of the kidnapping in Andorra. A copy was also sent to Francis.[144] Despite this information there is no proof of any trip to visit Andorra and the Pyrenees after 1883 to meet his uncle Isidro or his relatives. He also used to ask Aunt Harriet Butler – the 'official family historian' – about friends from years past, and, as always, Aunt Harriet's opinions about late celebrities were quite colourful: 'Sir William Hamilton was always entirely taken up with himself' or 'Darwin [Erasmus] was abused [she refers to his relationship with Miss Seward] – he was much praised too, but I think the abuse preponderated.'[145]

The Economic Circle

In 1884 Henry Fawcett died of pleurisy. During the summer of 1882 he had been ill with diphtheria and never completely recovered. With his death, Edgeworth lost an influential friend. He began attending the discussions of Wicksteed's 'Economic Circle' that had been meeting fortnightly at Beeton's Hampstead home since October 1884.

Wicksteed (1844–1927), a Unitarian minister, Dante scholar, active liberal and lecturer, became involved in political economy late in 1882 through Jevons, and for the next 45 years he considered himself Jevons's only disciple. Regular participants in the circle included the writer and dramatist G.B. Shaw, Foxwell, Cunynghame, Sydney Webb and Edgeworth, while Marshall seems to have attended only occasionally.[146] Despite the fact that Edgeworth held more conservative positions than most of those attending, he always kept his friends regardless of their political opinions because he never argued with conviction about anything.

In October 1886 he visited Sidgwick and then repeated the visit in March 1887. As Bonar remarked, 'partly from his strong sense of humour, Sidgwick came near to rivalling Marshall for first place in Edgeworth's admiration', but 'on the whole Marshall was the great Apollo, oracle or highest authority'.[147]

Family Visits and Observations of Wasps

During these years, Francis remained close to his sister Mary's family, the Sandersons, at Winchfield. Even though Oxford was more-or-less the same distance from Hampstead, he did not have the same close contact with the Butlers. By then, Harriet's family had four new members: Harold Edgeworth, born in 1878, Olive H., in 1880, Ruth F., in 1881, and Christina Violet, in 1884. That summer Francis went to Edgeworthstown and Kingstown to visit Eroles and his aunts Harriet and Lucy Jane.

In Edgeworthstown, Francis observed the movements of the wasps in a more systematic way than he had done during his childhood. He recorded the traffic rates at 8 am and at noon on 4 September 1884 and found the difference between them, half the modulus, to be 'insignificant'.[148]

This result was included as an illustration in his 1885 article 'Observations and Statistics'.[149] Edgeworth displayed these data about the traffic of wasps to show by analogy how to elucidate import and export statistics. Francis would revisit wasps and bees in September 1896 (Edgeworthstown), 1897 (Oxford), August 1906 (Edgeworthstown), September 1906 (Hampstead) and 1916–1920 (see Sections 3.19 and 4.10 below).

Sully's Activities

Edgeworth continued to meet Sully daily in their early morning athletics and served as an intellectual sounding board for the nearly three years it took Sully to write the textbook entitled *Outlines of Psychology*. It was published in England in 1884 and in America in 1886 under the title *The Teacher's Handbook of Psychology*. This textbook, addressed to educators, was also immediately translated into French and Spanish. During the 1880s, Sully also delivered a series of lectures at the new Women's Training College in Cambridge and wrote a successful novel, *Friendly Rivalry*. Then, in the second half of the 1880s, he lost his father, 'a former merchant and colliery proprietor and, also lost an excellent friend, Colonel Osborn'.

In 1887, Sully became close friends with Edvard (Alexander) Westermarck (1862–1939), a Finnish social anthropologist and scholar, who in that same year visited England, studied at the British Museum and lived most of the time in Surrey. They had met earlier in Norway by chance when walking in the mountains. In his *Memories of my Life*, published in 1929, Westermarck noted: 'This was the beginning of an acquaintance to which I have owed much in my life'. In England, Westermarck wrote his dissertation 'The Origins of Human Marriage', an immediate scientific success. Between 1897 and 1904, Westermarck made several trips to Morocco and developed field study methods in social anthropology.

Sully also forged a friendship with Richard Garnett, the keeper of printed books at the British Museum and grandfather of David Garnett, a member of the Bloomsbury Group. It is quite likely that Edgeworth also became acquainted with Sully's new friends.[150]

2.14 WORKING ON ACADEMIC PRESTIGE

King's College, London

The 1880s were advancing, and Edgeworth was still living from his inherited rents and a few pounds earned through poorly paid courses at King's College, London. In July 1885, Henry Wace and William Cunningham, Professors at King's College, London, offered Edgeworth the old Chair of Political Economy that had been occupied by Richard Jones and had been vacant since Nassau Senior's time, though 'its value must depend entirely upon such fees as may arise from'. [151]

Given these financial conditions, Edgeworth also kept lecturing on logic at King's, which he transformed into a course on 'Logic of Statistics'. His series of lectures on political economy had the following syllabus:

> Syllabus of Lectures on Political Economy by F.Y. Edgeworth, MA. In XI lectures: I Definition of Subject; II Division of the Subject; III Production; IV Land; V Labour; VI Laws of Progress, Increasing Returns; VII Increasing and Decreasing Returns; VIII Diffusion, Theory of Bargain; IX Theory of Bargain; X Theory of Bargain, Exchange; XI Exchange and Distribution.[152]

Reviews

Another activity entailed writing reviews. Between 1882 and 1890, Francis published reviews of twenty books in the journals *Mind* (*M.*), *The Academy* (*Ac.*), *Nature* (*N.*) and *Journal of Education* (*J. Ed.*):

1882, (*M.*): *The Science of Ethics* by L. Stephen.
1882, (*Ac.*) : *The State in Relation to Labour* by W.S. Jevons.
1883, (*Ac.*) : *Methods of Social Reform and Other Papers* by W.S. Jevons.
1884, (*Ac.*) : *Investigations in Currency and Finance* by W.S. Jevons.
1884, (*M.*) : *Studies and Exercises in Formal Logic* by J.N. Keynes.
1886, (*Ac.*) : *The Scope and Method of Economic Science* by H. Sidgwick.
1886, (*Ac.*) : *Journal and Letters of W.S. Jevons* edited by Harriet Jevons.
1886, (*Ac.*): *The Social Problem in its Economic, Moral and Political Aspects* by W. Graham.
1887, (*Ac.*) : *The Theory of International Trade* by C.F. Bastable.
1887,(*Ac.*) : *Inductive Political Economy* by W.L. Sargant.
1888, (*Ac.*) : *The Logic of Chance* by J. Venn.
1888, (*Ac.*) : *The Economic Interpretation of History* by J.E. Thorold Rogers.
1889, (*Ac.*) : *The Alphabet of Economic Science* by P.H. Wicksteed.
1889, (*Ac.*) : *Kapital und Kapitalzins* by E. Böhm-Bawerk.
1889, (*Ac.*) : *The Principles of Empirical or Inductive Logic* by J. Venn.
1889, (*Ac.*) : *Life of Sir W.R. Hamilton* by R.P. Graves (3 Volumes)
1890, (*N.*) and (*Ac.*) : *The Growth of Capital* by R. Giffen.
1890, (*N.*) and (*Ac.*) : *Principles of Economics* by A. Marshall.
1890, (*N.*) and (*Ac.*) : *Capital and Interest: A Critical History* by E. Böhm-Bawerk.
1890, (*J. E.*): *The Physiology of Industry* by J.A. Hobson and A.F. Mummery.

This information is indicative both of the type of books he was reading and of the high quality of the books chosen. During the 1880s, and being a newcomer among the reviewers, Edgeworth was very careful to restrain most of his ironic penchants, yet nevertheless his reviews do not hypocritically contained unlimited praise. For instance, in his review of the book *Studies and Exercises in Formal Logic* by John Neville Keynes, a book that was highly praised by his Cambridge friends, including Marshall, Edgeworth concluded: 'Keynes follows up the achievements of Aristotle; he relinquishes the aspirations of Boole.'[153]

The review of *The Alphabet of Economic Science* by P.H. Wicksteed contains a sample of Edgeworth's pungent prose: 'It appears to us that Jevons here goes to the very edge of a certain pitfall, and that Mr Wicksteed goes one step further.'[154]

However, the review that most interests us, in that it reveals part of Edgeworth's inner sentiments, is the one of Hamilton's biography by Graves:

> He was not merely the Irish Lagrange. . .it would not be extravagant to regard him as the Pascal or Descartes of his country. What Leibniz says of himself, that his mind could not be satisfied by one species of study, may be said with equal truth of the Irish polymath. 'If you had given your time to the practice of poetry, you would have succeeded' writes De Morgan to Hamilton; and a similar statement is, at least, equally true of metaphysics.

In praising Hamilton's poetry, Edgeworth neglected his father's opinion as he expressed it to Hamilton: 'After all, Hamilton, your poetry will not disgrace you.'[155] On the other hand, Edgeworth wrote: 'In Hamilton's speculations metaphysical refinements are accredited by mathematical results. We allude particularly to the wonderful paper on "Algebra considered as the Science of Pure Time"'. Francis Beaufort Edgeworth, as we have seen before, had been discussing the contents of this article with Hamilton and had clear reservations.[156] Edgeworth should have taken his father's opinion into account since his father was a much better philosopher than Hamilton.

This review of the book about Hamilton shows two things: first, that in those times great scientists were proud and not ashamed to be counted as philosophers and poets; and second, that Edgeworth felt highly rewarded by Hamilton's friendship with his family, although as a good reviewer he did not mention it.

2.15 ASCERTAINING THE VALUE OF MONEY

In the summer of 1886, Edgeworth was elected a member of the Council of the Statistical Society of London, later called the Royal Statistical Society, and Secretary of the Committee for Ascertaining and Measuring Variations in the Value of the Monetary Standard set up by Section F of the British Association for the Advancement of Science. Marshall, Foxwell, Nicholson, Palgrave, Sidgwick, J. Biddulph Martin and the statistician Stephen Bourne were the members of the Committee. Edgeworth obtained this position because of the influence of Marshall, who in 1885, after Fawcett's death, had been appointed to the Chair of Political Economy at Cambridge University.

The committee issued three memoranda written by the Secretary – Edgeworth – in 1887, 1888 and 1889.These memoranda were later published as two articles[157] with Francis cited as the sole author. In his report to the Council of King's College London, on 20 June 1888, when Edgeworth was applying for the Professorship of the Principles and Practice of Commerce, which had been left vacant by the death of Professor Leone Levi, he wrote:

> Among practical questions I have studied in particular the variation of prices, which now so seriously affects commercial interests. An essay written by me on that subject was published in the last *Report of the British Association.* Prof. J.S. Nicholson, of Edinburgh, in his recent treatise on Money, refers to this work as 'by far the best monograph on the whole subject'. . . . I have continued this subject in another paper, which was printed in the June number of the *Journal of the Royal Statistical Society.*
>
> I have made special studies on several other currency questions . . . such as the relation between the quantity of gold in circulation and the level of prices, upon which subject a memorandum has been furnished by me to the Royal Commission on the Precious Metals at their request. The methods of ascertaining the quantity of money in the country have been treated by me in some papers prepared for the British Association this year.[158]

Edgeworth's work on those committees was duly appreciated by his colleagues, who wrote several testimonials for him when he applied for the above-mentioned professorship at King's College, London, in 1888. Foxwell wrote:

> He [Edgeworth] holds the responsible post of secretary to two most important Committees recently appointed by the Economic Section of the British Association. The greater part of the work done on each Committee has devolved upon him, and has been performed with masterly thoroughness and laborious research. The Memorandum on the Measurement of the Variations in the Monetary Standard, which he prepared for one of these Committees, will always remain a standard work upon the difficult subject of which it treats.[159]

Giffen, the President of Section F of the British Association and a member of the Monetary Committee appointed by that body, of which Francis was Secretary, noticed Edgeworth's 'mathematical knowledge and skill to advance the study of these questions'. Nicholson admitted that 'as secretary, he has done the lion's share of the work of the Monetary Committee of the British Association'. Sir Rawson W. Rawson simply stated that Edgeworth's 'publications in the Journal of the Statistical Society and your papers read before Section F of the B.A. will speak for themselves'; Gonner conceded that he had been 'repeatedly impressed by the tenacity with which he follows a question to its very end' and Inglis Palgrave emphasised the 'industry and ingenuity which you have applied to the subjects in which we have had a common interest'.[160]

But the two testimonials that Edgeworth valued the most were the ones submitted by the professors he particularly admired, Sidgwick and Marshall, both of whom were also members of the committee. The former wrote that he had been impressed 'both with his scientific penetration, and the conscientious thoroughness that he has shown in dealing with this difficult problem'. Marshall also gave him high praise: 'He has a thorough knowledge of economic science; he is a very able man, with accurate habits of thought, and would discharge the duties of the Chair with great care and conscientiousness.'[161]

The Memoranda

The first and the third memoranda dealt with 'The Measurement of Change in Value of Money' while the second was about 'Tests of Accurate Measurement', as Edgeworth himself classified and titled them when he reissued them in 1925.[162] The memoranda contained surveys of the literature on those two subjects in order to give different types of prescriptions. In fact, Edgeworth maintained 'the desirability of prescribing separately for different interests'.[163] In his preliminary note in his 1925 reissue, Edgeworth alludes to separate prescriptions for different standards:

1. The 'Consumption Standard', which is designed 'to determine the change in the money value of the articles consumed by the population under consideration'. The simplest form of this standard is the comparative monetary cost of a fixed basket of commodities at two different moments in time or 'epochs'. When the basket portrays 'the quantities of commodities consumed at the initial epoch', then we get the method adopted by Sauerbeck in 1866, and by Giffen to study the period 1873–1883. If the basket depicts the quantities consumed at each subsequent period,we get the method used by Sidgwick as well as by Sauerbeck (after 1866), Mulhall and Giffen.[164] Means or combinations between these two types, Edgeworth states, had been used by Sidgwick, Marshall, Drobisch, Giffen and Palgrave. Another 'most refined form of this standard compares the amount of money required to procure the same *satisfaction* at different epochs' (as suggested by Sidgwick and Lehr). Edgeworth concludes by announcing that the Consumption Standard has been 'proposed by the Committee as the principal standard'.[165]
2. 'The *Production or Labour-Standard* compares the amounts of money procured by the same "real cost" in the sense of effort-and-sacrifice'.

In fact, this standard tries to reflect Adam Smith's labour-commanded theory, although 'quantity of labour may not be a very distinct idea' since according to Smith 'the greater part of people understand better what is meant by a quantity of a particular commodity'. At any rate, 'the *Production Standard* takes as the measure of appreciation or depreciation the change in the pecuniary remuneration of a certain set of services, namely, all (or the principal) which are rendered in the course of production, during a year'. This standard was referred to by Ricardo and used by Marshall, Giffen and Newcomb.

3. 'The *Capital Standard* takes for the measure of appreciation or depreciation the change in the monetary value of a certain set of articles, consisting of all purchasable things in existence in the community' (used by Nicholson).
4. 'The *Currency Standard* takes as the measure of appreciation or depreciation the change in the monetary value which changes hands in a certain set of sales, comprising all the commodities bought and sold yearly at the earlier epoch or at the later epoch, or some mean between those quantities' (as stated by Foxwell).
5. 'The *Income Standard* takes as the measure of appreciation or depreciation the change in the monetary value of the average consumption, or in the income per head, of the community'.
6. 'The *Indefinite Standard* takes as the measure of appreciation or depreciation a simple unweighted average of the ratios formed by dividing the price of each commodity at the later period by the price of the same commodity at the earlier period.' The average employed in this standard may be the arithmetic mean (used by Soetbeer), or the geometric mean (used by Jevons), or the median (recommended by Edgeworth).[166]

In a prior article, 'On the Method of Ascertaining a Change in the Value of Gold', (*JSS*, 1883),[167] Edgeworth defended, against Jevons's geometric unweighted mean, which, he claimed 'has not much ground to stand upon', the arithmetic weighted mean, just as he does here for the consumption standard. But after becoming more acquainted with Jevons, he became convinced by Jevons's theory that fluctuations in prices are the result of two distinct types of influences: 'changes on the side of money' which affect all prices in the same proportion, and 'changes on the side of the things', affecting pairs of prices relative to one another.[168] The 'changes on the side of money' were measured through the amount of the uniform residual movement, the indefinite standard. And here, in the memoranda, Edgeworth defends Jevons's unweighted procedure to measure this standard with the following argument:

> The case is, as if we wanted to discover the change in the length of shadows, due to the advance of day. If the objects casting shadows were unsteady – waving trees, for instance – a single measurement might be insufficient. We might have to take the mean of several shadows. Now for our purpose the *breadth* of the upright object casting the shadow would be unimportant.[169]

Nevertheless, Edgeworth later introduced some realistic considerations which led him to discard Jevons's geometric unweighted mean and replace it with a 'corrected median'.

Another important point of coincidence with Jevons was to stress the stochastic character of index-numbers due to sampling, which seems to have been the natural consequence of Edgeworth's immersion in statistics and also matches his conception mentioned at the end of Section 2.12 of the essentially stochastic nature of prices:

> If, like Jevons, we content ourselves with taking *samples* of commodities rather than all commodities – a perfect legitimate procedure, and justified alike by the theory of Laplace and the practice of statisticians . . . – then undoubtedly, the principles of *inverse probability* becoming applicable to this mode of measurement, greater weight should attach to the less fluctuating species of return.[170]

J.M. Keynes's Criticism

Edgeworth's memoranda for the British Association were received as sound doctrine not only by the members of the Commission but also by most prominent economists. In 1930, J.M. Keynes wrote in *A Treatise on Money* that Edgeworth's work was 'the first thorough Classification of Index-Numbers of Prices'. However, Keynes was against the 'Jevons-Edgeworth conception' of the two distinct types of influences on prices, namely 'changes on the side of money' and 'changes on the side of the things'. According to Keynes, he 'first endeavoured [himself] to deal with this point, though inadequately, in an Essay on Index-Numbers which gained the Adam Smith Prize in the University of Cambridge in the year 1907'.[171] One of the jury members for this prize was Edgeworth, and they both debated this point.

In *A Treatise on Money*, Keynes also explained that he had tried in this paper to persuade the academic world that 'the abstraction between the two sets of forces' was 'a false abstraction'. The reason was that 'the price-level, is itself a function of relative prices'. So, Keynes's intuition caught *avant la lettre* the econometric impossibility of identifying the 'level of prices' and the 'value of money' as distinct forces. And he added:

> I have long believed that this is a will-o'-the-wisp, a circle-squaring expedition which has given an elusive taint, difficult to touch or catch, to the treatment of

> the Theory of Price-Index Numbers traditional in England, [. . .through] Jevons, Edgeworth and Dr. Bowley. . . . I conclude, therefore, that the unweighted (or rather the randomly weighted) Index-Number of Prices – Edgeworth's 'Indefinite' Index-Number – which shall in some way measure the value of money 'as such' . . . has no place whatever in a rightly conceived discussion of the problems of Price-Levels.[172]

Not only the theoretical shadow of Jevons influenced the memoranda for the British Association; Giffen, Palgrave and Marshall might have also pressed for the commonsense declaration by which the Committee proposed the consumption standard as the 'principal standard'.

2.16 IN THE ORBIT AROUND MARSHALL

Alfred Marshall's aforementioned testimonial on Edgeworth came after their correspondence by post had become regular. For example, in June 1886, Marshall instructed Edgeworth on how Ricardian theories should be applied to Indian affairs, and he added as a recommendation a list of books and articles – by Hunters, Phear, Cossa, Ingram, Kautz, and Roscher.[173]

The letters sent by Marshall to Edgeworth in 1887 are all related to Edgeworth's work as Secretary of Section F at the British Association. Marshall suggested in January that he should read his own 'Remedies for Fluctuations of General Prices', an article about the purchasing power of money. In March, Marshall gave Francis his usual personal advice: 'The Report should I think be terse, judgematic, and err if at all on the side of being commonplace'. He also suggested to Francis 'the definition of a stable money as one in which the unit is procured by a certain amount of effort and sacrifice. Gold has behaved very well in keeping level with labour, rather than commodities; money wages remaining nearly constant while prices fell.'[174]

What Marshall was implicitly suggesting was the inclusion of an index based upon Smith's labour-commanded theory, advice followed by Edgeworth, who, upon Marshall's death, would complain that the theory Marshall proposed to him 'does not recur in Alfred Marshall's latest writing about Money'.[175] Two days later, Marshall urged Edgeworth, upon Giffen's instructions, to read Sir George Shuckburgh Evelyn on the 'tabular unit' in 'An Account of Some Endeavours to Ascertain a Standard of Weight and Measure', an article published in 1798.[176] The tabular standard of value had been studied and mentioned by Jevons (1875), based on the works of Joseph Lowe (1822) and Poulett Scrope (1833), who suggested, according to Jevons, that 'a standard might be formed by taking an average of the mass of commodities which, even if not employed as the legal standard, might serve to determine and correct the variations of the legal standard'. And Jevons imagines prophetically how the tabular standard of value would work:

> To carry Lowe's and Scrope's plans into effect, a permanent government commission would have to be created and endowed with a kind of judicial power. The officers of the department would collect the current prices of commodities in all the principal markets of the kingdom, and by a well-defined system of calculations, would compute from these data the average variations in the purchasing power of gold. The decisions of this commission would be published monthly.[177]

Edgeworth had alluded to 'the Jevonian tabular standard' in his aforementioned 1883 article.[178] Therefore, to reduce, as J.M. Keynes does in 1930, 'the Jevons-Edgeworth conception' to the 'Indefinite Standard', which aims to measure the value of money as distinct from the measure of the changes in the prices of commodities, is somehow unjust to both authors. But coming back to Marshall, we see that on this occasion his advice on the 'tabular unit' came too late.

In August, Marshall insisted on lightening up the final Report: 'I think it is very able and interesting: though I should prefer the omission of all about metaphysics.'[179] In 1925, Francis would justify him:

> In criticising the draft of these documents Marshall showed his characteristic concern for the 'general reader'. He would prune whatever was ambitious in mathematical expression or mechanical analogy. He would have approved, I think, of Dean Swift's advice to a young preacher, to omit philosophical terms and 'notions of the metaphysical or abstract kind'.[180]

Edgeworth maintained a sincere admiration for Marshall and usually showed a great reverence towards him, only interrupted from time to time by affectionate irony. Edgeworth expressed this irony mostly in his book reviews, and not even Marshall or Jevons were saved, as we have seen above and shall see below. Marshall perceived this duality – both reverence and humorous sarcasm – when he said jokingly that 'Francis is a charming fellow, but you must be careful with Ysidro!'[181]

2.17 *METRETIKE* AND A THEORY OF BANKING

Metretike: or the Method of Measuring Probability and Utility was a monograph by Edgeworth published by Temple in 1887. It is his third and last short book, shorter than most of his survey articles to come in the future.

From the first articles on probabilities and statistics dating from 1883 to 1884, Edgeworth was attempting 'to treat the mixed science of Probability and Utility: of what Laplace calls *espérance*, the product of probability upon utility'.[182] In *Metretike*, Edgeworth tries to give shape to his ideas about the relationship between probability and utility:

> The Relation of the Calculus of Probabilities to the Philosophy of Utility is partly (I) of Similarity; partly (II) of Identity.
> (I) Similarity exists both (A) as to the methods of calculation and (B) as to the problems proposed.
> (A) The similarity is contemplated under five headings, constituted each by the introduction of an attribute, which tends to render the calculation inexact. Under each of these headings it is argued that there have been granted to the Calculus of Probabilities postulates analogous to those which the calculation of Utility requires to be granted.
> (B) The canons of Belief and Conduct are of similar origin and authority.
> (II) The Theory of Errors, inasmuch as philosophically the *quæsitum* is rather the '*most* advantageous' than the *most probable* combination of the given observations, involves the Principle of Utility.[183]

In the process of finding similarities and partial identities between probability and utility, Edgeworth stresses that 'numerical accuracy [in the field of probabilities] is possible only in games of chance: or ideal problems about bags and balls'. He had previously accepted Boole's distinction between 'material probability', founded upon statistical fact, and 'intellectual probability', founded upon a more general sort of experience. For Edgeworth, 'the principal use of intellectual probabilities' is 'to afford an hypothesis which may serve as a starting point for further observation. This is a use more important than the only rôle which Cournot will allow to subjective probability, as he calls it: namely to regulate the conditions of a bet.'[184] A priori probabilities often have the nature of intellectual probabilities, and they cannot be measured with 'numerical accuracy'. And he continues:

> In this, the canons of Belief and Conduct may be compared: . . . The fond pretension to arithmetical calculation must be abandoned; but it has been made possible to estimate quantity of belief and quantity of happiness in terms which roughly indicate the more or less. To follow whose indications, in preference to custom and tradition, may well be 'more than half' the sum of wisdom.[185]

The final part of *Metretike* deals with the 'territory in common', the intersection between the 'two kingdoms'. As mentioned above in Section 2.12, Laplace, following Daniel Bernoulli, had distinguished between the *espérance mathématique* – the most probable value – and *espérance morale* – the most advantageous value – of a given variable.[186] This most advantageous value is found through observations distributed under a previously defined 'facility function', based on 'good sense', which depicts the utilities obtained:

> It may happen, when the law according to which observations of a physical quantity group themselves is not the normal symmetrical Law of Error, that the method of reducing (or taking the mean of) them which *most frequently* gives the true value, is not that which it is *best* to employ in the art of measurement. Good sense directed to utility, rather than mathematics, must dictate the proper correction.[187]

The joint analysis of probability and utility would be developed by other authors[188] only after Edgeworth's death; however, it is dubious whether *Metretike* exerted any influence in this development because it is one of Edgeworth's less acknowledged philosophical works. In 1926 Keynes wrote that it was 'a very disappointing volume and not much worth reading (a judgment with which I know that Edgeworth himself concurred)'.[189] This disappointment in definitely establishing the method of measuring probability and utility led Edgeworth to abandon his search for fully disclosing the nature of utility, though keeping alert to the scrutiny of all possible approaches to the subject.

On Banking

'The Mathematical Theory of Banking' appeared in the *Journal of the Royal Statistical Society* in 1888. In spite of its title, it is more an exercise in applied statistics than a paper on mathematical economics. Throughout the paper, Francis mentions Quetelet, Galton, Laplace and Poisson.[190] Let us read the initial scheme set forth by Edgeworth to identify its nature:

> Probability is the foundation of Banking. The solvency and profits of the banker depend upon the probability that he will not be called upon to meet at once more than a certain amount of its liabilities. There is involved not only the calculation of averages which is an affair of arithmetic, but also the doctrines of deviations from an average, the theory of errors, which has exercised the ablest mathematicians.

The problem that Edgeworth considers is determining the amount of monetary reserves a financial institution should have 'in order to meet (the possible excess of) its liabilities'. To show his line of reasoning, he displays a numerical exercise and concludes that 'the demands which a bank has to meet do not increase in proportion to its liabilities' but increase with the square root of its amount. At any rate, this proposition 'must be stated and applied with caution' when determining the monetary reserves because it is true only if creditors act independently. And he adds prudently that 'these speculations are at an immense height of abstraction above the affairs of earth. Appropriate general conceptions, rather than exact propositions, have been the object of the writer. He does not pretend to base any practical recommendations upon the theory.'[191]

2.18 1887–88: TRYING TO GET A GOOD ACADEMIC POSITION

Examiner of Political Economy

In March 1887 Edgeworth earned an Examinership in Political Economy for the College of Preceptors, also at London University, which was yet another poorly paid part-time job. Along with the rest of the examiners, he had to contribute examination questions, which had to be general enough to be answerable by the students in political economy from the entire university. Some sample examinations are still kept in Nuffield College's files. An examination paper stored in the file showed thirteen questions, of which a minimum of nine had to be answered within three hours. The following can serve as a sample to illustrate the nature of the questions: 'Describe and critically discuss the various forms in the XVIIIth century of what has been called the Mercantile Policy.'

Commerce Chair at King's College, London

Edgeworth applied for the Commerce Chair at King's College in the summer of 1888, and mentioned that he was 'Examiner of Political Economy for the College of Preceptors (London)' on his curriculum vitae and report to the Council.[192]

Edgeworth also added a posthumous testimonial from Jevons – written in 1881 and previously submitted in Liverpool – and another from the famous mathematician and statistician John Venn, author of *The Logic of Chance*, who described him as 'a diligent student of every branch of the theory of statistics'.[193] Before the end of 1888, as listed above, Edgeworth reviewed Venn's book and Venn did not at all accept the polite criticisms of the reviewer. This attitude made of him one of the *bêtes noires* – another was Karl Pearson – who tried to underrate Edgeworth's work in the field of statistics.

However, in spite of his extensive curriculum and all the aforementioned testimonials that he included, Edgeworth did not land the Commerce Chair at King's College, London. Yet he was not discouraged, and that summer Edgeworth also applied for the Drummond Chair of Political Economy in Oxford, as reported in *The Academy* on 21 July 1888. It was not yet his turn. The Chair was given to Thorold Rogers, who previously held the Tooke Chair of Economic Science and Statistics at King's College London.

2.19 AUNT HARRIET BUTLER'S LAST YEARS

It is rather touching to realise what respect and confidence Edgeworth had for his Aunt Harriet, to the point of sending her the galley proofs of his memoranda on the purchasing power of money, asking her for corrections. Her 1887 letters to Francis from Glasthule Lodge, Kingstown, reflect her broad interest in economic questions:

> Can a method untheoretically precise be 'practically very good?' . . . Here is your Monetary Standard British Association Essay – thank you. As it is, I suppose from its form, a Prof., I shall return it as soon as possible.
>
> Your additions and alterations in your Memoranda are great improvements. [H.B. discusses also some specific points of the Memoranda].

Her letters are also spiced with erudite wit, news about family and friends and suggestions about landscapes that Francis visited:

> 'Like a mistake in a fitted account' as Owen says in Rob Roy, as a blot in the character. Even bankers sometimes go wrong.
>
> Your letter is most interesting with your visits to the Cathedrals all so grand. Lincoln however the greatest. It and Ely have the advantage over Norwich in the nobleness of their situation presiding over such an extent.
>
> What Eroles says of the state of the country is melancholy [equivous (?)] and I fear hopeless.[194]

Aunt Harriet's last letter, stored at Nuffield College and dated 12 June 1888, answered some of Francis's questions about David Ricardo and his children:

> Mr Ricardo had three sons: Osman (so called because Mrs. Ricardo when they were courting wrote under the names of Osman and Priscilla) – who lived on the edge of Worcestershire, married Miss Mallory, had no children, died four years ago. Second son David, who inherited Gatcomb in Gloucestershire where

> Mr Ricardo lived, married Miss S. Quintin – David died some years ago and his son succeeded him. Third, Mortimer is also dead but left sons, one of whom has his uncle's Osman's property of Bromsberrow. There were four daughters – two married two brothers Austins, one was Mrs. Clutterbuck, beautiful, and Mary unmarried – all dead. . . . He belonged to the liberal party in Politics. He was undersized but his face handsome – his manners delightful. He died comparatively young – an impostation in his ear carried him off in two days.

This information about Ricardo's sons and daughters was quite valuable for Francis since, as we shall see below, it was the clue to finding the letters between Maria Edgeworth and David Ricardo that he published some years later.[195] Not long after she had provided this information, Aunt Harriet Butler died in 1889 at Kingstown in her 88th year, thus marking for Francis the severance of his main link with his enlightened Anglo-Irish ancestors.

2.20 FRIENDSHIP WITH PRICE AND BONAR

Langford Price

At the 1888 meeting of the British Association, Francis met Langford Lovell Frederick Price while they were jointly acting as secretaries of Section F.[196] The following year, Edgeworth was appointed president of that Section.

Price, who became one of Edgeworth's closest friends, was born in London in 1862 and had obtained firsts in Honours in Literæ Humaniores at Trinity College, Oxford, in 1885. That same year, Alfred Marshall lectured to Price at Oxford. Price held appointments as an extension lecturer at Oxford University until 1888, when he became a fellow and treasurer of Oriel College, where he held a fellowship until 1923. According to Petridis, in the *New Palgrave*, Price gave the Toynbee Trust lectures, in which he applied Edgeworth's scheme of bilateral monopoly to the labour market. The lectures were published in 1887 as *Industrial Peace* and served as the basis of a book published in 1888 which contains one of the earliest analysis of the bargaining problem in the short run. The model, though non-mathematical, 'postulated that in a bargaining situation there existed an upper and a lower limit for the wage, these limits being established by competitive forces. Between these limits the actual wage is determined by bargaining power'.[197] But, as Price himself confessed in 1926:

> Being no mathematician I could not follow with understanding the greater part of his [Edgeworth's] large contribution to the nice refinement of abstract

> economic theory jealously purged, as he tried to render it, of the corruption of concrete content. Nor could I dare appraise the meticulous scrutiny of the bases of statistical technique, his regard for which was religious in intensity and depth.

In spite of this handicap, once Price and Edgeworth were duly introduced in 1888, a friendship developed which became closer after 1891 in Oxford.

James Bonar

Another new friend in the academic arena in London, who lived with his wife Mary in Hampstead, was James Bonar, the son of Andrew Bonar, a famous preacher of the Church of Scotland in Glasgow. After graduating from Glasgow University, James went to Balliol College at Oxford. He then completed studies at Leipzig and Tübingen and taught in London's East End for three years, participating in the University Extension Movement. In 1881, Bonar entered the civil service, where he forged a career that culminated in 1907 as Deputy Master of the Royal Mint in Ottawa. He retired in 1919. Bonar combined his duties as a civil servant with studies of the classical economists, especially Malthus. He was the founder of the Adam Smith Club, published *Philosophy and Political Economy* and contributed seventy entries to the *Palgrave Dictionary on Political Economy*.[198]

Bonar was seven years younger than Edgeworth. They were introduced in 1889 on the steps of the British Museum by C.F. Keary, who according to Bonar was, with Sully, Edgeworth's best friend. Bonar, Price and John Maynard Keynes would be the authors of the three obituary articles written for Francis in 1926, and at that time they were considered Edgeworth's closest friends in the world of academia. Since Bonar regarded Keary as one of Edgeworth's close friends, and this is the only indication of a friendship not mentioned elsewhere, let us try to get some information about him from general sources. According to the British Library records, Charles Francis Keary (1848–1917) wrote *The Vikings in Western Christendom AD 789 to AD 888* (1891), and with Eliza Keary edited *The Francis Letters, with a Note of the Junius Controversy*, on the letters of Sir Philip Francis. Other records about Keary reveal that he is the author of the article 'Outlines of Primitive Belief among the Indo-European Races', published in *The New Englander* in September 1882, and the recipient of the Medal of the Royal Numismatic Society in 1894.

Returning to Bonar, he describes his friendship with Edgeworth in Hampstead and elsewhere in the following terms:

> As we lived those years – after 1889 – in Hampstead there were innumerable walks and talks of Sunday afternoons, or casual meetings in early morning, notes and messages in plenty, letters very few. There were (rarely) elaborate letters confidential and combustible. In private talks, it must be said, he dived where he chose and one could not always tell where he would come up again. When confronted with formal papers at a club or public meeting he wandered much less, he was a good listener and good-natured critic.

Bonar also refers to his travels with Edgeworth:

> Our friendship passed successively through the ordeal of travel. He was not in all senses a good traveller. . . . I remember only one occasion where a novel successively competed with nature. It was in the Gleckstein hut above Grindelwald. There was a brilliant display of lightning far below us in the valley. He kept saying 'Yes, very fine', without raising his eyes from Bulwer Lytton [novelist, playwight, essayist, poet and politician (1803–1873)], bequeathed to the hut by a previous traveller. His iron frame made climbing easy to him, and the strength seemed to most of us unaltered to the last. The head seemed unaffected too; always *mens sana in corpore sano*.[199]

Bonar's praise of Edgeworth's climbing abilities should be noted because Bonar was also a gifted mountaineer. In 1912 at the age of sixty he climbed the Wetterhorn (3708 m) in the Alps in the middle of a snowstorm.[200] Bonar does not specify the date when he alludes to another storm:

> [Edgeworth] applied the theory [of Probability]playfully when we were caught in a thunderstorm and took shelter (in face of folk-lore) under a tree: 'I know from the theory of probabilities it is unlikely we shall be struck, and I know from experience that if we stand out there it is quite certain we shall be wet and have a cold.' Another playful application was: 'Other things being equal the most unpleasant event is the most probable.'[201]

Bey Reshid

Another friend of Edgeworth during the late 1880s was Bey Reshid, a Turkish diplomat working in London. But in August 1890 he was reassigned to Rome and wrote to Edgeworth:

> It was my sincere desire to arrange an evening with you in order to take leave of you before my departure from England. In consequence of the order I received from the Government, demanding me to go to Rome without loss of time, I was obliged to deny myself this pleasure. I hope however I shall be able to make up this loss at a future date either in England or Rome.[202]

In spite of the distance, the friendship did not founder. In April 1891 Reshid wrote to Edgeworth to inform him that he had been elected a member of the Savile Club, 'an entirely unexpected pleasure to me for which you are alone responsible'.[203]

2.21 WAS THERE A BEATRICE POTTER AFFAIR?

Also during these years there was an episode involving Edgeworth and Beatrice Potter, a sociologist and economist who worked for the rich social reformer Charles Booth.[204] They met in 1889, when Francis was a 44-year-old bachelor and Miss Potter was a 31-year-old spinster. When she testified in the House of Lords on the subject of sweated labour, the *Pall Mall Gazette* described her as 'tall, supple, dark, with bright eyes and quite cool in the witness chair, who was fluent on coats and eloquent on breeches. Unfortunately, though her voice was a little shrill, it was very difficult to hear sentences, which were very sharply delivered.'[205]

Miss Potter was a very independent woman. In spite of having been intensely in love with Joseph Chamberlain, she forsook the relationship – 'of deep humiliation' – in July 1888 to save her professional career. However, she did not abandon the search for companionship, as she plainly put it in her diary on 7 March 1889: 'I long every day more for the restfulness of an abiding love – and yet I cannot sacrifice work for which all the horrible suffering of six years has fitted me.'[206]

In June 1889, after receiving a message from Miss Potter, Edgeworth sent her an invitation to meet him at the Reading Room of the British Museum:

> Your characteristic frankness was never exercised more delightfully than in the lines which I received this morning. This is a case in which speech is better than silver and your words seem gold to me and your 'confidence' and 'appreciation' the most precious things which I have on earth. I notice with interest that you speak of looking through authorities 'some day in the Museum'. Let me point out that summer days are peculiarly suited to that purpose both in account of the light and because air being artificially pumped in, the Reading Room of the Museum is in hot weather one of the coolest places in England. Marshall's suggestions deserve zealous prosecution.[207]

From this we can glean that Miss Potter had met Marshall before, and Marshall himself might well have introduced them to each other. On the other hand, we have to admit that the style of Edgeworth's letter is very gallant in a tone we have not found in any of his previous letters. However, he did not choose the right approach for a woman as independent as Beatrice Potter, since she wrote in pencil over the first page of the letter the words 'tiresome man'. And after her meeting with Francis she wrote the following in her diary:

> 4 June.
> Francis Edgeworth (nephew to Maria Edgeworth). Professor of Political Economy at King's College, London: eminent statistician of the mathematical type, a gentle-natured and intellectual man with queer cramped nature, excessively polite and diffident in manner, formal and exact in phraseology – an old admirer and a present lover! . . . This man is pathetic: his somewhat pedantic and thoroughly conventional conversation, his starved affections, the furtive glance of unsatisfied desire peering through the old bachelor habits and appearance of forty-four years of celibate continence, the little stories from *Punch*, the correct literary reminiscences, the greenness in all that appertains to emotional experience – and yet, withal, a hidden fire burning within, which might burst forth and burn down conventional barriers and dry sticks and leave the ground free for a new spring growth of thoughts and feelings. Poor fellow! He bores me.[208]

With all this information in hand, we can say that the romance had ended before it had started. But Francis Edgeworth, the 'old admirer and present lover',[209] did not perceive the discouraging signals sent by Beatrice Potter, and just like Dante, he descended to the Inferno. In this case, the Inferno was the Annual Co-operative Congress, which was held a few days afterwards, on 7 June at Whitsuntide in Ipswich.

Potter reports in her diary that 'the unfortunate lover has followed me hither', to the White Horse Inn, where forty congress participants were installed, and she is proud to write that 'they drank whisky and smoked tobacco. At the Co-operative Congress there is an absolute equality. All live together on the freest terms – food, excursions and business are all given and taken under the democratic co-operative system.'

At this Co-operative Congress of June 1889, Marshall had been elected to deliver the inaugural address. Miss Potter described him as 'a small, slight man with bushy moustache and long hair, nervous movements, sensitive and unhealthy pallid complexion' and refers to him as the 'little professor'. Beatrice Potter further added a few words about Edgeworth's presence:

> On the whole the Ipswich Congress is to me personally unsatisfactory. There is the queer, pedantic, unhappy statistician [Edgeworth] who dogs my steps, makes elaborate speeches on formal matters, and jerks out, every now and again, agonized expressions of romantic regard. I hesitate to dismiss the man with rude coldness, for I feel that intimate friendship with an outspoken sympathetic woman might transform him from a statistical measuring machine into a human observer. But a lover has an evil effect and makes me self-conscious.[210]

The impossible 'romance' with Edgeworth ended in Ipswich. This time Francis understood Beatrice Potter's message and did not press further. Potter did not assign any importance to her contact with Edgeworth, who is not even mentioned in her 1926 book *My Apprenticeship*.

She often informed Marshall about her work. In July 1889, Marshall wrote to her that he was looking forward to her study of the Co-operative movement, but a few months later Marshall tried to persuade Beatrice Potter to quit working on a history of Co-operation and focus her research on the field of female labour.[211]

Notwithstanding Marshall's opinion, Beatrice Potter succeeded in writing a book on Co-operation that was published in 1891. Ten years after its publication, the 'little professor' Marshall said to one of his pupils, Charles Rye Fay, 'all right, work on Co-operation, but try not to be influenced by a pernicious book written by Beatrice Potter on this subject'.

In January 1892, Miss Potter accepted a marriage proposal from Mr Sidney Webb (1859–1947), one of the founders of the Fabian Society, who had been described by her as 'a remarkable little man with a huge head on a very tiny body'. Their relationship had started in around February 1890, and she did not make the courtship easy. In November she wrote to him that she did not love him and that 'all the misery of this relationship arises from this . . . there is no change in my feeling except a growing certainty that I cannot love you'. However, Sidney changed her mind by helping her in her work, even as her amanuensis to answer her correspondence.[212]

2.22 SOPHIE BRYANT

In February 1891, Edgeworth was offering a 'country walk' to a 41-year-old Irish widow, Sophie Bryant, who replied that 'the walk you outline in your letter seems excellent. We must hope that the weather will be favourable . . . the prospect of a country walk so soon seems very pleasant.'[213]

Mrs Bryant was no ordinary woman. She was on the staff of the North London Collegiate School in Camden, in 1882 had been elected a member of the London Mathematical Society, and in 1884 was the first woman to become a Doctor of Science in Great Britain. She was one of the first women to own a bicycle and was also an enthusiastic alpinist.

Throughout her life, she was involved in several academic areas which were also of interest to Edgeworth. Her doctoral subject was mental and moral philosophy, and her paper for the London Mathematical Society, also published in 1884, was on the honeycomb as the most efficient natural form, in line with Francis's everlasting interest in bees.

Yet before February 1891 ended, Edgeworth had new prospects – to be described in the next chapter – that obliged him to leave London during the academic year. There is another letter from Mrs Bryant congratulating him on these new prospects and suggesting that she would try to see him 'on Sunday and if you have time for Essex Hall and still go with me there so much the better'.[214] There are no more letters from Mrs Bryant in the Nuffield files.

In October 1901, Bryant joined forces with Sully, who called a meeting at University College London with eight more people, to found the British Psychological Society. In 1904, she received an honorary doctorate from Trinity College Dublin. Thereafter, she became a supporter of the libertarian feminist movement and was involved in the issue of women's education and a member of The Society for Promoting Employment of Women. Bryant was interested in Irish politics, was a convinced Protestant Irish nationalist and helped to found a Home Rule pressure group.

Thus, Sophie Bryant and Francis Edgeworth had a lot in common in their daily activities and interests. However, we have no information about further friendly encounters between them after February 1891 and before August 1922, when she died due to an accidental fall while hiking in the Alps, near Chamonix.

Indeed, F.Y. Edgeworth's private archives contain no further letters with invitations for walks or museum visits to prospective companions, either male or female. With regard to this absence of sentimental affairs, Keynes offered the following conjecture: 'His difficult nature, not his conception of life, cut him off from a full intimacy in any direction. He did not have as much happiness as he might have had.'[215]

Nevertheless, Keynes did not provide any detail about what he meant by 'difficult nature'.

NOTES

1. Creedy (1986), p. 13.
2. H.J.B.'s report, probably dating from 1938 (EP NUFFIELD C UO, Box E 3, 9/1).
3. See Jaffé (1965), Letter 544.
4. From 'Diogenes at the Savile Club'. Stevenson's *Complete Works* (Skerryvore edition, London, Heinemann, 1924–6, Vol. XIV, p. 349).
5. C.E. to H.B., November 1874. Quoted by Creedy (1986), p. 12.
6. 15 November 1875 (EP NUFFIELD C UO, Box D 5).
7. October 1881 (EP NUFFIELD C UO, Box D 7/14). Quoted by Creedy (1986), p. 9.
8. EP NUFFIELD C UO, Box D 5.
9. Mirowski (1994), p. 8.
10. Sully (1918), p. 164.
11. Creedy (1986) and Newman (1987) identify Sully's 'bachelor chum' as Francis Edgeworth.
12. Sully (1918), p. 164.
13. Sully (1918), p. 164.
14. Eldest son of George Henry Lewes, partner of Mary Ann Evans (George Eliot).
15. *Mind*, 1876, pp. 570–71.
16. Whitaker (1996), I, p. 125; Newman (1987), p. 40.

17. Newman (2003).
18. Alexander Pope's *An Essay on Man,* Second Epistle (1733). Edgeworth included these verses on p. 60 of *New and Old Methods of Ethics* and also in *Mathematical Psychics*, p. 139. Verses quoted also by Creedy, (1986) p. 15, who was the first to provide the sources.
19. Keynes (1926), in Vol. X of Keynes (1971–1989), p. 257. Quoted by Newman (1987) and (2003) p. xxii.
20. Barratt (1869) and Sidgwick (1874).
21. Barratt (1869), pp. 238, 241 and 290, quoted by Edgeworth (1877), pp. 2–3.
22. Edgeworth (1877), pp. 4, 9–10, 22.
23. Edgeworth (1877), p. 38. Quoted by Creedy (1986), pp. 28–29 and Newman (1987) and (2003), p. xxix.
24. Wundt, W. (1874/1893), *Principles of Physiological Psychology*, trans. E.B. Titchener, London, Allen, p. 432.
25. Edgeworth (1877), pp. 40 and 42. Noted by Creedy in 1980. See Creedy (1998), 4, pp. 371–6, esp. p.373. Quoted by Newman (1987) and (2003), p. xxix.
26. Edgeworth (1877), p. 36. In fact, he anticipated the Bergsonian (1938) welfare function.
27. Edgeworth (1877), p. 38. See also Creedy (1998), 4, pp. 371–6.
28. Edgeworth (1877), pp. 43–4. Quoted and commented on by Creedy (1998), 4, pp. 371–6 and Newman (1987) and (2003), p. xxx.
29. Noted by Newman (1987) and (2003), p. xxxiv.
30. Edgeworth (1877), p. 55. Quoted by Mirowski (1994), p. 18.
31. Bain to F.Y.E., 26 January 1878 (EP NUFFIELD C UO, Box D 2 /1).
32. Sully (1877).
33. Keynes (1926), in Vol. X of Keynes (1971–1989), p. 264.
34. Butler (1972), p. 17, n. 1. Mentioned by Creedy (1986), p. 9.
35. Halsey (1976), pp. 43 and 45.
36. *Mind*, July 1879, pp. 394–408. Mentioned in Whitaker (1996), I, p. 124.
37. See Baccini (2007).
38. Edgeworth (1881), pp. 61 and 68–9.
39. Edgeworth (1925), II, p. 149, mentioned by Creedy (1986), p. 29. On Edgeworth's attitude towards 'authority', see Creedy (1986), pp. 17–18 and 29–30.
40. Jevons and Jevons (1934), p. 230. Nowadays, at The Chestnuts – a Youth Hostel – there is no blue plaque indicating that Jevons lived there but there is one for Paul Robeson, the Afro-American singer and actor. Neither is there any indication at 5, Mount Vernon about Edgeworth, but there is one at no. 7 about Stevenson.
41. The articles were published in the *Journal of the Statistical Society of London.* See Jevons and Jevons (1934) and Keynes (1936a).
42. Sully (1918), p. 180.
43. And in 1880 Jevons wrote a letter to his friend Broadfield 'from the Athenæum Club'. Jevons (1886). The letters were dated 14 November 1878 and April 1880.
44. EP NUFFIELD C UO, Box D 2 /1/9.
45. Jevons (1886), 25 April 1879.
46. Jevons (1905), p. 94.
47. In Jevons's 'Richard Cantillon and the nationality of Political Economy'. *Contemporary Review*, January 1881. Also included in Henry Higgs's 1905 edition of *The Principles of Economics*.
48. Sully (1918), p. 208.
49. H.B. to F.Y.E., 18 May 1880 (EP NUFFIELD C UO, Box D 2 /1/9).
50. In 1879 Jevons went to Norway with his cousin Arthur Jevons, in 1880 with his brother Tom and in 1881 with his cousin William Jevons.
51. Annan (1984), pp. 97–98.
52. Mountaineering became fashionable among intellectuals. Alfred Marshall in the Black Forest, John Maynard Keynes in the Pyrenees and Arthur Cecil Pigou in the

Alps are other examples of excellent mountaineers. Pigou contributed to *The Alpine Journal* with the article 'Some Alpine Expeditions in 1922'.
53. Pigou (1925), p. 66.
54. Groenewegen (1995), pp. 176–179. See also D.P. O'Brien's introduction to the 1994 reprint.
55. Whitaker (1996) I, pp. 124–5, 134–6.
56. During these years, Edgeworth sent all his writings on ethics and political economy to Marshall. According to Peter Groenewegen, Marshall's recent biographer, much of Edgeworth's material 'was relevant to the content of what became Book III ["On Wants and their Satisfaction"] of the [Marshall's] *Principles* that would appear in 1890'.
57. Whitaker (1996) I, Marshall to F.Y.E., 8 February and 28 March 1880.
58. Whitaker (1996) I, Marshall to F.Y.E., 28 March 1881. Of course, neither Marshall nor Edgeworth were yet aware of the identification problem concerning these curves.
59. Pigou (1925), p. 66.
60. EP NUFFIELD C UO, Box D 7/14.
61. Rev. Alfred Barry, D.D., 22 March and 1 October 1881 (EP NUFFIELD C UO, Box D 7/14).
62. Edgeworth, June 1888, 'To the Council of King's College' (EP NUFFIELD C UO, Box D 9/10/1).
63. Edgeworth (1881), p. 1.
64. Edgeworth (1881), p. 15.
65. Edgeworth (1881), pp. 1–2.
66. Edgeworth (1881), p. 9.
67. Edgeworth (1881), p. 11.
68. Edgeworth (1881), p. 7.
69. Edgeworth (1881), pp. 8–9. The expression 'innumerable smile of seas of love' has been ascribed to Aeschylus (*Prometheus Bound,* line 90) and 'gold sands' appears in a poem by Tennyson, a favourite of Edgeworth (*Locksley Hall,* II, pp. 31–2). The poets are not explicitly mentioned by Edgeworth and were researched by Newman. See Newman (2003), pp. 158–9.
70. Edgeworth (1881), p. 11.
71. Edgeworth (1881), pp. 16–17.
72. Edgeworth (1881), pp. 17–19.
73. Edgeworth (1881), pp. 20 and 40.
74. See Jevons (1871), pp. 136–9 and Walras (1874–77), Art. 50. 'Uniformity of price' is the terminology adopted by Edgeworth.
75. Edgeworth reinforces his argumentation by quoting 'an able critic' in the *Saturday Review* dated 11 November 1871: Edgeworth (1881), p. 110.
76. Edgeworth (1881), pp. 26–28. As we have seen above in Section 2.3, Edgeworth had first used Lagrangian Multipliers in his book of 1877. See Creedy (1986), pp. 54–60 and Creedy (1998), 4, pp. 371–6, especially p. 373.
77. Edgeworth (1881), pp. 21–5.
78. Edgeworth (1881), p. 28, Fig. 1. With regard to graphical representations, Edgeworth is also prone to use polar coordinates, a representation that he had learned from Hamilton. See, for instance Edgeworth (1881), pp. 39–41. See also below in Section 4.12 Edgeworth's review of Bowley's *The Mathematical Groundwork of Economics*.
79. Edgeworth (1881), pp. 22–4. As Creedy (1986) points out, this result leads to what we now call the 'first fundamental theorem of welfare economics': the price-taking solution is Pareto efficient.
80. Edgeworth (1881), p. 42. Therefore, the solution corresponds to a price-taking equilibrium in a competitive market and so it cannot be 'blocked' by a coalition, a result that the first users of game theory – including Shubik, Scarf, Debreu, Aumann, Arrow and Vind – arrived at in the late 1950s, as they also applied the coalitional argument used by Edgeworth. And, after realising his precedence, some of them referred to the solution

point as 'the Edgeworth allocation point of a competitive exchange economy' or, more briefly, 'the core of a competitive economy'. See also Creedy (1986), Ch. 4, pp. 63–77.

81. Edgeworth (1881), pp. 50–51.
82. Edgeworth (1881), pp. 51–4. For a thorough analysis of Edgeworth's Economical and Utilitarian Calculus, see Creedy (1986), Chs 3 to 5.
83. Edgeworth (1881), p. 56.
84. Presidential address delivered to Section F of the British Association in 1889, see Section 3.1 below. Included in Edgeworth (1925), Vol. II, p. 284. Noted by Creedy (1986), p. 81. See also Creedy (1998), 7, pp. 609–18 and 22, pp. 34–52.
85. Edgeworth (1881), p. 55. Noted by Creedy (1986), p. 82.
86. Edgeworth (1881), pp. 53 and 128. Creedy (1986), pp. 82–3, stresses this contribution and points out that the utilitarian part of *Mathematical Psychics* 'was, however, ignored for many years' and that more recently, in 1971, had been only partially 'acknowledged by Rawls'.
87. Edgeworth (1881), p. 109. W. Jaffé has questioned 'Edgeworth's erudite interpretation' based on Berkeley. See Jaffé (1973), pp. 294–95. On the other hand, Creedy stands for Edgeworth's defence of Jevons and attributes the origin of his concept of trading body to Jevons's discussions with Fleeming Jenkin. See Creedy (1986), pp. 44–7 and 74.
88. Edgeworth (1881), pp. 126–34.
89. Edgeworth (1881), pp. 128–9 and p. 138.
90. Edgeworth (1881), p. 134.
91. H.S. Foxwell to F.Y. Edgeworth, from 12, Ryder Street, London, no date [1881] (EP NUFFIELD C UO, Box D 2).
92. *The Academy*, 1881. Quoted by Keynes (1924), p. 25, and also by Bowley (1934), p. 115.
93. *Mind* (1881), 6, pp. 581–3.
94. Francis Galton to F.Y.E., 28 October 1881. (EP NUFFIELD C UO, Box D 1/19).
95. W.S. Jevons to F.Y.E., from 2, The Chestnuts, Branch Hill, Hampstead, 12 November 1881 (EP NUFFIELD C UO, Box D 1/16).
96. These three families were fine illustrations of what Noel Annan would refer to as the 'intellectual aristocracy', mostly the result of genetic transmission, a phenomenon studied by Galton. See Annan (1984), p. 5.
97. W.S. Jevons to F.Y.E., 26 December 1881. In Jevons (1886), p. 439.
98. See R.D. Collison Black's 'Introduction' to Jevons (1871), p. 28.
99. *The Academy*, Vol. XXXVIII, July–December 1890.
100. Keynes (1926), in Vol. X of Keynes (1971–1989), p. 262. Quoted also by Skidelsky, 1992, II, p. 413.
101. Sully (1918), p. 182.
102. Report to the Council of University College, Liverpool, October 1881 (EP NUFFIELD C UO, Box D 7).
103. He presented testimonials from George Francis Armstrong (Professor of History and English Literature, Cork), Alexander Bain (Aberdeen), William Stanley Jevons (University College, London), James Sully (by then also at King's College, London), and Alfred Barry (Principal of King's College, London) (EP NUFFIELD C UO, Box D 6).
104. He submitted testimonials from Jevons, Marshall (University College, Bristol), Alfred Barry and G. Robertson (King's College, London). Marshall also submitted testimonials for John Neville Keynes and Herbert Somerton Foxwell. See Whitaker (1996), I, p. 137 (EP NUFFIELD C UO, Box D 6).
105. Edgeworth printed and submitted testimonials signed by Marshall, Jevons, Thomas Fowler (Professor of Logic at the University of Oxford), Green (Professor of Moral Philosophy at the University of Oxford), Alexander Bain (Professor of Logic at the University of Aberdeen), Foxwell (Professor of Political Economy at University College, London; Examiner in Political Economy at the University of London), Alfred Barry (Principal of King's College, London), and Walter Wren (as mentioned above) (EP NUFFIELD C UO, Box D 7).

106. Professorship of Political Economy at University College, London (Manuscript letter at EP NUFFIELD C UO, Box D2/ 1/10).
107. EP NUFFIELD C UO, Box D 7/14. Included in Whitaker (1996), I, p. 146.
108. Jevons to F.Y.E., 21 September 1881.
109. Mirowski (1994), pp. 66–67, fn. 24. Information supplied by M. White.
110. EP BODLEIAN UO, Ms. Eng. lett. c730.
111. *Palgrave*, Vol. 2, Eatwell et al. (1987): 'Fawcett, Henry (1833–1884)'.
112. H.B. to F.Y.E., 9 September 1881, from Armagh (EP NUFFIELD C UO, Box D 2/1/12 & 13).
113. H.B. to F.Y.E., 23 September 1881, from Armagh (EP NUFFIELD C UO, Box D 2/1/14).
114. H.B. to F.Y.E., 14 January 1882 (EP NUFFIELD C UO, Box D 2/1/17).
115. Nowadays, Dún Laoghaire.
116. Jevons (1886), p. 450.
117. F.Y.E. to Harriet Jevons, 16 August 1882 (Black (1977), Vol. 5, p. 201).
118. *The Academy*, no. 538 (26 August 1882), pp. 151–152.
119. Sully (1918), pp. 223.
120. Keynes (1936a), in Vol. X of Keynes (1971–1989), pp. 137 and 148 (n. 4).
121. Marshall to F.Y.E., 12 September 1882, from Bournemouth. In Whitaker (1996), I, p. 150.
122. Palgrave (1925–26), II, p. 475, 'Stanley Jevons'.
123. Leslie Stephen to F.Y.E., 15 October 1882 (EP NUFFIELD C UO, Box D 2/1/18).
124. Four of these articles were published in 1883: 'The Law of Error' in *Philosophical Magazine*; 'The Method of Least Squares' in *Phil. Mag.*; 'The Physical Basis of Probability' in *Phil. Mag.*; and 'On the Method of Ascertaining a Change in the Value of Gold' in *Journal of the Royal Statistical Society*. Four articles were published in 1884: 'The Philosophy of Chance' in *Mind*; 'On the Reduction of Observations' in *Phil. Mag.*; 'A Priori Probabilities' in *Phil. Mag.*; and 'Chance and Law' in *Hermathena*. In 1885, three more articles appeared: 'Methods of Statistics' in the *Jubilee Volume of the R.S.S*; the first part of 'The Calculus of Probabilities Applied to Psychic Research' in *Proceedings of the Society for Psychic Research* (the second part appeared in the same journal in 1886); and 'On Methods of Ascertaining Variations in the Rate of Births, Deaths and Marriages' in *JRSS*.
125. In 1876 John Venn had published the second edition of *The Logic of Chance*, and in fact two of Edgeworth's articles were a reaction against some of Venn's frequentist views. George Boole's *An Investigation into the Laws of Thought*, 1854, and his articles on the theory of probabilities dating from 1851 to 1862 were also, along with the chapter on probability in Augustus de Morgan's *Formal Logic*, common material for scholars in the period 1880 to 1910. The works on probabilities and statistics by William Fishburn Donkin (1814–1869) and James Whitbread Lee Glaisher (1848–1928) were also in vogue, as Donkin was professor of Astronomy at Oxford and Glaisher lectured in mathematics at Cambridge.
126. A.A. Cournot, *Exposition de la Théorie des Chances et des Probabilités*, 1843; J.S. Mill, *A System of Logic, ratiocinative and inductive*, 1843; W.S. Jevons, *Principles of Science*, 1874.
127. *Mind* (1884), pp. 233–4.
128. 'A Priori Probabilities', *Philosophical Magazine*, 1884. In Mirowski (1994), p. 165.
129. Edgeworth (1881), p. 55. Stressed by Creedy (1986), p. 82.
130. 'The Philosophy of Chance', *Mind* (1884). In Mirowski (1994), pp. 152–4.
131. 'The Method of Least Squares', *Philosophical Magazine* 1883. In Mirowski (1994), p. 135.
132. As noted by B. L. Welch in ''Student' and small sample theory', *Journal of the American Statistical Association*, 1958, pp. 777–88. Mentioned by E.S. Pearson, 'Some reflections on continuity in the development of mathematical statistics', 1885–1920,

Biometrika, 1967, reprinted in Pearson and Kendall (1970), p. 343. Mentioned also by Stigler (1978), p. 294.
133. 'Chance and Law', *Hermathena*, 1884. In Mirowski (1994), p. 173. This article appears among the references listed by J.M. Keynes in *A Treatise on Probability* (1922).
134. Edited by H.S. Foxwell.
135. Stigler (1978), p. 295.
136. Pearson and Kendall (1970), p. 342; Stigler (1978), p. 295.
137. Keynes (1921), in Vol. VIII of Keynes (1971–1989), p. 437.
138. See Stigler (1978), pp. 295 and 300.
139. Bowley (1934). See below, Section 3.16.
140. Jevons's works are *Money and the Mechanism of Exchange,* International Scientific Series, London, 1881, and *A Serious Fall in the Value of Gold Ascertained, and Social Effects Set Forth*, Edward Stanford, London, 1863.
141. *Journal of the Royal Statistical Society*, no. 47, 1884, pp. 164–6.
142. Graves (l882, 1885 and 1889).
143. H.B. to F.Y.E., 2 February 1883 (EP NUFFIELD C UO, Box D 2/1).
144. This copy is now at EP NUFFIELD C UO, Box E 3/7. The content of this letter was discussed in Section 1.7.
145. HB to F.Y.E., 2 and 20 February 1883 and 2 October 1886 (EP NUFFIELD C UO, Box D 2/1 & 1/24).
146. Whitaker (1996), I, p. 255, n. 6, p. 276, no. 4, Reisman (1990), p. 159. On 4 January 1994 Sydney Webb and G.B. Shaw had participated in founding the Fabian Society – thus named in honour of Quintus Fabius Cunctator – with the objective 'to help on the reconstruction of the Society in accordance with the highest moral possibilities'.
147. Bonar (1926), pp. 649–50. Quoted by Newman (1987), p. 48.
148. Reported by Stigler (1978), p. 297.
149. 'Observations and statistics: An essay on the theory of errors of observation and the first principles of statistics' in *Transactions of the Cambridge Philosophical Society*, 14.
150. Sully (1918), pp. 223–4.
151. H.Wace and W. Cunningham, King's College, London, 13 and 16 July 1885 (EP NUFFIELD C UO, Box D 2/1/32 & D 1).
152. EP NUFFIELD C UO, Box D 2 /9 & 10.
153. Noted by Skidelsky in his biography of J. Maynard Keynes. And Skidelsky concludes, 'this is now the generally accepted verdict'. See Skidelsky (1983), p. 59.
154. Newman included this sentence in the list of samples of what he called 'Edgeworth's lapidary wit'. Newman (2003), p. 515.
155. Mentioned above. See Graves (1882).
156. See F.B. Edgeworth-Hamilton's correspondence 1835–1836 (EP BODLEIAN UO, Ms. Eng. lett. c746).
157. 'Measurement of Change in Value of Money' and 'Tests of Accurate Measurement' in Edgeworth (1925), I, pp. 195–297 and pp. 298–343.
158. The full name of the Royal Commission mentioned by Edgeworth was 'The Royal Commission appointed to enquire into the recent changes in the relative values in the precious metals', it was also known in short as 'The Royal Commission on Gold and Silver'. Report kept at EP NUFFIELD C UO., Box D 9/10/1.
159. H.S. Foxwell, 25 May 1888 (EP NUFFIELD C UO, Box D 9/10).
160. R.Giffen, 18 June; J.S.Nicholson, 22 May; Sir Rawson W. Rawson, 25 May; E.C.K. Gonner, 11 June; J. Biddulph Martin, 18 June; and R.H. Inglis Palgrave, 31 May 1888 (EP NUFFIELD C UO, Box D 9/10).
161. A. Marshall, 21 May and H. Sidgwick, May, 1888 (EP NUFFIELD C UO, Box D 9/10).
162. Edgeworth (1925), pp. 195–297 and 298–343. Also included in Vol. 6 of Darnell (1991).

163. Edgeworth (1925), p. 201.
164. Nowadays these two methods are respectively called 'Laspeyres' and 'Paasche' methods.
165. Edgeworth (1925), I, pp. 212–14.
166. Edgeworth (1925), I, pp. 295–7. Newmarch is quoted by Jevons in *Money and the Mechanisms of Exchange* (1875).
167. Included in Vol. 4 of Darnell (1991).
168. See Jevons (1863); (1865).
169. Edgeworth (1925), I, p. 247.
170. Edgeworth (1925), I, p. 217.
171. Keynes (1930) I, p. 81, n. 2.
172. Keynes (1930), I, pp. 80–81.
173. Marshall to F.Y.E., 17 June 1886, in Whitaker (1996), I, p. 209.
174. Marshall to F.Y.E., 11 January 1887. In Whitaker (1996), I, p. 223.
175. Edgeworth refers to *Money, Credit and Commerce* (1922): Pigou (1925), p. 68.
176. *Proceedings of the Royal Society of London*, pp. 133–82, especially 175–6. Marshall to F.Y.E., 11 and 13 March 1887. In Whitaker (1996), I, pp. 233–4.
177. Lowe, Joseph (1822), *The Present State of England in regard to Agriculture, Trade and Finance*; Scrope, G. Poulett, (1833) *An Examination of the Bank Charter Question, with an Inquiry into the Nature of a Just Standard of Value*; Jevons, W.S. (1875), *Money and the Mechanisms of Exchange*, Ch. XXV, p. 330. Marshall to F.Y.E., 11 and 13 March 1887. In Whitaker (1996), I, pp. 233–4.
178. Edgeworth (1883), 'On the Method of Ascertaining a Change in the Value of Gold', *Journal of the Statistical Society*, pp. 714–718, included in Mirowski (1994). Mentioned in Section 2.12.
179. Marshall to F.Y.E., August 1887. In Whitaker (1996), I, p. 235.
180. Pigou (1925), pp. 67–68.
181. Quoted by Keynes (1926), in Vol. X of Keynes (1971–1989), p. 265.
182. 'The Philosophy of Chance' *Mind* (1884), in Mirowski (1994), p. 158.
183. *Metretike* (1887), in Mirowski (1994), p. 81.
184. 'The Philosophy of Chance', *Mind* (1884), in Mirowski (1994), p. 157.
185. *Metretike* (1887), in Mirowski (1994), p. 105. This assertion gives a hint of ordinal utilities, and explains Edgeworth's receptivity for work in this direction by Fisher (1892), Voigt (1893) and Pareto (1898). See below, sections 3.7, 3.14 and 4.6.
186. M. Le Comte Laplace, *Essai Philosophique sur les Probabilités* (1814), p. 53; D. Bernoulli, 'Specimen theoriæ novæ de mensura sortis' (1738). (*Econometrica*, 1954).
187. *Metretike* (1887), in Mirowski (1994), p. 106.
188. Such as Ramsey and Von Neumann and Morgenstern.
189. Keynes (1926), in Vol. X of Keynes (1971–1989), p. 261.
190. This last author is quite significant since Poisson's distribution depicts the distribution of the time intervals between two entries used in queuing theory and inventory theory, branches of operations research that have been developed since World War II. In this case, Edgeworth's hints predated subsequent work on the subject by more than 60 years.
191. 'The Theory of Banking' (1888), in Mirowski (1994), pp. 199, 207 and 208. Included also in Vol. 4 of Darnell (1991).
192. He presented testimonials from Sidgwick, Foxwell, Jowett, Wace (Principal of King's College, London) and Rawson (EP NUFFIELD C UO, Box E 2).
193. EP NUFFIELD C UO, Box D 9/10/1.
194. H.B. to F.Y.E., 29 May, 27 July and 29 November 1887 (EP NUFFIELD C UO, Box D 1).
195. In *Economic Journal*, XVII, 1907. See Sections 1.4 above and 4.3 below.
196. Price (1926).
197. Petridis (1987).
198. *New Palgrave* Eatwell et al. (1987), 'Bonar, James'.

199. Bonar, *EJ* (1926), p. 647.
200. *New Palgrave*, Eatwell et al. (1987), 'Bonar, James'.
201. Bonar, *EJ* (1926), p. 647.
202. Reshid to F.Y.E., 6 August 1890, from 'Ambassade de Turquie', Rome (EP NUFFIELD C UO, Box D 1).
203. Reshid to F.Y.E., 3 April 1891, from 'Ambassade de Turquie', Rome (EP NUFFIELD C UO, Box D 3).
204. This episode has been reported by Creedy (1986) and Newman (1987). Booth published *Life and Labour of the People in London,* a huge 17-volume work (1891–1903) with many statistical tables and maps on poverty, crime and related matters.
205. From *The Diary of Beatrice Webb (1873–1943)*, MacKenzie (1982).
206. MacKenzie (1977), p. 134.
207. F.Y.E. to Beatrice Potter, June 1889 (EP LSE Archives).
208. According to Creedy (1986), 'her diary says more about her than about F.Y. Edgeworth'.
209. 'Lover' in the old English meaning of 'somebody in love' which does not imply necessarily any physical liaison.
210. MacKenzie (1982), pp. 283–7 and also Webb (1926), pp. 300 and 318–19.
211. A. Marshall to Beatrice Potter, 2 July 1889 (Passfield Papers, LSE Archives).
212. MacKenzie (1982), p. 344.
213. Sophie Bryant to F.Y. Edgeworth, 9 February 1891 (EP NUFFIELD C UO, Box D 1).
214. Sophie Bryant to F.Y. Edgeworth, 23 February 1891 (EP NUFFIELD C UO, Box D 3).
215. Keynes (1926), in Vol. X of Keynes (1971–1989), p. 278.

Prof. F.Y. Edgeworth in about 1895

Reproduced with permission from Dr David Edgeworth Butler

3. Professor F.Y. Edgeworth

When from their game of dice men separate,
He who hath lost remains in sadness fix'd,
Revolving in his mind what luckless throws
He cast: but meanwhile, all the company
Go with the other; one before him runs,
And one behind his mantle twitches, one
Fast by his side bids him remember him.
He stops not; and each one, to whom his hand
Is stretch'd, well knows he bids him stand aside;
And thus he from the press defends himself.
(Purgatory, Canto VI, vv. 1–9).
Dante Alighieri, *The Divine Comedy*.

3.1 PRESIDENT OF SECTION F OF THE BRITISH ASSOCIATION

Edgeworth's efforts on the committees of the British Association for the Advancement of Science were finally rewarded by his being elected president of Section F of this Association in 1889 – a presidency that he held again in 1922. As his first presidential address, he chose one of his favourite topics: 'On the Application of Mathematics to Political Economy'.[1]

The address was an explicit homage to Jevons, who had 'submitted to this section a general mathematical theory of Political Economy', which, as he himself stated, was 'received without a word of interest or belief'. Along with this tribute to Jevons, Edgeworth also confirmed his own conception of 'our mathematical method rightly understood':

> It is concerned with quantity, indeed, but not necessarily with number. It is not so much a political arithmetic as a sort of economic algebra, in which the problem is not to find x and y in terms of given quantities, but rather to discover loose quantitative relations of the form: x is greater or less than y; and increases or decreases with the increase of z.

In fact, he develops the first appendix of his *Mathematical Psychics* in a more rhetorical fashion, where he attributes his point of view to that of

Cournot, discovered by Edgeworth through Jevons; Cournot holds that 'one of the most important functions of [mathematical] analysis just consists in assigning determined relations among quantities whose numerical values, and even the algebraic forms, are absolutely unassignable'.[2]

In the presidential address, Edgeworth stresses the fact that Cournot had already presented the demand and supply functions as an interdependent system and had drawn both curves in a diagram, just as Marshall would do in his *Principles of Economics* which would appear in 1890, only several months after Edgeworth's address.[3] Marshall and his view of the preferably hidden use of mathematics in economics are duly and respectfully quoted. Edgeworth explains that 'upon this view, mathematic discipline might be compared to grammar or to the study of classical literature, which it is profitable to have learnt thoroughly, while it is pedantic to obtrude one's learning'.

Of course, Edgeworth himself did not mind appearing to be a pedant, such as when he refers in his address to the quaternions of his beloved Sir William Rowan Hamilton, by claiming that 'algebra and geometry are to ordinary language in Political Economy somewhat as quaternions are to ordinary algebraic geometry in mathematical physics'. And emphasising Clerk-Maxwell's distinction between ideas on the one hand and operations and methods on the other, so adroitly applied by Hamilton, Edgeworth proclaims that 'this is the spirit in which the economist should employ mathematics, "the ideas as distinguished from the operations and methods"'.

Finally, it is worth highlighting that in political economy, Edgeworth assigns a prime role to the law of large numbers and the calculus of probabilities in general, when he points out in the address that 'the advantage of what has been called the law of large numbers may equally be enjoyed by a theory which deals with markets and combinations'. However, he warns that 'great caution is required in transferring the Theory of Errors to human affairs; and the Calculus of Probabilities may be easily made, in Mill's phrase, the "opprobium of mathematics"'.

One of the great mathematical economists of the age, Léon Walras, is presented by Edgeworth as 'the Helvetian Jevons', although some of his contributions are mildly and politely criticised in his address. The criticisms are the same that Edgeworth made in his review of Walras's *Éléments d'Économie Politique Pure*; the presidential address was given on 12 September 1889, just one week after the review was published in *Nature*. However, before we explain the content of this review, we should first examine the complex academic relationship between Walras and the English school – Jevons, Marshall, Edgeworth and Wicksteed – in general, and between Walras and Edgeworth in particular.

3.2 CORRESPONDENCE WITH WALRAS (AUGUST 1888–JANUARY 1891)

Walras and Jevons

Léon Walras was eager to have his mathematical theory of exchange, based on Cournot and his own father Auguste Walras, recognised in England. On 1 May 1874, he sent Jevons his communication presented to the Institut de France 'Principe d'une Théorie Mathématique de l'Échange', and later, on 29 July, he wrote to him that 'in reading your work, the slight disappointment that I have felt in realising that you were already installed in my territory has been largely compensated by the vivid satisfaction of seeing the confirmation of my results'.[4]

This letter was followed by the exchange of several new works, opening a not very intense yet cordial correspondence. Walras always had an excellent relationship with Jevons. After Jevons's death in August 1882, Walras wrote to Foxwell inquiring about British economists with mathematical skills, and Foxwell replied on 30 December 1882 that 'the ablest of our living economists is Prof. Alfred Marshall (Bristol). He is a first rate mathematician but in weak health and with little leisure and much interested in Mathematical Economics.'[5]

Walras and Marshall

Early in 1883, Walras sent some of his articles to Marshall, and in July he dispatched 'my memoirs collected in one volume'. Marshall reciprocated by sending him his *The Economics of Industry*. After reading it, Walras commented on 28 October 1883, 'you have accepted Jevons's theory on the proportionality of values to final utilities in the exchange'. To that, Marshall readily replied on 1 November:

> I cannot be said to have accepted Mr. Jevons doctrine of 'final utility'. For I had taught it publicly at Cambridge before his book appeared. . . . Following the lead of Cournot I had anticipated all the central points of Jevons book, and had in many respects gone beyond him. . . . I read on Oct. 30, 1873, before the Cambridge Philosophical Society a short paper anticipating incidentally your doctrine of unstable equilibrium.

This was too much for Walras. After that, they would only exchange a few papers between 1884 and 1886, and Marshall would summarise his position on 28 November 1886 by stating that 'as to Jevons and myself, I do not claim to have anticipated him as to the doctrine of final utility, but what I have learnt from anybody I have learnt from Cournot and not from him'.

After that, there is only one letter from Marshall (9 September 1989) thanking Walras for sending him the newly released second edition of *Éléments d'Économie Politique Pure*, and commenting that 'the right place for mathematics in a treatise on Economics is the background. But I think it is most desirable that different seekers after truth should take different routes.'[6]

Walras and Edgeworth

Since his relationship with Marshall was unsatisfactory, Walras approached Edgeworth with the idea of testing him in order to find out if he was the right man for him in England. To this end, in August and November 1888 Walras sent some of his papers to Edgeworth, with the first edition of his *Éléments d'Économie Politique Pure* and the new chapter on markets with capital goods ready for the second edition. Then, in March 1889, Walras also sent him the new chapter as an independent article for publication in an English journal. Edgeworth replied and asked him a few questions about points he did not understand. Walras answered them but Edgeworth was not convinced. Moreover, he remarked that the chapter could not be published separately from the book since it could not be understood properly on its own, and on 13 July he proposed that he write a review of the second edition of *Éléments*. After receiving the second edition from Walras, Edgeworth invited him to the meeting of the British Association in Newcastle on 4 August, but Walras declined.[7]

In September 1889, Edgeworth published the article 'The Mathematical Theory of Political Economy: Review of Léon Walras, *Éléments d'Économie Politique Pure*' in *Nature*, where he introduced Walras as 'one of the favoured few to whom belongs the honours of having made a discovery in Political Economy, . . . the last of a small band of original thinkers'. Edgeworth considered that Walras's 'principal achievement is the copious exposition of the one fundamental theorem that value in exchange . . . corresponds to the utility of the last, the least useful, [. . . in Walras's terminology] *rareté*'. Edgeworth also praised Walras's conception of the entrepreneur: 'Prof. Walras is one of the first who correctly conceived the *entrepreneur* as buying agencies of production (use of land, labour and capital), and selling finished products in four markets, which thus become interdependent'.[8]

However, Edgeworth rejected Walras's view on the rate of profit tending towards zero through competition; according to Edgeworth, this 'could be amended by introducing the "disutility of labour" of the entrepreneur as a factor of economic equilibrium'.[9] He also criticised his 'lesson on capitalization' (new in the second edition), the *tâtonnement* process, as unrealistic and the 'use of symbols in excess of the modest requirements of elementary mathematical reasoning'. Specifically, he claimed that 'the exuberance of algebraic foliage rather than the fruit of economic truth is the outcome of science thus cultivated'.[10] Moreover, Edgeworth complained about Walras's extremely severe criticisms of the English school. Nonetheless, he ended his review by quoting: 'one may say as Napoleon of his victories: *"Il y a là du solide que la dent de l'envie ne peut ronger"*.'[11]

In his letter to Walras dated 20 August, announcing his 'notice' of the second edition of the *Éléments*, Edgeworth added that he had 'acted on the Shakespearean canon of criticism: "Nothing extenuate / Nor set down aught in malice".'[12] This review marked a turning point in their relationship. Walras felt offended by Edgeworth's criticisms and wrote to his friend Perozzo on 13 October 1889:

> I am anxious to see Pure Political Economy appearing as a science in itself, and I think that this will happen when a mathematical economist, after thoroughly reading me, explains my system by apprehending it and testing its accuracy. I had hoped for a while that Edgeworth would render me this service, but he has his reservations. I think him to be a little infeudated to Marshall. Wicksteed is much more independent and he may well be my man in England.

Wicksteed, the new favourite, would also fall short after expressing some differences with Walras. And to Charles Gide, editor of *Revue d'Économie Politique*, Walras wrote on 3 November 1889:

> I regretfully realise that there is a mathematic phraseology and charlatanry and that Edgeworth has become master of them. His review of my work is exactly as I had figured. He has not read it carefully and he rebukes me without any argument on the points where I feel most assured: the *tâtonnement* theory and . . . the theory of new capital. His *Opening Address* [as President of Section F of the British Association, 1889] is also rich in errors of appreciation and doctrine; moreover, it contains serious factual errors.

Walras also wrote to Perozzo in March 1890 that he was amazed by the fact that:

> the economists who are mediocre mathematicians, like Jevons, have produced very serious theories, while the mathematicians who are not good enough economists, such as Edgeworth, Auspitz and Lieben [Austrian mathematical economists], say a lot of nonsense. I have concluded that it is essential to lay the foundations of Pure Political Economy very solidly before building on these foundations the mathematical constructions which they have to support.[13]

Meanwhile, on 10 October 1889 and 22 March 1890, Walras quite hypocritically gave instructions to Ladislaus von Bortkiewicz, then a student in Saint Petersburg, to respond to Edgeworth's objections. He also included some derogatory words about Edgeworth, saying that he built his mathematical models on an 'insufficient economic basis'. Walras also urged Bortkiewicz on by saying that Edgeworth's criticisms had to be rebutted without delay since it meant 'at the same time a rebuttal of Marshall himself'.[14] In *Revue d'Économie Politique*, January–February 1890, Bortkiewicz's article appeared as a book review of the *Éléments*, 2nd edition, an article that was '*corrigé*' and dispatched to the editor C. Gide by Walras personally.[15] On 20 February 1890, Walras sent Bortkiewicz's review to Edgeworth with a letter that reflected Walras's devious behaviour. In the letter, Walras applauded himself for 'having followed in this circumstance' his 'ordinary rule of conduct by which he waited for a third party to take charge' of the reply. Unaware of Walras's game, Edgeworth modestly replied:

> You will hardly expect me to agree with the writer. I must admit however that he hits one weak point, namely that I ought not in criticising an author of such eminence to have expressed myself so succinctly and without a full array of proof. . . . I appreciate your good feeling in the matter.

Walras's answer came immediately, on 8 March 1890 and was extremely cynical, as he told Edgeworth: 'Mr. Bortkiewicz has offered to me in the most complete manner the satisfaction that I hoped for in seeing a competent man wholly appraise my system . . . To tell you the truth, all the points of your criticism, without exception, look extremely weak to me.' Walras was proud of his manoeuvres and wrote to his friend Maffeo Pantaleoni on 13 March 1890:

> It looked to me that Edgeworth, with whom I largely and uselessly corresponded last year, offered an easy target in his articles in *Nature*; therefore, I have encouraged Mr Bortkiewicz, who was very well disposed and prepared, to fall upon him.

Pantaleoni, by then Professor of Political Economy in Rome, had candidly interpreted Bortkiewicz's review as an attack, not only on Edgeworth, but on Walras himself, and was somewhat taken aback at Walras's confidence.

Nine months later, on 8 January 1891, Edgeworth wrote his last direct letter to Walras. It was merely to send him an application form for membership of the British Economic Association and a subscription to the *Economic Journal* (see below), which Walras duly filled out and submitted.[16]

In answer to Bortkiewicz's criticisms, Edgeworth gave a 'fuller statement' of his 'matured views' in the January 1891 issue of *Revue d'Économie Politique*, 'La Théorie Mathématique de l'Offre et de la Demande et le Coût de Production'. According to Jaffé, 'Edgeworth disclaimed repeatedly any want of respect for Walras, but he insisted that Bortkiewicz's resentment against any criticism of the master was a disservice both to Walras and to the science of Economics'. Walras once again tried to encourage Bortkiewicz, and on 9 January 1991 surprised him with the following offensive generalisation about the English people:

> English people are more and more gentle as we are more and more firm with them. One can apply to them, up to a certain point, our ancient French proverb: '*Oignez vilain, il vous poindra. Poignez vilain, il vous oindra*'.[17]

But this time, Bortkiewicz was becoming reluctant to continue the attack because of a lack of time and interest, as he had been advised by his mentor Professor Jahnson that he should prepare himself to become a professor of statistics. In view of Bortkiewicz's mood, on 27 February 1891 Walras decided not to pursue further debate with Edgeworth, arguing that he was feeling 'terribly tired'.[18]

3.3 1890: THE TOOKE CHAIR

In February 1890, Edgeworth was elected as the successor to the economist and historian Thorold Rogers in the Tooke Chair of Economic Science and Statistics at King's College, London. At 65 years of age, Rogers had applied successfully, in July 1888, for the Drummond Chair of Political Economy at Oxford, a position that he had occupied in the past. For Edgeworth, the Tooke Chair, though rather poorly paid, was much better than the old Chair in Political Economy that he had been holding since 1885, also at King's.[19]

At the end of February, Edgeworth received a letter from L.R. Phelps, an Oxford friend and competitor (they had both vied for the Drummond Chair in 1888), who congratulated him on his appointment and prophetically bid him 'welcome back again to Oxford'. After some years, Phelps became Provost of Oriel College, where he was famous among the students for saying, 'Never worry after five o'clock'.[20]

Two Political Economy Clubs

In July, Marshall wrote to Edgeworth to remind him that the Tooke Professorship also carried honorary membership of the Political Economy

Club in London.[21] Marshall was anxious to find forums where his new book, *Principles of Economics*, could be discussed. Edgeworth wrote a review of Marshall's new book for *The Academy*, where he remarked that '*Natura non facit saltum* [nature does not move by leaps], the motto which Professor Marshall has adopted, is not altogether appropriate to a treatise which advances the position of science as it were by leaps and bounds'. But Edgeworth also criticised Marshall's propensity to treat a given subject in different parts of the book; thus, 'the student may complain that he is not sufficiently assisted by the index in bringing together all the remarks relating to the same topic'.[22]

Another club that met monthly, The Junior Economic Club, was based at University College, London. It had been founded in 1890 by H. Higgs, H.R. Beeton and C.E. Collet, who at the time were young postgraduates under Foxwell's influence. Foxwell had held the professorship in Political Economy there since 1881, when he succeeded Jevons. Edgeworth participated in and presided over several meetings where, according to Miss Clara Collet, 'Edgeworth criticised Marshall's *Principles* and Marshall answered or rather gave his own criticisms of his work'.[23] At that time, Miss Collet, like Beatrice Potter, worked as a sociologist with Charles Booth; afterwards, she became an honorary fellow of University College London and was finally appointed Senior Investigator of Women's Industries at the Board of Trade (Labour Department).

Finally, when Edgeworth was just three months short of celebrating his 46th birthday, he obtained – thanks again to Marshall's influence – the two positions that would occupy him during the rest of his life: the editorship of the *Economic Journal* and the Drummond Chair of Political Economy at Oxford University.

3.4 EDITOR OF THE *ECONOMIC JOURNAL*

The British Economic Association

In March 1890, while Marshall was serving as President of Section F of the British Association for the Advancement of Science, he was pressed by Foxwell, Palgrave and Gonner to send round a circular to the members of the committee of this section in order to found 'an English Economic Journal somewhat similar in character to the American Quarterly [*Quarterly Journal of Economics*]' and 'an English Economic Society or Association [. . . for] the encouragement of research and discussion, the publication of monographs, the translation of foreign works and the republication of English works that are out of print'.[24]

As early as 1887 Foxwell had already indicated to the Harvard *Quarterly Journal of Economics* that an economic society was to be formed 'with the hearty co-operation of the leading English economists', issuing a quarterly economic journal 'probably edited from Cambridge under the direction of Professor Marshall . . . following the lead so ably set up on your side of the Atlantic' (by the founding of the American Economic Association in September 1885). But Marshall's indecision and the possibility of broadening the Statistical Society's membership, which was finally discarded, delayed the project until 1890. Meanwhile, a group of economic historians at Oxford, who were mostly clergymen, went ahead – led by W. Cunningham – with their plans to publish a quarterly *Economic Review* with the support of the Christian Social Union. Those were the times in which the historicist approach to political economy was in open fight in Great Britain with the marginalist approach to economics in the sense advocated by Marshall. The *Economic Review* appeared only two months before the *Economic Journal* was issued for the first time and the editors of the former remarked that the 'fight between the two schools still rages somewhat fiercely'.[25]

Finally, after several preparatory meetings, on 20 November 1890, the British Economic Association (which would change its name to the Royal Economic Society in 1902) was founded at a meeting with 200 participants held at University College, London. The appointed council members included Marshall, Sidgwick, Ingram, Foxwell, Giffen, J.N. Keynes, Price, Bonar and Wicksteed. Edgeworth was named Secretary of the Council of the Association.[26]

Francis Ysidro Edgeworth is described in the members' list as M.A. (Master of Arts) and D.C.L. (Doctor of Civil Law), while the name of his brother Antonio Eroles Edgeworth is accompanied by the initials J.P. (Justice of the Peace) and D.L. (Doctor of Letters). This means that Francis had obtained recognition as Doctor of Civil Law – from the University of Durham, as we shall discover in the next section – and that Eroles did not limit his activity to merely serving as landlord of Edgeworthstown; in fact, just like his grandfather Richard Lovell Edgeworth, he was also a Justice of the Peace in the best feudal tradition.[27]

With regard to the nature of the association, Edgeworth, Bonar and Foxwell defended a 'closed' association by imposing entry barriers of academic merit. Marshall, however, opposed this and was in favour of 'openness', like the American Economic Association and the London Statistical Society. Marshall prevailed, and 606 people initially applied for membership, each of them paying the individual fee of one guinea per year, which included a free subscription to the *Economic Journal*. One year later, at the end of December 1891, the number of members had risen to 710. Out of these 710 members, 115 had foreign addresses (including the ones mentioned in note 26, as well as Gide and Pantaleoni), 168 were bankers, businessmen or insurers, 56 were civil servants or worked in the government, 66 were in accounting and law and 99 were teachers at all levels.[28]

Immediately after the inaugural meeting of the British Economic Association on 20 November 1890, when Edgeworth was appointed secretary of the association, he was also appointed Editor of the *Economic Journal*. This post, which carried a yearly salary of £100, fitted Edgeworth's talents like a glove. Moreover, he was the potential candidate who had published the most academic articles and reviews and who knew more languages, both modern and classical. Marshall had in mind his Cambridge colleague John Neville Keynes for the new post, but Keynes declined to run for the job. The council had to choose between Edgeworth and his Oxford friend, Langford Price.[29]

The *Economic Journal's* Editorial Policy

Edgeworth outlined the editorial policy adopted by the council in his editorial appearing in the first issue of the *Economic Journal* in March 1891:

> The British Economic Association is open to all schools and parties; no person is excluded because of his opinions. The *Economic Journal*, issued under the authority of the Association, will be conducted in a similar spirit of toleration. It will be open to writers of different schools. The most opposite doctrines may meet here as on a fair field. . . . Nor will it be attempted to prescribe the method, any more than the result, of scientific investigation.
>
> Is it extravagant to hope that this toleration of the differences between the votaries of economic science may tend to produce agreement between them? . . . The Association is to be not only 'British' in its love of fair play and free speech but also 'Economic' in the character which the term suggests of special knowledge and scientific accuracy.

In order to get a regular flow of suitable papers, Edgeworth invited a group of 80 economists, who agreed to appear in a 'list of writers who have undertaken to contribute to the Journal'. We find professors, such as Bastable, Foxwell, Gonner, Ingram, Marshall, Nicholson, Cannan and Cunningham, as well as some professional economists and social scientists including Giffen, J.N. Keynes, Palgrave, Phelps, Webb and Wicksteed. The list also contains foreign professors, such as Böhm-Bawerk, Brentano, Wieser, Bauer, Hasbachs, Gide and Pantaleoni from continental Europe; Dunbar, Seligman, Taussig, Walker and Mayo-Smith from America; and Professor Haslam from New Zealand. A few of Edgeworth's personal friends also appear on the list, including James Bonar, Langford Price,

Clara Collet, Beatrice Potter and Fawcett's widow, Millicent, leader of the constitutional faction of the British suffragettes.

John Rae, who had recently published a book on socialism, had the honour of being the author of the first article published by the *Economic Journal*, 'The Eight Hours Day in Victoria'. Rae became a regular contributor of notes and reviews during the first three years until he had a public argument with Edgeworth that we shall explain in detail below.

An openness to foreign economists and social scientists was obvious from the first issue: the second article was 'The Eleventh Census of the United States' by the American demographer Richmond Mayo-Smith. One 1891 issue included 'The Austrian School and the Theory of Value' by F. von Wieser, a disciple of Menger and friend of Marshall. Moreover, the council appointed several geographical correspondents: Taussig (USA), Ashley (Canada), Duckworth (New South Wales), Harrison (Calcutta), Bauer (Austria-Hungary), Mahaim (Belgium), Greven (Holland), Cohn (Germany), Castelot (Paris), Gide (rest of France) and Bey Reshid (Turkey).

Edgeworth established cooperative ties with the editors of the most important foreign journals in the field of economics. Just a few prominent examples include his contact with Charles Gide (1847–1932), founder of the *Revue d'Économie Politique* in 1887. At that time he was teaching political economy at the Faculty of Law in Montpellier, and from there he went on to Paris. He was also in touch with Maffeo Pantaleoni (1857–1924), Professor of Political Economy at the University of Bari, later professor at Naples and finally at Rome, who together with several friends had bought the title rights and served as editor of the *Giornale degli economisti.* In 1889, Pantaleoni had published *Principii di Economia Pura* and declared himself an admirer of the English (Marginalist) School, and particularly of Marshall. Indeed, Pantaleoni wrote that he considered Edgeworth 'the closest approximation of a match for Marshall living in England. You know that to my mind, Marshall is simply a new Ricardo who has appeared in the field.'[30] In return, Marshall found Pantaleoni's work quite interesting. In 1889 he wrote to J. Neville Keynes that Pantaleoni seemed 'to have much truer mathematical instincts than Jevons, Walras, Laundhardt & Co. and I may now add Wicksteed. But I have not told him that.'[31] But let us return to the *Economic Journal.*

With regard to the reviews, Edgeworth did not leave them exclusively to chance spontaneous contributions but complemented them with reviews commissioned from a team initially made up of Bonar, Price and himself. Accordingly, after 1891, he published all the reviews he wrote, with just one exception,[32] in this journal; therefore, he no longer submitted reviews to *The Academy*, *Mind* or *Nature*.

In spite of Edgeworth's concern for contributions by women, they were very rare: during the first 3 years of the *Journal*, only 2 out of the 73 articles were submitted by women, and both of them were about subjects which were regarded as being somewhat feminine in nature: 'Women's Work in Leeds' by Clara E. Collet and 'Fashion' by Caroline A. Foley. Furthermore, only 7 out of 174 reviews were signed by women, all of them in the third year of the journal, and of the 69 notes, only 2 were signed by women, namely Collet and Fawcett.

Having been appointed by Marshall, Edgeworth often asked him for advice about how to handle the job. As he explained some years later: 'New to that sort of work I wrote to Marshall asking for advice on every small difficulty which arose, until he protested that, if the correspondence was to go on at that rate, he would have to use envelopes with my address printed on them'.[33] Nonetheless, Marshall still wrote to him on one occasion that 'by the aid of your letter I came to see more fully the dry and caustic humour of your notice: but I don't think the ordinary readers will; and I dislike jokes in an economic journal'.[34] To Edgeworth, leaving aside his personal irony was quite difficult, especially in his reviews and notes.

3.5 DRUMMOND PROFESSOR AT OXFORD

In late 1890, Edgeworth applied for a Professorship of Political Economy known as the Drummond Chair at the University of Oxford, which had been left vacant by the death of Thorold Rogers on 12 October 1890. Edgeworth submitted his personal curriculum vitae to the electors, which is included below as Appendix H. Through this report we learn that the Doctorate in Civil Law that he had achieved was from the University of Durham and was honorary in nature. Although no year is stated, we can deduce that this doctorate must have been recent, from the second half of 1888, 1889 or 1890, as Edgeworth had not mentioned it in his applications for professorships prior to July 1888.

He accompanied this report with testimonials written by Bastable (Trinity College, Dublin), Foxwell (University College), Giffen (Board of Trade, Whitehall), Jowett (Balliol), Marshall (Cambridge), J.B. Martin (Secretary of the Royal Statistical Society), J. Munro (Owen's College, Manchester), Nicholson (Edinburgh), Sidgwick (Cambridge) and Wace (Principal of King's College, London). Edgeworth also added letters from Böhm-Bawerk and Pantaleoni.[35]

This time, the candidates vying with him were L.R. Phelps (the editor designate of the *Economic Review*), W.J. Ashley (Professor of Economic History at the University of Toronto), H. Llewellyn Smith and R. Inglis

Palgrave, both historical relativists. Ashley, one of Edgeworth's main competitors, complained in a letter to Seligman that he was afraid 'they will elect Edgeworth and Oxford will be doomed for its sins to unlimited psycho-mathematical economics'.[36] Due to the fight between 'historicists' and 'marginalists', Edgeworth's election has been presented, even by some contemporary authors, as a result, not of his outstanding academic merits, but of Marshall's unquestionable authority.[37]

Edgeworth was officially appointed to the Drummond Chair on 21 February 1891, just thirteen days after his forty-sixth birthday. This chair also implied being elected Fellow of All Souls College and, being unmarried, to have rooms there. He lived at All Souls for the rest of his long life.

Congratulations

During the last week of February, Edgeworth received many letters of congratulation from his friends, former professors and colleagues: A. Marshall, J.A. Bryce, L. Langford Price, B. Jowett, H. Higgs, J. Sully, H. Wall, J.N. Keynes, J.S. Nicholson, G. Griffith, V. Perin, J. Gould, L. Stephen ('I hope that your marginal utility in that capacity will be a maximum – i.e., as I interpret the formula that you will do as much good with as little labour as possible'), R.I. Palgrave and H. Llewellyn Smith (two of his competitors), J. Venn (also at Oxford), F. Watson, C.L. Graves, G.A. Amilta (from London), G.T. Savage (from Cork County), Miss C.E. Collet, Mrs S. Bryant, Mrs C.A. Foley and last but not least, Miss B. Potter. In March and April, Edgeworth received the last two congratulatory letters, from C. Limerick and M. Pantaleoni, who wrote from Bari. He gave Edgeworth humorous advice on how to survive in the rarefied world of academia.[38]

Among all these letters, the two that deserve a special attention are the ones from Alfred Marshall and Beatrice Potter, since they give us hints about these two personalities. Marshall, after an introductory 'Hurrah! Hurrah!! Hurrah!!!' advised Edgeworth to retain the Secretaryship of the British Economic Association and the Editorship of the *Economic Journal* by paying a separate secretary.[39] Beatrice Potter's letter, is rather difficult to decipher. After congratulating him she wondered about the composition of the Electors' Commission : 'is it not an important coincidence that out of the five Presidents [for the Electors] two are "orthodox" Economists and three are not Economists at all but only *Political Individualists*. And the two outsiders in the council happen to be representative of non-State-interference.'[40]

One year before becoming Mrs Webb, Miss Potter was mockingly referring to the new theoretical trends as 'Political Individualism' instead of 'Economics', the new name bestowed by 'the little professor', as she called Marshall. We do not know what Edgeworth's immediate reaction to Miss Potter's derisive lines was; however, in August of that year, while he was spending part of his long vacation with the family of his sister Mary Sanderson at the rectory in Winchfield, he sent Beatrice Potter a very professional letter as editor of the *Economic Journal*:

> It appears advisable to have a discussion in the *EJ* on the regulation of female labour and the prohibition of home work, of the use of heavy implements and [?] a question recently raised. It is easy to obtain able advocates of the extreme view of female liberty. But for an opposite or at least different view I know not where to look unless to you. . . .
>
> I am sorry that the number of the [*Economic*] *Journal* which is just forthcoming does not contain a review of your truly *magnum opus* [*The Cooperative Movement in Great Britain*]. I entrusted it to a particularly competent reviewer. But it is the nature of great artists not to hurry themselves.[41]

The contrast in tone with his former letter of June 1889, in which Francis gallantly invited Beatrice Potter to visit the British Museum is evident. The competent reviewer that Edgeworth mentions is James Bonar. The review came together with that of another book on the same subject, *The Cooperative Movement To-day*, by G.J. Holyoake. Bonar presented Potter's book as the socialist vision as opposed to Holyoake's individualist vision in such simplistic terms that it very probably enraged both authors.

Inaugural Lecture

Edgeworth's inaugural lecture as Drummond Professor was 'The Object and Methods of Political Economy', a general subject, as the occasion required.[42] Obviously, Edgeworth tried to be eclectic and defended both the abstract method, which is similar to that of mathematical physics, and the historical method, which is based, according to Professor (and recent competitor) Ashley, on 'direct observation and generalisation from facts past or present'. He also used a couple of significative quotations: the first from Malthus, who wrote that 'many of the questions both in morals and in politics seem to be of the nature of the problems *de maximis et minimis* in fluxions';[43] the second from Mill when he warned that 'against the danger of *overlooking* something, neither strength of understanding nor intellectual cultivation can be more than a very imperfect protection'. And Edgeworth added that against this danger, 'no remedy can be prescribed except to cultivate open-mindedness and candour, and above all sympathy, the absence of which has aggravated the most serious mistakes which have been committed in Political Economy'.

After declaring, to his own advantage, that 'statistics are an indispensable part of the equipment of the modern publicist', he closed the address with a well-intentioned quotation from his friend Price: 'The quarrel between "old" and "new" economists seems to be giving way on all sides to a hearty desire to recognise good work wherever it is to be found, and to an honest endeavour to seek for grounds of agreement rather than reasons for difference.' [44]

The Tooke Professorship vacated by Edgeworth was assigned to Dr William Cunningham, author of the much-praised three-volume work on economic history, *Growth of English Industry and Commerce* (1882).[45]

3.6 DAILY LIFE AT OXFORD, ACCORDING TO PRICE AND BONAR

Langford Price's Description

Being a professor at Oxford also implied membership of the Oxford University Political Economy Club. Edgeworth was a regular diner there, and according to Price, he 'was more happy at critical participation than at expository introduction'.

Price also provides the best account of Edgeworth's routines at Oxford over the next 30 years, such as walking and talking 'once a week during Term. . . . His physical vigour was such that it was not easy at times to keep pace with him, and I suspect that often I returned more tired than he, despite the disparity in our ages'. After walking fast at the start,

> to get soon into air more stimulating than that which circulated at the low level of the Oxford marsh . . . before we had gone very far I would notice him feeling in his pocket for a slip of paper, furtively produced, to refresh his memory upon items noted through the week for consultation (like a tiny detail in the working of the *Economic Journal* or a small problem of daily life).

His talk was sometimes:

> connected with a picture or a play which we had seen, or a film which he had witnessed (at one period he was much interested in visiting cinemas), or a book in general literature which he had been reading or re-reading. But most of the interrogation aimed at me was on Economics, and generally it was about theory rather than practice.[46]

Just as in his daily exercises in Hampstead, every day Francis bathed in the waters of some appropriate spot, even in the coldest days of winter.

He chose Parson's Pleasure for his plunges and he probably contributed to establishing an Oxford tradition that links Parson's Pleasure with venerable Oxford University dons swimming naked.[47]

But let us go back to Price's description of Edgeworth's usual manners and mood at Oxford. Price claims that 'it was impossible, so kindly affectionate that he was, to quarrel with him, and differences of view were induced to shade so delicately and imperceptibly into one another that they did not warrant any *casus belli*'. This shading was the consequence of the extreme difficulty he had in reaching conclusions. In controversies, Edgeworth used to be 'non-committal though alert to see and eager to expose illogical argument'. According to Price, the reason behind his inconclusiveness was that:

> he was obsessed by the intrinsic hardness and intertwined complexity of economic reasoning as he conceived and practised it. Again and again, in the course of our weekly walks, do I remember him concluding some discussion with the characteristic commentary that our subject was so difficult and complicated.

Price also describes 'the contents of the rooms he occupied at All Souls' since they 'seemed to me indicative. The few books were full of paper slips, inserted, I suppose to facilitate quick and constant reference. Of adornment and comfort there was little sign.' This Spartan setting was not induced by stinginess since, at the same time, 'he was, too, hospitable to a degree, and his eagerness to entertain his friends was only curbed by limiting space at the All Souls high table'.

Price had the privilege of travelling with Francis to Italy, where 'Edgeworth took a genuine pleasure', though 'he obviously considered it an imperative duty, after frivolously spending time like any ordinary tourist, such as myself, upon fashionable sight-seeing, to resort to the corrective of serious attendance at a formal lecture on a learned subject'.

Price also explains that once, upon entering Oriel College with Edgeworth,

> as he passed into the second quadrangle, [he] remarked on the beauty of the creeper then growing on the wall of the Common Room. I concurred, adding however, that a member of our body had lately urged its removal. Instantaneously came the complimentary observation 'a very just opinion' and I do not think that he cared for, if he noticed, any inconsistency with what he had been saying.

On the other hand, Price stresses the contrast between Thorold Rogers, who had always been 'positive and sure', and his successor at Oxford, who in contrast was

> hesitating and tentative, was always seeking shelter behind reference to multiplied authority and yet was not displeased to find, and was punctilious in exhibiting, minute discrepancies in the numerous texts consulted, ending as a result, to all appearance, in more, rather than less, unstable ambiguity than that with which he started.

And Price follows the description of Edgeworth as professor:

> As a lecturer, it must be granted, he travelled soon beyond the ken or grasp of ordinary students, as in the discussion in articles or reviews of burning questions of the day he was wont to introduce recondite argument and nice consideration that to average readers could only seem far-fetched. . . . His more useful work as Professor . . . was in 'informal instruction' . . . there he was strikingly felicitous in overcoming the initial inertia of arousing and maintaining a brisk, enlightening discussion and deftly compelled his class to debate among themselves rather than listen merely to a conversational discourse from him. . . . He was fitted, in short, for the better and the best, and not for the worse or mediocre pupil. He was however no unskilled examiner.[48]

Edgeworth lectured on general theory and on specific topics of political economy, including trade unions, wages and bimetallism, and also gave an advanced course to deal with the more complicated questions arising in the general lectures. He also gave a course on statistics and brought external scholars such as Acworth, Hewins, Giffen and Flux to lecture on their subjects: railway rates, state regulation of wages, the use of statistics and the economic effects of internal migration, respectively.[49]

Afterwards, in the 1940s, Langford Price complained about Edgeworth:

> [one] could hardly ever obtain from 'my balancing friend Edgeworth' a definite conclusion on any matter, except occasionally the damnation of an outrageous fallacy. . . . Economics at Oxford looked like slumbering quietly or in effect at least must languish comparatively as it rested, so to say, inert in Edgeworth's keeping. There was no active stir of a resonant hive of busy students gathering honey under his helping regime.[50]

In fact, the number of students of first research degrees established after 1895 in Oxford – the B.Litt. and the B.Sc. – who asked Edgeworth for advanced supervision in economics or statistics was always very low.[51] The same thing happened in Cambridge.

James Bonar's Description

Another close friend, James Bonar, also reported on some of Edgeworth's traits as Professor of Political Economy.[52] Bonar thought, as Price did, that

Edgeworth 'often over-estimated the power of the ordinary economist, still more of the mere ordinary reader, to follow him'. Bonar describes his style in front of an audience, how Edgeworth smiled

> when he proposed to clear up an argument by a diagram on the blackboard. He went calmly on his way and let people talk. He made up his own mind whenever the matter was within his own ken. Where the matter was out of his province his humility was profound. He was fonder, too, of maintaining his argument than of converting others to his own conclusions.

And Bonar goes on with his portrait of Edgeworth:

> Though he was always learning, he unlearnt little; his views of life were in essentials unchanged; he remained in philosophy naturalistic, in economics Ricardian, in ethics Utilitarian of the old school, with the slogan of 'Greatest Happiness *sans phrase*'. To the end his early heroes remained on their pedestals. . . . He had a gift of expression not given by any system of education and not imitable or easily parodied, working slowly and surely rather than fluently. He was content to enjoy literature, without contributing to it (as far as I know), even the conventional Latin verses of the sixties at Oxford. He never read so exclusively in his own subject that he had not time for excursions into Dante and Æschylus, Milton or Gibson, and into lighter literature. It was against his nature to inflict an insult, and against his habits to be conscious of receiving one.

Bonar stresses that Edgeworth was extremely cautious and diffident. As we have already mentioned, Bonar quotes a pessimistic principle of maximum likelihood ironically espoused by Edgeworth: 'Other things being equal, the most unpleasant event is the most probable.' On the other hand, Bonar's descriptive remembrances of Edgeworth are quite specific:

> He collected his friend's opinions on the subject of matrimony and told me he was disappointed: 'They were all so happily married'. But here as usual he had formed his own opinion first, not to be much modified by any answers to his inquiries. [. . .In] Edgeworth's admiration . . . Marshall was the great Apollo, oracle or highest authority. 'Marshall was at the Council today; it was as if Achilles had come back.' No one was, in another sense of the word humour, more good-humoured than he [FYE] and farther from the 'gruffness fatal to friendship' (Aristotle, *Ethics*, VIII, 5). He was provokingly modest; he deferred to many as being far above him – not only to Marshall and Sidgwick, but to Giffen and Goschen, Mrs. Bryant and Mrs. Webb.[53]

Edgeworth's admiration for and subjugation to Marshall has been stressed by all of his biographers and has become a cliché. Even Roy Harrod, then an Oxford student, tells in his biography of John Maynard Keynes that 'at Oxford, Professor F.Y. Edgeworth, an original economist of notable achievements, was his [Marshall's] unqualified admirer'.[54]

But even though Edgeworth's praise of Marshall may have been 'unqualified' to his students, we do not know if Bonar realised that Edgeworth had compared Marshall with Achilles, who had his own secret flaw, as opposed to Apollo. The comparison with Apollo is, as far as we know, Bonar's alone, and perhaps he missed this sparkle of irony from Ysidro – using Marshall's dichotomy between the 'candid' Francis and the 'mischievous' Ysidro. On the other hand, it is interesting to note that Bonar extends Edgeworth's admiration not only to Sidgwick, Giffen and Goschen, but also to Sophie Bryant and Beatrice Webb. Bonar also describes several characteristics of Francis that we are already aware of through his aunts and family:

> He was a bad sailor even on Swiss lakes . . . He was sometimes 'at sea' on land. He has been known to mislay his railway tickets, and to let a train pass at a junction when he was deep in a book. He bowed before the Arts and made it a duty to take his friends to the yearly exhibitions of the Royal Academy. Music appealed to him little, and the Church not much more [like most Edgeworths before him. . .] The consequent peace of mind was a greater reward to him than his wide reputation and honors. . . . His personality was of a type of its own, unlike any we had ever known or shall know elsewhere.[55]

3.7 THE FIRST STAGE OF THE *ECONOMIC JOURNAL*: 1891–95

In June 1891, Marshall suggested names to Foxwell in order to ease Edgeworth's burden in London, namely for the editorship of the *Economic Journal (EJ)* and for secretaries of the British Economic Association and the Royal Statistical Society. The candidate that they both found suitable was Henry Higgs, a 28-year-old graduate from University College London and a student of Foxwell. Higgs had participated in his master's early efforts in 1888 with respect to the Association and was one of the founders of the Junior Economic Club. He had entered the civil service at the age of eighteen as a Lower Division Clerk, and two years later, in 1884, he passed the competition for Class I and was assigned a clerkship in the Secretary's Office of the General Post Office. He then started to study law, and in January 1885 he attended lectures at University College London. He earned honours and was awarded a £40 University Exhibition per annum for two years, and later won the Joseph Hume Scholar of £120 per annum for three years. In 1890, he earned his Bachelor of Laws and became a Barrister of the Middle Temple. Higgs was also a close friend of James Bonar, the civil servant with academic interests in economics who was also a friend of Edgeworth. Attracted by Bonar's friendship, Higgs went to live in Hampstead and was thereafter in close contact with Francis.[56]

Higgs was named secretary of the British Economic Association in 1892. Edgeworth retained the editorship of the *Economic Journal*, and in 1896 Higgs was named Assistant Editor. Therefore, there is an initial five-year period from 1891 to 1895, in which Edgeworth was the sole editor of the *Economic Journal*. During this first stage, Edgeworth faced just one incident, when John Rae, a regular contributor of the *Journal* accused him of inefficiency and failure to pay the amounts due to him. The Council of the British Economic Association acquitted Edgeworth of discourtesy.[57] John Rae's contributions were an article, fourteen notes and three reviews during the first three years of the *Journal*. One year after this incident, in 1895, Rae published his most successful work, *Life of Adam Smith*.

During 1891 to 1895, Edgeworth contributed a vast number of reviews and articles as an author. We have tallied 35 reviews of books written in four languages: English, French, Italian and German. The complete list is offered as Appendix I. In the years to follow, Edgeworth would also review books in Spanish and modern Greek. With the obvious exception of the international contributions, Edgeworth published all his writings on political economy in the *Journal*. The list of the articles and notes he published in this period also appears in Appendix I.

Among them, the most important (to Edgeworth as well, who selected them for the 1925 edition) were 'The Theory of International Values' and several notes connected with index numbers and the debate on bimetallism, totalling 103 pages altogether.[58]

As editor, following the policy set forth by the British Economic Association in the first issue of the *EJ*, Edgeworth tried to open the journal to the full spectrum of political trends, as well as to foreign professors and women. We have mentioned some early instances of this editorial policy in Section 3.4, and there are further examples in this period. In 1891 there was a review by Pantaleoni of Schullern's *Die theoretische Nationalökonomie Italiens in neuster Zeit*. In 1892, Pareto contributed a note on 'State Expenditures in Italy as Compared with the National Wealth'. In 1894 there was a review by Sidney Webb (Fabian Society) and later, in 1898, there was a review by the heterodox writer Hobson. In 1895 we find a review by Charles Gide, editor of *Journal des Économistes*, another by Mary P. Marshall and an article by M. Pantaleoni.

New Contributions to the Theory of Utility: Irving Fisher and Andreas Voigt

After Edgeworth's attempt to measure utility and probability in his 1887 essay *Metretike*, it is particularly interesting to see how he kept open to new advances in the theory of utility. In this regard, we find in the 1892 *EJ* his review of Irving Fisher's book, *Mathematical Investigations in the Theory of Value and Prices*, and two allusions to the German economist Andreas Voigt in the 1894 *EJ*.

Though Fisher introduces a 'unit of utility' – 'the utility of the hundredth loaf per year' – and arrives at similar results as Jevons and Marshall – 'price, production and consumption, are determined by *the equality of marginal utility and marginal cost of production*' – , Fisher affirms that 'unless certain mathematical conditions are fulfilled by the "maximum directions" [the normal to the indifference surfaces] and the prices which act along them, integration is impossible and there *is no such quantity as total utility or gain*'.

For Edgeworth, 'the importance of this theorem' lies in Fisher's conclusion:

> if we seek only the causation of the *objective facts of prices and commodity distribution* certain attributes of utility as a quantity are entirely unessential. . . . We may dispense with the total utility density and conceive the economic world to be filled merely with lines of force or maximum directions and we may also dispense with the attribute that one man's utility can be compared to another's.

In other words, Fisher shows and Edgeworth notices that demand functions do not entail cardinal utility but only ordinal utility. Neither Fisher in the book nor Edgeworth in the review use the term 'ordinal'.

Edgeworth accepts this dispensations of cardinal utility for the Economic Calculus but reminds Fisher of 'the great gulf which separates economics from moral philosophy' and the exigency of interpersonal comparisons of utility in the Utilitarian Calculus, 'to regulate the utilitarian distribution'. Finally, Edgeworth ends the review with laudatory terms, by predicting 'to Dr. Fisher the degree of immortality which belongs to one who has deepened the foundations of the pure theory of Economics'.[59]

Two years later, we find in the 1894 *EJ* an article by Edgeworth – 'Professor Nicholson on Consumer's Rent' which includes the following footnote on p. 155: 'On the measurement of sensation consider Dr. Voigt's proposal to use only *ordinal* – not cardinal – numbers, referred to on p. 202 of the present nr. of the Journal'.

Then, at that p. 202, at the beginning of the section about recent periodicals, we find a reference to an article by Dr. A. Voigt, 'Zahl und Mass in der Œkonomic' published in *Zeitschrift für die gesamte Staatswissenschaft*, 1893, vol. 4, which includes the sentence: 'Economics deal with quantities, such as utility, which not being expressible in units, are measured only by *ordinal* numbers.'[60]

Edgeworth would, in the years to come, be the main diffuser through the

EJ, of the distinction beteween cardinal and ordinal utility. We shall come back to it in Section 4.6 where we shall follow a further stage of the *EJ*.

The Theory of International Values

This was the first of the comprehensive surveys that Edgeworth wrote for the *EJ*. He initially framed the subject 'on classical lines' by stating that 'in technical usage international is distinguished from home trade by the existence of barriers that prevent owners of the means of production in one . . . sphere of industry from employing those means in another sphere'.[61] Or more briefly, in Bastable's words, the main peculiarity is 'the immobility of industrial agents'.[62]

Nonetheless, Edgeworth holds the view that, in essence, 'International Trade is that general theory which Jevons called the Theory of Exchange and Professor Marshall describes as "an inquiry into the balancing of the forces of Demand and Supply".' In fact, the pure theory of international trade by Smith, Ricardo and J.S. Mill may be considered the classical theory of exchange, insofar as 'it is a corollary of the general theory that all the parties to a bargain look to gain by it'. And he supports his point of view by arguing:

> Foreign trade would not go on unless it seemed less costly to each of the parties to it to obtain imports in exchange for exports than to produce them at home. This is the generalised statement of the Principle of Comparative Cost, with respect to its positive part at least. The negative part that the value of articles in the international market is not proportioned to the cost – the 'efforts and sacrifice' – incurred by the respective producers, is superfluous if the definition here proposed is adopted.[63]

Therefore, Edgeworth brings the pure theory of international trade of the classic economists to his own terrain of the bilateral exchange from *Mathematical Psychics*. By doing so, he was able to develop the analysis of bilateral barter between nations in terms of 'native indifference curves' and 'foreign indifference curves' which come out of their respective 'collective utility curves', defined – as in the case of the joint happiness from his previous 'exact utilitarianism' – through a functional combination of individual utilities.[64]

Edgeworth then analyses the works of Ricardo, J.S. Mill, Cairnes and Sidgwick on this topic and tries to develop them in diagrammatical and analytical terms, with added allusions to Pareto, Marshall and Cunynghame. Finally, he examines some precedents for the mathematical treatment of international trade by Cournot (1838), Hagen (1844), von Mangoldt (1863) and Auspitz and Lieben (1889).

Some Works outside the *Economic Journal*

In February 1891, Edgeworth published an 'enthusiastic review' – not exempt from some reservations – of J. Neville Keynes's *Scope and Method* in *Nature*. Edgeworth reviewed it again in the *Economic Journal* dated June 1891.[65] He also published an article in the *Giornale degli Economisti*, 'Osservazione sulla Teoria Matematica dell'Economia Politica, con Riguardo Speciale ai Principi di Economia di Alfredo Marshall'.[66] Some of Edgeworth's criticisms in 'Osservazione sulla . . .', centred on the *Principles'* scarce attention to the problem of indeterminacy in barter, were not at all to Marshall's liking, and he expressed his wounded feelings to Edgeworth:

> I now throw myself on your kind and generous forbearance, and ask you to listen without anger . . . What I want to say is that I do not think you at all appreciate the deadly and enduring injury that A does to B, if he reads rapidly a piece of hard argument on which B has spent an immense deal of work; and then believing that argument to be wrong, writes an article full of the most polite phrases, in which a caricature of that argument is held up to the most refined, but deadly scorn. I fancy you think that the polite phrases diminish the mischief. Really it is they that cause the most harm. Their effect, though *certainly not* their intention, is that of a white flag under which one ship approaches close to another and rams or torpedoes it.[67]

3.8 GALTON AND EDGEWORTH

We have seen in Section 2.8 that in 1881 Galton had defended Edgeworth's *Mathematical Psychics* against Jevons's criticisms. Three years earlier, Edgeworth had written to Galton after accompanying Sully to the Anthropological Society to hear Galton lecture on the 'power of visualisation', in which the eminent biostatistician suggested that the visualisation of numbers varies among different people. In a letter dated 15 March 1880, Francis sent him his own 'number-curve':

> It is shaped like a hunting whip. There is a point of discontinuity at 12. [He adds a drawing of the whip with a straight vertical whip-handle and a 0 at the base and a 12 at the top. Then he draws a leftward semicircular lash from the top with the numbers 20, 30, 40, 50 . . . 90 descending at diminishing intervals].
>
> I have had this presentation as long as I can remember, certainly at the age of four when it used to be my amusement trying how high I could count. The hundreds I think of as along the digit line (as the whip-handle).[68]

In September 1885, the British Association for the Advancement of Science held its academic meeting in Aberdeen. On the 12th, Galton presented his celebrated paper on 'Regression Towards Mediocrity in Heredity Stature', in which, under Jevons's influence, he used the term 'correlation'. Two days later, Edgeworth read the article 'On Methods of Ascertaining Variations in the Rate of Births, Deaths and Marriages',[69] in which he performed the aforementioned analysis of variance for a two-way classification, thus predating R.A. Fisher's work on this topic by 40 years.[70]

This encounter with Galton was a source of inspiration for Edgeworth, who keenly followed all of Galton's works on correlation and in the early 1890s published several important papers on this subject.

In 1889, 1894 and 1896 Galton tried unsuccessfully to get Edgeworth elected Fellow of the Royal Society, which after being founded in 1660 had been the meeting point of Britain's most famous scientists. Galton was elected as a fellow in 1860. Most of the 25 letters that Edgeworth wrote to him and that are stored in the Galton Papers, University College London, date from these years, and Francis politely thanked him for 'labouring for my admission to the Royal Society' and for 'writing about the result of my candidature for the Royal Society'. The correspondence with Galton is quite distant in tone, perhaps because Galton was 23 years older, and only in the last letter remaining, dated 27 September 1896, does Edgeworth discuss specific statistical inference problems with him.

At any rate, Galton could not convince other scientists in the field, such as John Venn, that Edgeworth should be admitted to the Royal Society at the 1896 election. Venn, a fellow since 1883, replied to Galton in April 1896:

> I am sorry to hear what you say about Edgeworth, but I think that your remarks exactly hit the point. I have a very high opinion of the aggregate of his contributions to the Theory of Statistics, and have felt the help of his work in a number of directions; but it is difficult to point to any single essay of his which, like K. Pearson's, is decisive at once in the way of power and originality. I feel strongly that Edgeworth ought to be on the Royal Society.[71]

Karl Pearson[72] was elected as a fellow of the Royal Society in June 1896 instead of Edgeworth.

The file of Edgeworth's letters addressed to Galton also contains a report on his most 'noteworthy' relatives as a contribution to Galton's project on 'noteworthy families' to examine the hereditary influence of intellectual genius. After publishing a first volume in 1906 in conjunction with B. Schuster, Galton collected information about more families and asked Edgeworth about his own family for a second volume that was ultimately never published. Francis mentioned, as expected, his grandfather Richard Lovell Edgeworth, his aunt Maria, his uncles Charles Sneyd and Michael Pakenham Edgeworth, his cousin Thomas Lovell Beddoes and even the Abbé Edgeworth of Firmont. And on the Beaufort side he referred to his great-great-grandfather 'Daniel Cornelius Beaufort (1700–88), provost and archdeacon of Tuam, a French refugee'; his great-grandfather 'Daniel Augustus Beaufort (1739–1821), hon. LL.D., geographer, published a map of Ireland, 1792'; his father's uncle and direct uncle by marriage, 'Sir Francis Beaufort (1774–1857), K.C.B., F.R.S., Rear Admiral & Cartographer. Surveyed coast of Karamania 1811–12, hydrographer to the Navy 1829–55', and his father's cousin, the son of the aforementioned, 'Francis Lestock Beaufort (1815–79), author of *Digest of Criminal Law Procedure in Bengal*'. Francis did not mention any intellectual achievements from his mother's side.

This time Francis also suggested that Galton could find more details about the Edgeworth family, if necessary, by asking his cousin Harriet, who was married to Arthur Butler, the brother of Louisa Butler, Galton's wife.

3.9 PEARSON AND EDGEWORTH

The letters from Edgeworth kept in the Pearson Papers, University College, London,[73] prove that he was the main inducer of Pearson's decision to enter the arena of statistical inference. In contrast to the colder, more formal style used with Galton, Edgeworth was much warmer and more spontaneous with Pearson, probably because Pearson was twelve years younger than him.

In April 1891, around the time when Edgeworth sent Pearson his first letter, the former had just obtained the Oxford chair and the editorship of the *Economic Journal*. Pearson, who had been appointed to the Chair of Applied Mathematics at University College London in 1885 and had married Maria Sharpe in 1890, was living at number 7, Well Road, Hampstead, quite close to Edgeworth's quarters. In this first letter, he invited Pearson to write an article for the *EJ* about 'the applicability of Mathematics to Political Economy'.

Pearson did not accept the invitation. Nevertheless, their Hampstead conversations continued, as Edgeworth's letters of those years prove: two letters in 1891, two in 1892, thirteen in 1893, five in 1894, eight in 1895 and eleven in 1896. But the topic of statistical inference, which was Edgeworth's main academic theoretical concern from 1891 to 1896, does not appear until 1893.

In May and June of 1892, Edgeworth gave six Newmarch Lectures at University College, London, 'On the Uses and methods of Statistics', where he displayed his theoretical work over ten years – 1883–1892 – in statistics, which included the work in multiple correlations and tests of fit. The lectures took place nearly five months before the first public discussion of any of these topics by Karl Pearson.[74]

In August of 1892 Edgeworth, following in Galton's footsteps, published part of his statistical research on the topic of correlation in the *Philosophical Magazine*, in articles entitled 'Correlated Averages' and 'The Law of Error and Correlated Averages'.[75] In these articles, where he introduced the term 'coefficient of correlation', Edgeworth worked on the analysis of correlation and succeeded in furnishing the first thorough mathematical analysis of correlation and in revealing its relation with the multivariate normal distribution.[76]

Indeed, Edgeworth's role as a pioneer in the field of correlation was initially acknowledged by Pearson. In 1893, Pearson, was working on this topic and trying to publish the first article of a series of *Mathematical Contributions to the Theory of Evolution*.[77] Edgeworth, who was ten years ahead in the subject of statistics through his published works, wrote patiently, in a postcard dated October 1893, that 'you need not apprehend my taking the wind out of your sails. We are not in the same hemisphere, in the lines of my article of 1886 [on asymmetric curves]. I do not borrow anything from our recent discussions.' And five days later, Edgeworth reassured him yet again:

> I will put off publishing my paper for any length of time rather than that you should have any feeling of dissatisfaction about the matter. I would do so even if I were proud of the paper which is far from being the case. . . . For in justice to myself I must repeat that it is not in the slightest indebted to you.[78]

When Pearson's article appeared, Edgeworth wrote to him, saying, 'I don't know whether to admire more the literary elegance or the mathematical force of the article.'[79] After 1893, most of Edgeworth's letters to Pearson were quite long and dealt with specific statistical problems. Every time Pearson published an article, Edgeworth sent him his compliments, such as 'I have read your great paper with great interest and admiration. It is certainly a most important discovery.'[80]

In the period 1893 to 1895, Edgeworth published seven articles and notes in the *Philosophical Magazine* and the *Journal of the Royal Statistical Society* on correlation, its application to social phenomena, calculation of errors and a summary of 'Recent Contributions to the Theory of Statistics' which culminated in his four important works from 1896: 'Statistics on Unprogressive Communities', 'The Asymmetrical Probability Curve', 'The Compound Law of Error' and 'Supplementary Notes on Statistics'. In the latter, he criticised Pareto's Distribution Law and favoured instead a corresponding formula by Pearson, 'the author who has made the greatest advance in the science of Probabilities since the era of Poisson'.[81]

Despite Edgeworth's repeated praise, Pearson's aggressive personality manifested itself on multiple occasions. In a letter dated 5 November 1895, from Oxford, Edgeworth complained that Pearson is charging him with '(1) misinterpretation, (2) mathematical error, (3) logical fallacy & (4) unjustifiable tone'. However, it appears that Pearson tried to apologise, since ten days later Edgeworth replied that he 'was much gratified by the tone of your last'.[82]

The eleven letters from the following year, 1896, still reflect a smidgen of humorous affection from Edgeworth's side, but after that the correspondence was limited to academic politeness and years of silence.

3.10 1890S: FAMILY LIFE IN OXFORD

After establishing himself at All Souls College in Oxford, Edgeworth kept up a close relationship with his sister Mary Sanderson and her family. Winchfield moved nearer to Oxford, and Francis became fond of his nieces Mia, Fanny, Helen, Felicia and Rosa. However, in 1893, when she was 54 years old, Mary died. Back in 5 October 1891, their aunt Lucy Jane Robinson had written to Mary, reminding her that this date was both her father's birthday and the day he died. Aunt Lucy Jane called her 'Miss Minnie' as when Mary was a child, and confessed that how 'I survived all the healthy strong ones is a mystery, . . . recollecting as I do some moments of my early life when they were standing, leaning over me and thinking I was quite gone'.[83] Aunt Lucy survived Mary for four more years. She died in 1897 at the age of 92 at Kingstown, 153 years after the birth of her father.

Now that Edgeworthstown was renovated, Francis spent some of his holidays there with his brother Eroles and Françoise, his French wife. He was there in the summers of 1895 and 1896. Through the return addresses on his letters from August 1895 to Karl Pearson, we learn that Francis was in Edgeworthstown, and from there he went on a tour of Northwest Ireland, namely to Clifden (Connemara), Athlone and Achill Island (Connaught). And in an article published in 1907, Edgeworth reported that on 9 September 1896 he was recording observations on wasp movements (*Vespa germanica*) in Edgeworthstown. More observations of a nest of bumblebees (*Bombus hortorum*) followed at Oxford in 1897.

Cousin Harriet Jessie Butler

Edgeworth's relationship with his cousin Harriet Jessie Butler, who lived in Oxford, developed quite slowly. In the time that had elapsed, she had become the perfect housewife. According to a source based on her daughter

Violet, she 'was on the Committee of a home for unmarried mothers, took a strong interest in moral welfare causes, ran the Mother's Union at the family's country home at Birdip in Gloucestershire, and as a matter of course visited old people in the workhouse once a week for fifty years'.[84] Harriet wrote many years later that Francis 'was always rather odd, and when he first came to live in Oxford at All Souls [1891] was somewhat aloof'.[85] In a letter to her son, she reported that Mia Sanderson, Francis's eldest niece, had sent a photo of him 'dated I should fancy at least 30 years ago – anyway it represents to me the cousin I knew but slightly, and did *not* like! in the 70s and 80s. He changed and *steadied* and mellowed greatly as years went on: tho' I must say "a charming old gentleman" does not appear to me to describe him.'[86]

Harriet's children had an altogether better view, as Christina Violet Butler stated to Maynard Keynes in 1926 that 'he was really extraordinarily kind as an uncle and "quasi-uncle" to his younger relations [. . . kindness expressed] in little things such as elaborately planned and strenuous river picnics [in Oxford] replete with chocolate boxes, when we were all children'.[87] Harold Butler also reported a good relationship with Edgeworth:

> I have very pleasant memories of his charming hospitality at All Souls and on many rides on horseback and cycling together. He was a wonderful man in his own peculiar way. For sheer cleverness I have never met anyone his equal. And he was a most brilliant and delightful talker. He perhaps amused people more than he quite realised, but he had an extraordinary sparkle, and inspired real affection too. He must have been quite unique in his gifts as well as his idiosyncrasies. I know nothing of his earlier life, but his Oxford years were, I think, very happy.[88]

In those years, Harriet recalled:

> By this time Francis was settled in Oxford, and Eroles visited him, and came to see us, looking comparatively respectable. He [Eroles] amused A.G. Butler immensely, dining with him in College and telling endless stories of dubious veracity. He came to us one evening when Fanny Fox [a cousin of Francis and Harriet] was with us and talked without ceasing. When he was gone, Fanny remarked quietly, 'Of course there is not a word of truth in all those stories'. However he managed to live respectably at Edgeworthstown and do his duty as landlord until his death in 1917 [in fact, 1911].

A little further on in her report, Harriet criticised both Eroles and Francis Edgeworth for another incident:

> It was there [in Oxford] that he [Francis] had Eroles to visit him, and he appeared to support him in various strange acts, including allowing the use and

> partial republication of the private Memoir of Maria Edgeworth in an inaccurate book by Augustus Hare. I protested, as in duty bound, against this publication (to which he had no possible right), since the original work was written nominally by my Grandmother, but really by Aunt Butler, as entirely private . . . Eroles however simply handed the Memoir over to Hare when he asked for it, telling nobody and refusing even to apologise or express any regret to me. Francis said little about it. It was a pity; for Hare made many small mistakes in it, which he would have saved, if he had been dealt with properly.[89]

During the spring of 1892, Edgeworth took part in the celebration of the tercentenary of Trinity College Dublin, accompanied by Sully. Sully mentioned this visit in his 1917 book of memoirs, when he explained that in the spring of 1892:

> Croom Robertson resigned the Grote Chair of the Philosophy of Mind and Logic at University College, [London], I substituted him, stood for the vacant chair and was elected. The news of my appointment came when I was in Dublin, taking part in the celebration of the tercentenary of Trinity College.[90]

In 1892, Sully also published his lengthiest work, *The Human Mind*. Afterwards, he wrote *Studies of Childhood* (1895) and *Essay on Laughter* (1902). In 1903, Sully resigned from his professorship. Yet before that, in 1897 he had left Hampstead and abruptly disappeared from Edgeworth's life for a period of more than fifteen years.

3.11 *PALGRAVE'S DICTIONARY OF POLITICAL ECONOMY*

Around 1891, Edgeworth had suggested to the economist R.H. Inglis Palgrave that he produce a dictionary of political economy, along the lines of the one that Leslie Stephen had been developing since 1882 with the *Dictionary of National Biography*. Palgrave followed Edgeworth's advice and, assisted by a group of friends, he was able to construct this remarkable collective work within a reasonable period of time. The first volume was published in 1894, the second appeared in 1896 and the third in 1899. Edgeworth was one of the main contributors to the *Dictionary*, which is commonly called *Palgrave*. He contributed 115 entries, which are listed in Appendix J.

Most of Edgeworth's entries give the reader a survey of the relevant literature on each topic. In the entries about political economy, Edgeworth very often quoted Marshall, Sidgwick, Jevons, J.S. Mill, De Quincey, Cournot, Walras, Auspitz and Lieben, and Walker. Other authors mentioned are Ricardo, Cairnes, Wicksteed, Giffen, Taussig, Wagner, Fisher,

Locke and Hume. In the entries about statistical inference, he often quoted Laplace, Quetelet, Galton, Venn and Pearson.

Some of Edgeworth's entries offer unexpected contents: the entry 'Demand Curves' shows his statistical pessimism by mentioning that 'Jevons's hope of obtaining demand curves by statistical observation . . . may appear chimerical'.[91]

Another entry, 'Absentee', is an implicit homage to his aunt Maria Edgeworth, who in 1812 published a series of *Tales of Fashionable Life*. In the second series, there was a 'moral tale', widely recognised as one of her best, entitled *The Absentee*, in which Maria attacked the Irish landlords who leave their property to establish themselves at London.

Finally, one of the most curious entries written by Edgeworth is the article on 'Doctrinaire', in which he proposes literary humour to counteract doctrinairism. After defining doctrinaire as 'one who applies theory without due regard to the facts', he writes: 'Literary humour is a potent antidote against the crotchets of doctrinaires. Take as examples: Voltaire's *L'homme aux quarante écus*, Scott's *Malachi Malagrowther*; some of Thomas L. Peacock's stories, especially *Crotchet Castle*; part of Ruskin's *Unto this Last*.'[92]

3.12 ACADEMIC EVENTS IN THE PERIOD 1895–97

The London School of Economics and Political Science (LSE), founded by members of the Fabian Society – which included Beatrice and Sidney Webb and George Bernard Shaw – opened in October 1895 with W.A.S. Hewins as its first Director. Following the model of the *École des Sciences Politiques* in Paris, the LSE laid more stress on historical, statistical and applied training than in reasoning through theoretical abstractions as guidance for future social scientists. Hewins invited many external professors to give lectures, from Edgeworth and Giffen to William Cunningham and Foxwell. LSE became more and more theoretical after the turn of the century.[93]

During these years, the friendship between Marshall and Edgeworth was flowing quite smoothly, in spite of some turbulence that can be detected through their correspondence. In 1896, Marshall was informed by Edgeworth that Edwin Cannan was a great cyclist. This was a surprise for Marshall, who had to write to Cannan apologising for a previous conversation between them. Cannan (1861–1935), a professor at the LSE from its founding, was considered to be a 'Marshall man', even though he had both personal and professional difficulties with Marshall.

The event that involved the most gossip in the closed world of academia also took place in 1896: the Cambridge referendum on granting degrees to women. After a flurry of opinions both in favour and against, the votes went against the proposal. In spite of his learned and supportive wife, his experience in women's colleges such as Newnham and mixed colleges like the University College of Bristol and all his previous statements, Marshall voted against it. In 1925, after Marshall's death, Edgeworth would comment:

> Concern for the practice of family duties was the ground of Marshall's opposition to the granting of degrees to women (1896). Without offering an opinion on this issue, I may point out that his arguments were deduced from principles which with general approbation he applied to another issue, that which is raised by Socialism. Again and again he has expressed sympathy with the generous aspirations of the Socialists, while declining to follow them far on untried abrupt paths. In a similar spirit he urges the Cambridge Senate to begin with half measures, to wait for experience before taking a step of doubtful policy but great magnitude.[94]

Then in January 1897, Marshall declined Edgeworth's invitation to write the obituary notice for the American economist F.A. Walker, Professor at Yale University, for the *EJ*. The obituary was ultimately written by L.L. Price. Afterwards, in October 1897, Marshall wrote to J.N. Keynes about organising the Moral Science Tripos set of examiners, stating that he wanted to have in it 'any Oxford man, other than Edgeworth'.

When in 1897 Edgeworth published a survey in Italian on the theory of monopoly, entitled 'Teoria Pura del Monopolio', in the *Giornale degli Economisti*, he quoted generously from Marshall among other quotations from Cournot, Bertrand and Amoroso. Apparently Marshall's reprimand to Edgeworth, when he published his 1891 article in the same Italian journal, was quite effective, since this time Marshall did not raise any objection. An English version of this article appeared in the *Economic Journal* – see below.[95]

3.13 ECHOES FROM WALRAS AFTER 1891

Walras never relaxed his antagonistic treatment of Marshall and Edgeworth, not only in his letters to Bortkiewicz but also in his correspondence with Maffeo Pantaleoni (1857–1924), Vilfredo Pareto (1848–1923) and Enrico Barone (1859–1924). Pantaleoni, then Professor of Political Economy in Naples, had introduced Walras to Pareto through a letter dated June 1891. Shortly thereafter, Pareto made a brief visit to Lausanne which was not especially remarkable on either side, nothing like the first encounter of kindred spirits, such as the first meeting of Marx and

Engels in Paris. Nevertheless, Walras wrote to Pareto in a confidential tone on 12 March 1892:

> The price of producing services can be deduced from the price of the products, but the price of products cannot be deduced from the price of producing services . . . This is what Jevons has managed to discern . . . and it is deplorable to see Marshall, *this great white elephant of political economy* and Edgeworth, through impotence and jealousy, hurrying to rehabilitate Ricardo and Mills's theory on the price of products. The Austrians have had the merit of using their *Grenznutzen* to avoid this nonsense. (Emphasis in the original).

Walras also accused Marshall and his followers of confusing the demand curve with the marginal utility (*rareté*) curve.

Several months later, Pareto offered himself, via Pantaleoni, to succeed Walras in Lausanne when the latter decided to retire, even though he was only 58 years old. The French economist accepted Pareto's offer and wrote back to Pantaleoni on 4 November 1892 that 'your idea was already mine'. Before the year was over, Walras had retired and Pareto had taken his place.

In 1895 Walras also corresponded with Barone, a major in the Italian army who was interested in mathematical economics. Edgeworth did not accept Barone's article entitled 'Essay on the Coordination of the Laws of Distribution' for the *EJ*. Barone complained to Walras in a letter dated 26 October 1895:

> After Mr Edgeworth has begged me to examine the work of Mr Wicksteed, and to develop the observations that I had made in one of my letters addressed to Mr Edgeworth himself into an article; after having written to me that my note was, according to him, a new ray of light on a very important question, one fine day he let me know that it was impossible for him to publish it. Obviously, my note has shocked those gentlemen, since they have not understood either your theory or the role of your entrepreneur; it is evident that my note has not earned the privilege of being accepted by the pontiff of the English economists, Mr Marshall, and poor Mr Edgeworth has found himself in an embarrassing situation!.

Walras answered Barone on 24 November 1895:

> You cannot get anything from the Englishmen unless you put them in their place and refuse to be bothered by them. It would not be implausible that some people were shocked by the manners of Marshall and Edgeworth and would be tempted to avoid any association with them. I know by experience that to have the Englishmen against you attracts the Americans, who are generally very well-disposed in our favour.[96]

Barone aimed to befriend Knut Wicksell and wrote to him on 3 November 1895: 'I fully and profoundly regret the controversies among my master [L. Walras], Prof. Marshall and Prof. Edgeworth, since, as you have observed, my scientific work has been so far only partly addressed to making peace among them.'[97]

3.14 EDGEWORTH, PANTALEONI AND PARETO

Pantaleoni

Edgeworth maintained a friendly academic correspondence with Pantaleoni and Vilfredo Pareto. Pantaleoni was named professor in Geneva in 1897 thanks to Walras; later, in 1900, he would be appointed professor in Pavia, and in 1902 in Rome. Nevertheless, he was always sympathetic towards Edgeworth whenever he mentioned him to Walras. In fact, quite independently of Walras, Pantaleoni had quoted Edgeworth four times[98] in his 1889 *Manuale di Economia Pura* and invited him to write for the *Giornale degli Economisti*. Edgeworth accepted and in 1891 he published 'Osservazione sulla Teoria Matematica dell'Economia Politica', the article that occasioned Marshall's rebuke. Curiously, Pantaleoni's *Manuale* did not incorporate any Edgeworthian concepts, but was very Marshallian. Despite the fact that Marshall's *Principles of Economics* had not yet been published and the only available works were *The Pure Theory of Foreign Trade and Domestic Values* (1879) in its unfinished version, circulated by Sidgwick, and *The Economics of Industry* (also from 1879), these two works were quoted 25 times. Pantaleoni worshipped Marshall, as he wrote flatteringly to Edgeworth in a letter dated 15 November 1890: 'I think and say that you are the closest approximation of a match for Marshall living in England. You know that to my mind, Marshall is simply a new Ricardo who has appeared in the field.'[99]

Pantaleoni also wrote a testimonial letter about Edgeworth for the Drummond Chair and congratulated him when he was appointed.[100]

Pareto

Vilfredo Pareto did not behave in such an upright manner as Pantaleoni; on 12 August 1893 he sent Walras a letter that he had received from Edgeworth and belittled it by adding sarcastic remarks:

> Here I enclose a letter from Prof. Edgeworth that I have understood but little, since without mentioning the handwriting, it is written with red ink on transparent paper!
>
> I think that Prof. Edgeworth returns to the question of the utility curve and the price curve. I do not quite conceive how such a question may be raised by persons who know mathematics, and I still think that on this subject you are completely right. Prof. Edgeworth also deals with the imperfection of

competition, and this I regard as another question entirely. I openly admit that this imperfection may exist.

In fact, Edgeworth's letter to Pareto was quite friendly:

Your interpretations of Walras are most interesting and flattering for him. You do not, however, quite persuade me that he is free from the defect which I have ventured to indicate in my article related to him in the *Revue d'Économie Politique* (Jan. 1891?), which does not sufficiently express the quality of a net advantage in different occupations, the essence of the great Ricardian theory of *cost of production*.[101]

Pareto's mean-spiritedness becomes more evident when we compare it with Edgeworth's tone, as it is shown in his letter to Pareto from January 1895:

Can you refer me to passages in Prof. Walras's book which give an answer to my difficulty [on net advantages in different industries in the Marshallian sense]? I hope that in the next *Economic Journal* there will appear some account of your contribution to mathematical economics which I have gathered with great interest and admiration.

After receiving a new note from Edgeworth in June 1895, Pareto jestingly reported to Walras: 'If you want to have fun, read this letter from Edgeworth as well as the note that he sends me. You have also my answer.'[102]

At the end of 1896, the academic relationship between Edgeworth and Pareto openly deteriorated when they had an acrimonious discussion on the curve of the distribution of wealth, which was an empirical law enunciated by Pareto with no theoretical support.[103] This retrogression in their relationship is not strange if we take into account that Pareto also privately criticised Walras when he published his more applied works. For instance, after Walras issued 'Théorie de la Propriété',[104] Pareto wrote to Pantaleoni: 'Walras publishes incredible things. You are younger than me. When I write such things I beg you to give me notice to stop writing.'

Pareto's radical liberalism and his 1900 shift to anti-democratic positions which condemned all kinds of socialism shocked Walras, and ultimately it undermined their friendship. Barone attempted to secure a reconciliation between them, to no avail.[105]

Returning now to the scholarly relationship between Edgeworth and Pareto, we have to add that this bitter discussion proved no obstacle to Pareto relying upon some of the most distinctive features of Edgeworth's mathematical economics. For instance, in his article published in October 1893 in the *Giornale degli Economisti*, Pareto considered his *ophélimité élémentaire*, namely utility, to be a single function for each individual of several variables (the quantities of the different goods), just as Edgeworth did in 1881 in *Mathematical Psychics*. This is in contrast to Jevons, Walras and Marshall, who for each individual needed as many functions of just one variable, the quantity of one specific good, as there were different goods present at the exchange. This formalisation led Pareto, in his *Cours d'Économie Politique* of 1896 and 1897, as well as in his *Manuale d'Economia Politica* of 1906, to accept Edgeworth's indifference curves and the contract curve in isolated bilateral exchange and to follow the original geometrical argument displayed in the diagrams of *Mathematical Psychics*.[106]

Pareto also took Edgeworth's indifference curves and the Cournot conception of the 'mathematical method' as set forth by Edgeworth in 1881 and also clearly enunciated by Fisher in 1892, to build his own ophelimity indexes in two articles of 1898 and 1900, in a contribution of 1903 to *The German Encyclopædia of Mathematical Sciences* and in his *Manuale* of 1906. These ophelimity indexes constitute a decisive step towards the ordinalisation of utility completed later by Hicks and Allen, though Pareto never used the words 'cardinal' and 'ordinal':[107]

> The notions of indifference lines and preference lines have been introduced by Professor Edgeworth. He came to them from the notion of utility (*ophelimité*), which he supposed to be a known quantity, and he deduced from there the definition of these lines. We have reversed the problem. . . . By coming from the notion of indifference lines . . . one can arrive at the determination of the economic equilibrium and finally get certain functions containing the *ophelimité*, if it exists. In any case, we obtain indexes of *ophelimité*.[108]

Edgeworth had acknowledged this ordinalisation in a 1903 review of Pareto's 'Anwendungen der Mathematik auf Nationalökonomie'.[109] On the other side, Pareto also mentioned Edgeworth's work on the 1894 theory of international trade in the *Cours* and he quoted his work on the 1897 theory of monopoly in the *Manuale*.[110]

Moreover, as Knut Wicksell points out, Pareto's theory about the purchasing power of money, based upon the abstract marginal utility that can be procured with one extra unit of money, is taken from Edgeworth's 1887 work on the 'best method of ascertaining and measuring variations in the monetary standard'.[111]

3.15 1896–1905: SECOND STAGE OF THE *ECONOMIC JOURNAL*

In 1896, Henry Higgs, the secretary of the British Economic Association, was appointed Assistant Editor of the *Economic Journal*. Edgeworth was assisted by Higgs in the editorial task during the period 1896–1905. As Keynes explained:

> In 1896 [Higgs] joined him [Edgeworth] on the editorial side as Assistant Editor. His tenure of both offices and his close association with Edgeworth continued until 1905, when, on his becoming the Prime Minister's Private Secretary, the pressure of higher official duties led to his retirement from them. He remained on the Council until the end of his life. Edgeworth's diffidence (though far from incapacity) in administrative things gained much needed and greatly valued support from Higgs, whom Edgeworth regarded as a man of affairs in touch with the great world, who would keep him straight on worldly and official matters; though, between them, they had sometimes a marvellous capacity for magnifying matters of form! [112]

The Edgeworth-Higgs editorial duo worked efficiently together. In 1902, under Higgs's initiative, the association earned a Royal Charter and changed its name to the Royal Economic Society. But in 1905, Higgs was named private secretary to Sir Henry Campbell-Bannerman, the liberal prime minister, and felt forced to resign due to 'the pressure of higher official duties'.[113] After this editing joint venture, Edgeworth saw a good deal of Higgs in Hampstead. Higgs married Miss Winifred South in 1908 and immediately after their honeymoon in Picardy he wrote from Hampstead to Clara Collet that his wife 'is devoting herself to my old father, who is in his 84th year and has recently had a stroke'.[114] He also kept up his close friendship with the Bonars.

Reviews by Edgeworth

The quality of Edgeworth's choices, the variety of languages and the number of books he reviewed during this period – forty of them – justify his prestige as the most erudite economist of his time. A detailed list is provided in Appendix K.

In spite of Marshall's recommendations, Edgeworth could be ironic and even sarcastic in his criticisms. For instance, in his 1899 review of Macfarlane's *Value and Distribution* he says: 'Upon the whole we are disposed to say of the author's own theories what he has said of Prof. Böhm-Bawerk's: "This entire discussion is not only without any real profit but is actually misleading".'

In his 1899 review of Vidaurre's *Economía Política*, Edgeworth writes: '[The book offers] a sort of jet of laws of nature [. . .and] the author's beliefs are unaltered, like the Laws of Nature which he formulates they have been, are and always will be, the same.'

In his 1905 review of H. Cunynghame's *A Geometrical Political Economy* he states: 'It is given to few to unite like Mr. Cunynghame the powers of popular exposition and scientific investigation. Apparently it is not given even to him to apply both powers at the same point.'[115]

Sometimes Edgeworth's verdict is apparently very flattering but is actually veiled in irony, as in the review of Plunkett's *Ireland in the New Century*: 'If author's words prove as effectual as his work, the name of Horace Plunkett will be remembered enshrined in History with the names of Arthur Young and Thomas Drummond.'

On other occasions, Edgeworth uses the words of a third party to spotlight a critical opinion. This is the case with the review of Cournot's *Recherches* on the occasion of the 1898 translation into English edited by Irving Fisher, *Researches into the Mathematical Principles of the Theory of Wealth*. In this review, Edgeworth remarks on 'Fisher's criticisms that not all parts of Cournot's main work are of equal merit'. The last two chapters, are labelled by Fisher 'an ambitious but erroneous theory'. Edgeworth had previously pointed out – in 1894 – particular errors relating to these chapters on trade between regions.[116]

Edgeworth's Articles and Notes

With regard to Edgeworth's own articles and notes, we have a total of around 230 pages published by him in the *Economic Journal* during this period, as detailed in Appendix K. The most important articles are on the theory of monopoly, the pure theory of taxation and the incidence of urban rates.

Because of the regularity of Edgeworth's scientific output, it may seem surprising that during the four years 1902 to 1905 he did not publish a single article or note in the *Economic Journal*. This gap does not correspond to any personal crisis, since we find that during this period not only did he publish the aforementioned reviews in the *EJ*, he published three important articles on statistical inference and one on political economy elsewhere. The last article was published in the *Quarterly Journal of Economics* as a consequence of his journey to America. It is interesting to realise that these were the years during which two absorbing stories filled the minds of Edgeworth, Marshall and their fellow economists, and also filled the pages of the *Economic Journal*: the establishment of the Economic Tripos at Cambridge and the *Manifesto* on the Tariff Reform of 1903.[117] However, first let us look at Edgeworth's most important articles from this period.

The Pure Theory of Monopoly

In 1897 Edgeworth published two large-scale, interconnected surveys, 'Teoria Pura del Monopolio', which exceptionally appeared in Italian in the *Giornale degli Economisti*, and 'The Pure Theory of Taxation', which appeared in the *Economic Journal*. Both articles were selected and discussed by Edgeworth for the 1925 reissue of his works on political economy. Let us look at the contents of the first article:

> Section I. – The theory of taxation in the simpler cases of a single monopolist dealing with a group (or groups) of competitive individuals.
> Section II. – Proof of the proposition that in the case of two or more monopolists dealing with competitive groups, economic equilibrium is indeterminate.
> Section III. – On the effects of taxation (and other kinds of governmental regulations) in the more complicated cases of a single monopolist dealing with groups of competitors – cases of correlation in respect of production and consumption.
> Section IV. – Summary in simple language of the theses maintained in the preceding sections.

Through the first three sections Edgeworth often refers to the theories of monopoly put forth by Cournot, Mill and Marshall. Thus, in the second section he dismisses 'the proposition clearly stated by Cournot, and to all appearances generally admitted, except by Bertrand and Marshall', that in the case of a duopoly or oligopoly 'the action of economic forces would tend to a definite position of equilibrium, a determinate set of values, – this plausible proposition is proved to be unfounded'.[118]

He then concedes that 'this theory is less evident, the opinion of Cournot is more plausible, in cases where the competing monopolists are dealing not in "rival" but "complementary" articles'. Edgeworth takes the concepts of 'rival' – substitutive – and 'complementary' objects of demand from Auspitz and Lieben's *Über die Theorie des Preises* (1889).[119] And, long before Hicks and Allen or Mosak or Samuelson, he links them to the sign of the second cross partial derivatives of the utility function in money terms: 'complementary' when the sign is positive and 'rival' when the sign is negative.[120] Edgeworth then studies the specific cases where the behaviour of each of the duopolists can be foreseen, and he analyses the stability of the solutions found. In this analysis, he reaches and anticipates the 'Stackelberg solution' and the collusive solution, the 'Bertrand solution' to duopoly.[121]

A second thesis follows in Marshall's footsteps when 'Marshall, by means of his curves demonstrates that it might be advisable to tax one kind of commodity and employ the proceeds in bountying another kind.' Edgeworth proves the paradoxical result that 'when the supply of two or more *correlated* [that, is, rival or complementary] commodities is in the hands of a single monopolist, a tax on one of the articles may prove advantageous to the consumers as a whole'. From this, Edgeworth suggests 'the control of monopolies by governmental interposition'. In his analysis, he generalises the concept of correlated commodities from both the demand and the supply points of view by defining '*complementary* and *rival* articles in production':

> (a) *complementary* products in the case of which the production of one article becomes less difficult and expensive by the increased production of the other article; (b) *rival* products in the case of which the production of one article becomes more costly according as the production of the other is increased.[122]

At the end of 'Teoria Pura del Monopolio', Edgeworth analyses the effects of prices on several different fiscal policies affecting monopolies: (a) fixing a maximum tariff, (b) applying a fixed tax, (c) applying a specific tax (so much per unit of commodity) and (d) a tax proportional to the profits or progressive with them, which falls entirely upon the monopolist, as shown by Cournot, Marshall and Wicksell. The analysis is conducted bearing in mind whether or not the articles are complementary or rival in production and in demand, a dichotomy that leads to different conclusions about fiscal policy.[123]

The Pure Theory of Taxation

All the conclusions in Edgeworth's 1897 Italian article referring to the taxing of monopolies were included in his 1897 *Economic Journal* article, 'The Pure Theory of Taxation'. In the summary added in 1925, Edgeworth explains that this long paper deals with two topics, 'the incidence of taxes' which affect transactions, and 'the criterion of a good tax system, the rule according to which the burden of taxation ought to be distributed among the tax payers'. For the first question, Edgeworth introduces a classification of four dichotomous cross divisions:

> Either (A) all the transactions under consideration are exposed to competition; or (a) among the parties with whom we are concerned there is at least one monopolist.
> Either (B) all the products with which we are concerned obey the law of increasing cost; or (b) some do not.
> Either (C) the mobility of capital and labour is not taken account of, or (c) it exists and is taken account of.
> Either (D) the taxation considered varies with the quantities of articles, . . . and so may be described as a tax on *margin*; or (d) it does not so vary, . . . and so may be described as a tax on *surplus*.[124]

Then Edgeworth considers the 'more important cases formed by the combination of these attributes', starting with (A)(B)(C)(D), which is the case of international trade. He then deals only with several special cases, namely: 1) the case of perfect inelasticity on one side of the market, such as the one examined by Fleeming Jenkin in 1871;[125] 2) other 'peculiar cases', such as those which arise when commodities are correlated – rivals or complementary – with respect to production and consumption, and Edgeworth explains 'in less technical terms' the taxation paradox mentioned in his former Italian article about monopolies. However, his 1925 introductory summary of the paradox refers to broader terms:

> It is shown that (even in a regime of competition) when demand and production are complementary, a tax on one may cause the price of *either* article to fall, with advantage to the consumers as a whole. It might have been added that if the commodities are rivals both in production and consumption, a tax on one may cause the price of both to fall. This *curiosum* does not depend on a change in the marginal value of money.[126]

Edgeworth's taxation paradox was acutely criticised by Seligman, who used it to proclaim the evil consequences of the mathematical method, which 'sometimes leads to results which are likely to divorce still more the Economics of the closet from the Economics of the market-place' as 'may be illustrated by a slip of Mr. Edgeworth himself'. So, Edgeworth replied to him by writing 'Professor Seligman on the Theory of Monopoly', in which he tried to explain in plain English, with just a little arithmetic and three simple diagrams, the mathematical arguments that support the taxation paradox.[127]

The second topic, the 'criterion of good taxation', occupies the latter part of the article. Here, through the application of the utilitarian principle, 'there is advocated the minimum aggregate sacrifice, distinguished from equal sacrifice or proportional sacrifice, as the peculiar conception of Professor Seligman and Cohen-Stuart may be called'. This minimum aggregate sacrifice entails equal marginal sacrifice. In 1925, Edgeworth was pleased to announce that 'the principle of *minimum* sacrifice is now very generally accepted, praised by Cannan, and used by Marshall. Professor T.N. Carver, who was the first to propound this doctrine, has exhibited its application with convincing clearness in his last work, *National Economy*.'[128]

This principle had previously been mentioned by J.S. Mill: 'Whatever sacrifices it [a government] requires from them, [persons or classes] should be made to bear as nearly as possible with the same pressure upon all, which it must be observed, is the mode by which least sacrifice is occasioned on the whole. . . . Equality of taxation . . . means equality of sacrifice.'

Carver remarked that 'Mill affirms the two divergent principles [minimum sacrifice and equal sacrifice] in the same passage'. And Edgeworth, trying to solve Mill's apparent contradiction, wonders, 'what if equal sacrifice is but a corrupt reading for *equimarginal sacrifice*, the condition of minimum disutility?' As usual, Edgeworth was eager to refer to an authority – Mill, Sidgwick, Jevons and Marshall were his favourites – and tried to provide a convincing interpretation of his quotations in advance of possible criticisms.

In contrast with Jevons, who hated most of Mill's work, Edgeworth was especially devoted to Mill's authority, and on this occasion he tried to save that 'classical paragraph which has influenced the most influential of his successors. Both Mr. N.G. Pierson and Sir Robert Giffen profess to follow Mill.'[129]

Thirteen years later, in 1910, Edgeworth would write 'The Subjective Element in the First Principles of Taxation' for the *Quarterly Journal of Economics*, an article that in the 1925 collected *Papers Relating to Political Economy*, was retitled 'Minimum Sacrifice *versus* Equal Sacrifice', insisting on the 'defence of minimum aggregate sacrifice as the criterion of good taxation'.[130] Here Edgeworth's main argument was that the criterion of minimum aggregate sacrifice automatically implies progressive taxation, while equal or proportional sacrifice needs further hypotheses to avoid regressive taxation.

Hobson in the *Economic Journal*

With regard to other authors publishing in the *Economic Journal* between 1896 and 1905, it is interesting to take note of an 1898 review of Dr Ludwig Stein's *Die sociale Frage im Lichte der Philosophie*, by John A. Hobson. It is interesting because there has been rather ill-founded speculation that Edgeworth had blocked Hobson's chances of being appointed an Extension Lecturer of Political Economy at London after Hobson had published, in conjunction with A.F. Mummery, *The Physiology of Industry*. Hobson attributed the exclusion to 'the intervention of an economic Professor who had read my book', but he did not name the professor. Since in 1890 Edgeworth had reviewed the book in the *Journal of Education*, in 1953 he was made the only suspect by some scholars.[131] But this suspicion not only does not match Edgeworth's nature, depicted by Bonar as 'entirely incapable of intentional discourtesy' and that 'it was against his nature to inflict an insult'; neither does it dovetail with the fact that in 1898 Hobson published the aforementioned review in the *EJ*.[132]

Hobson, who held a bachelor's degree in classics and mathematics – informally named Greats – from Lincoln College Oxford, had applied to lecture in literature and political economy at both London and Oxford in the Extension Programme. Curiously enough, his application to teach literature was accepted at both places, but only Oxford accepted him to teach political economy. The reason for this discrimination may lie in the fact that a more suitable candidate according to the standards of the electors was found for London.[133]

3.16 MARSHALL'S CAMPAIGN FOR A TRIPOS IN ECONOMICS

During the period from 1899 to 1902, Marshall was busy campaigning for a specific curriculum in economics at the undergraduate level, leading to an Economics Tripos. In March 1899, he had been induced by George Howard Darwin, the second son of Charles Darwin, who was an astronomer and mathematician, and who was by then Secretary of the Cambridge Association, to state the needs of Cambridge teaching in the field of economics. This area in 1899 amounted to Marshall's Chair in Political Economy, Cunningham's lectureship at Trinity College in Economic History and Foxwell's lectureship at St John's in Political Economy. He realised that the best way to secure better funding from the academic authorities was to have a curriculum in economics at the undergraduate level. There were already examples of this, including the London School of Economics and some American universities. With regard to the LSE, Marshall also wrote to Foxwell in March 1899:

> I am afraid of talking about London. I don't want to seem to attack them. But the comparisons which Hewins is *constantly* making to our disadvantage are rather riling. He boasts of his fifty courses. But what is a course? I don't believe that many more *lectures* are given in London school on economics and economic history in a year than in Cambridge.[134]

Notwithstanding Marshall's claim of not wanting to attack the LSE's schemes, in April 1902 he convinced J.N. Keynes, Edgeworth, Bonar, Foxwell, Gonner, Price and Sanger to sign a petition to the Senate of another London institution, the University of London, requesting reconsideration of the scheme for internal degrees for the Faculty of Economics that had been promulgated in the *London University Gazette* on 21 December 1901. The petitioners complained that the subjects were too numerous and too widely and imprecisely defined, that they encouraged rote memorisation, and that no serious study of economics was being provided.[135]

Concerning the much better status of American universities, Marshall wrote in his request to Darwin dated March 1899:

> We need one additional Professorship or Readership in Economics, and at least one University Lectureship . . . Even so, Cambridge would be less well equipped not only than Harvard and Yale, but also than some of the younger Universities of America. To take a strong instance: The Faculty of Political Science at Columbia consists of nineteen Professors and Lecturers; and of these, seven belong to the department of Economics and Social Science.[136]

After this petition, Marshall began discussing the idea of the Economics Tripos mainly with Foxwell. What should the profile of economics students be like? Here are his reflections, in a letter to Foxwell dated March 1899, on what the students should not be like:

> But to state that those people who are studying economic history as a mere series of facts without any scientific analysis, are students of economics, would I believe be a falsehood. It would I think be misinterpreted by Schmoller's students just as much as by Edgeworth's.[137]

Marshall also mentioned Edgeworth in a long letter to Arthur Lyon Bowley of March 1901, about the nature of economic analysis:

> Surely the thing to do is to build the basis of our economic structure soundly and not to put a varnish of mathematical accuracy to many places of decimals on results the premises of which are not established within 20 or 50 per cent . . . Surely the thing to do is to seek the Many in the One, the One in the Many. And who is to do it? . . .
>
> Edgeworth might have done something great at it: but he has crushed his instincts between the cog wheels of his mathematical machinery: and I doubt whether there are many other men from the impatient Oxford who have the patience to do it. It must for the greater part be done by Cambridge men, or left undone.[138]

After several years as a schoolmaster, Bowley (1869–1957), tenth wrangler at Trinity College, Cambridge, from 1888 to 1891, became a lecturer in mathematics and Professor of Mathematics and Economics at University College, Reading, from 1900–1919. Bowley also taught at the London School of Economics from 1895, rising successively from lecturer to reader to professor. In 1902, Bowley, in conjunction with Edgeworth, published the article 'Methods of Representing Statistics of Wages and Other Groups Not Fulfilling the Normal Law of Error'. This was Edgeworth's only shared publication. In 1919, Bowley was the first holder of the Chair of Statistics at the University of London. In 1924, he published his famous *Mathematical Groundwork*, where he dealt with Edgeworth's theory of barter and popularised its diagrammatic presentation. This diagram is known nowadays as the 'Edgeworth box'.[139] But let us go back to Marshall.

In May 1901, Marshall had been placed on a committee formed of the Master of Peterhouse and G.L. Dickinson, Professor of Greek History, with the purpose of advising 'as to how best to extend the study of modern economics and politics in the University'.[140] In spite of the apparent progress of his project, in February 1902 Marshall had not yet convinced J.N. Keynes, yet at the same time he was trying to involve Oxford as a whole to exert pressure on Edgeworth. To this end, Marshall wrote to J.N. Keynes: 'I think you are the only correspondent who has found much fault with my scheme. Oxford blesses it heartily and an address to their V.C. [Vice-Chancellor] is being got up, which Phelps tells me is practically a request for Part II of my scheme.'[141] Concerning Marshall's indirect pressure on Edgeworth, Langford Price wrote in a letter to Marshall that 'Edgeworth is not an easy burden to propel . . . I tried desperately but in vain the whole afternoon to get him to write the letter to the V.C. and I really think that he is not seriously discontented with the present deplorable situation'.[142]

Edgeworth responded to the pressure by writing, on the 26 June 1902, a memorandum about the status of Oxford's economic studies, presenting a proposal for the committee appointed by the Hebdomadal Council of Oxford University. It was not the proposal that he preferred, but he accepted it in deference to his friend Langford Price, who was indeed strongly in favour of it. As a consequence of the memorandum, a one-year diploma in economics was established in 1903 at a postgraduate level, differing from Cambridge, where the economics curriculum was planned by Marshall to be established at the undergraduate level. Theoretical economics at this undergraduate level was kept in Oxford in the curriculum of Literæ Humaniores, as Edgeworth and Price had found when they studied. Edgeworth argued that 'modern economics is the one branch of knowledge outside mathematics and mathematical physics, which has realised in any considerable degree the idea of a demonstrative science to which Greek philosophy aspired in vain.'[143]

In the summer of 1902, Marshall was still campaigning for the new Tripos and wrote 'A plea for the Creation of a Curriculum in Economics and Associated Branches of Political Science', which was widely circulated. On 7 May of the following year, there was a first discussion about it in the Senate House, and afterwards Marshall had to send two reports to its members and the new Tripos with a new Board of Studies was finally approved.[144]

Once Oxford had taken the decision to offer the diploma in economics, Edgeworth wrote to Marshall on academic issues, sending him some comments on Marshall's article 'Distribution and Exchange', published in the *Economic Journal*, vol. VIII (1898). Marshall answered on 28 August 1902 from Wolkenstein, in the Southern Tyrol, where he was spending the long vacation: 'You know I never apply curves or mathematics to market values. For I don't think they help much. And market values are, I think, either absolutely abstract or terribly concrete and full of ever-varying (though individually vital) side-issues.'[145]

Despite this exchange of letters in the summer of 1902, Marshall considered that none of Edgeworth's articles and books was suitable for the students of economics in the new Cambridge Tripos and did not include any in the list of readings. The Economics Board introduced several changes, but none of Edgeworth's works were included in the final list.[146]

3.17 HARVARD AND YALE IN THE AUTUMN OF 1902: IRVING FISHER

As we have seen in Section 3.7, Edgeworth's review of Irving Fisher's *Mathematical Investigations in the Theory of Value and Prices* ended with sincere praise, by predicting that he would reach a 'degree of immortality'.

Fisher had declared in the preface to his PhD thesis, read on April 1892, that 'three days after Part II was finished I received and saw for the first time Professor Edgeworth's *Mathematical Psychics*'. Nevertheless, Fisher had been clever enough to discover by himself 'that in exchange, utility is a function of both commodities (not only one as assumed by Jevons)' and, like Edgeworth too, he mentioned a certain curve 'which may be called *indifference curve*'. Another author would have been suspicious about these coincidences, but in his aforementioned review Edgeworth was quite generous with the young American PhD.

His words probably stimulated Fisher to visit England in 1894, when he travelled to Europe for a sabbatical year accompanied by his wife Margaret Hazard, after they were married in 1893. There, he concerned himself with the controversy over bimetallism and met Edgeworth, who encouraged him to write an article for the *EJ*. When Fisher visited the continent, he attended lectures by the eminent mathematicians Frobenius and Poincaré.[147]

Back at Yale, after teaching mathematics, he entered the economics department in 1895. The following year, Fisher published 'What is Capital?' in the 1896 volume of the *EJ*, and in 1898 he became a full professor at Yale University. In that year, he was struck by consumption, confirming his inherited propensity to this disease, since his father had died from tuberculosis. By conscientiously adopting a healthy lifestyle and nutritious diet that he would preach throughout his whole life, Fisher recovered and after a three-year leave, he returned to Yale.

In 1901, in the first year after his health crisis, Fisher invited Edgeworth to travel to Yale to deliver a series of lectures. Harvard then followed suit, so he had the chance to lecture at the American Cambridge as well.

The invitation was very tempting for Edgeworth. There was a certain tradition started by John Stuart Mill and carried on by Alfred Marshall, which held that a proper economist should visit the United States. However, unlike Mill and Marshall, Edgeworth was to go not just to visit institutions like a learned tourist but to give lectures at leading universities. The topic that Edgeworth chose may be interpreted as an homage to the American economist John Bates Clark, who in 1899 had published *The Distribution of Wealth: A Theory of Wages, Interest and Profits*. Perhaps in tribute, Edgeworth prepared a series of lectures under the heading 'The Theory of Distribution', and in spite of his dislike of seafaring journeys, he delivered these lectures at Harvard University and at Yale's Political Science Club during the autumn of 1902. The lectures were later published in the February 1904 *Quarterly Journal of Economics*.[148]

However, Edgeworth's 'Theory of Distribution' was not in any way a detailed review of Clark's book. It is more a defence of Marshall's positions set forth in his *Principles of Economics*, especially against Walras and Barone on the theory of the entrepreneur with no gain or loss, than an analysis of Clark's work, which is mentioned incidentally only in a few paragraphs scattered over four pages out of a total of 48 pages.[149]

In December 1902, once he had returned to Oxford, Edgeworth received a typewritten letter from New Haven, Connecticut; Fisher was sending him ten dollars, a lecture fee that Edgeworth had not solicited.[150]

After Edgeworth's visit to America, his friendship with Irving Fisher was sustained by both of them, as they defended their mutual academic positions even though they would never meet again. In 1915, Fisher published the bestseller, *How to Live: Rules for Healthful Living Based on Modern Science*, which was translated into ten languages. After the First World War, he also entered a field in which Edgeworth had worked: that of index numbers. In 1922, he published *The Making of Index Numbers*, which he dedicated to 'Prof. Francis Edgeworth and Mr. Correa Walsh', the latter being another author on the subject, also reviewed by Edgeworth. In the 1920s, by designing an index card system, he made a reputation as a forecaster and a fortune which he later lost in the 1929 Wall Street crash. His reputation as a forecaster also vanished, as a few days before the crash he had reassured investors that stock prices were at their proper level with no danger of recession. He retired from Yale in 1935.[151]

All this happened long after Edgeworth had left the American scene. Edgeworth would never go back to the United States.

3.18 THE *MANIFESTO* ON THE TARIFF REFORM OF 1903

In 1902, the Colonial Secretary in the government of Arthur Balfour, Joseph Chamberlain, was convinced that the British Empire should be viewed as a united trading block: the colonies should benefit from the same economic advantages as the metropolis, and this entailed a tariff reform in order to create an imperial preference.

Chamberlain's chance to introduce this imperial preference fiscal policy came when the Chancellor of the Exchequer, C.T. Ritchie, proposed the suppression of an import tariff of one shilling per quarter of corn without taking into account the frontiers of the Empire. Despite all Balfour's efforts to preserve the cabinet's unity, disagreement prevailed and Chamberlain resigned from office in September 1903 in order to free himself to campaign in favour of his plan.

Because of the economic nature of the matter, academic opinion was sought by both Ritchie and Chamberlain. The debate reached the newspapers. On 15 August *The Times* published a letter, which became known as the *Manifesto*, signed by fourteen professors and followed by a series of rejoinders and replies.[152] The *Manifesto*, which came out decidedly in favour of global free trade, held that imperial preference would 'most probably lead to protection', and that instead of uniting the empire, it would 'engender irritating controversies between the different members of the empire'.[153] Through a letter that Marshall wrote from the Tyrol to his German economist friend Lujo Brentano, we know some details about the writing of the *Manifesto*:

> It was mainly drafted by Edgeworth in consultation with Bastable and Nicholson: I having declined to draft it, because when I was asked to do so I thought there was nothing sufficiently definite to kick against. Afterwards, when Chamberlain and his [Tariff Reform] League committed themselves to the most glaring economic falsities, I changed my mind and suggested that one should be drafted in England . . . and now I think that on the whole we may be proud of it.[154]

'The most glaring falsities' which Marshall referred to are those set out by Hewins in his series of articles in *The Times* in support of Chamberlain's imperial preference. Hewins had resigned from the London School of Economics and was the Secretary of the Commission appointed by the Tariff Reform League.[155]

The main supporters of the League were the aforementioned Hewins, who was Chamberlain's main economic adviser, Ashley, from the LSE, and Foxwell from Cambridge. Ashley and Foxwell were advising Lord

Balfour and had a more flexible position than Hewins in the defence of the tariff reform.

So, through the press the public realised that there was a profound division of opinion among the economists, and sarcastic allusions to the free-traders appeared in the newspapers. Besides, Price accused free-trade economists of propounding 'an *odium theologicum* which could discern no via media between economic salvation and damnation'. Price had not only refused to sign the *Manifesto* but his letter of refusal was published in *The Times* immediately below the fourteen professors' letter.[156]

Moreover, Marshall, who was accused of 'addressing the layman in condescending terms', regretted having signed the *Manifesto*[157] and addressed his ire to Foxwell, who wrote:

> It was just because Marshall felt that his signature was a blunder that he was so angry with me for attacking it. He came to my room scolding violently for an hour and finally said 'You were lying, and you knew it'. . . . The Manifesto has had the effect which so many of us foresaw at the time of putting economists out of court altogether. We are now hopelessly discredited.[158]

The breach between Foxwell and Marshall became insurmountable, and, as we shall see – in Section 4.1 – it was crudely disclosed some years later. Chamberlain's tariff reform and imperial preference policy was finally implemented in 1919, after the Great War.

3.19 1902–10: PUBLICATIONS, MAINLY ON STATISTICS

As mentioned before, Edgeworth's works on statistics became more important than his works on economics after 1902. This is notably so in the period 1902–10. During those years Edgeworth's most important work on economics was the aforementioned critical survey, 'The Theory of Distribution', for the American *Quarterly Journal of Economics (QJE)* (1904). To this survey we should add a long article on the effects of different tax policies on urban land values and a survey of mathematical theories for the *Economic Journal* (1906 and 1907–1908); a mathematical article 'On the Use of the Differential Calculus in Economics to Determine Conditions of Maximum Advantage' for *Scientia* (1909); and the short essay 'The Subjective Element in the First Principles of Taxation', also for the *QJE* (1910).[159]

In 1902 he published in the *Journal of the Royal Statistical Society*, jointly with Arthur L. Bowley, the article 'Methods of Representing Statistics of Wages and Other Groups Not Fulfilling the Normal Law of Error'.

Working on Skew Curves

Also in 1902, Edgeworth contributed to the *Encyclopædia Britannica* with his article 'The Law of Error'. The same title he applied to the article published in *Transactions of Cambridge Society* of 1905. In the next year he published 'The Generalised Law of Error, or Law of Great Numbers', and in 1907, 'On the Representation of Statistical Frequency by a Series', both in the *Journal of the Royal Statistical Society*. This is an important group of papers in which Edgeworth succeeded in completely developing methods to model skew data. He had worked intensely on these methods, starting in 1895, when Pearson competed with him on the derivation of asymmetrical frequency curves, in other words, of skew curves.

Edgeworth considered skew distributions found in nature to be the result of aggregating relatively small numbers of non-normal components. One of his approaches – the method nowadays called the 'method of the Edgeworth series' – consisted of a generalisation of the central limit theorem by introducing correction terms which became negligible if the number of components in the aggregation was large.[160] This method to manage skew data had been tentatively proposed by Edgeworth as early as 1883 in 'The Method of Least Squares', and became a full-fledged approach in 1905. The 'Edgeworth series' expansion and the 'Edgeworth box' diagram are the two instances in which his name is still explicitly honoured.

Another method of devising skew curves, designed by Edgeworth in 1898 and further developed in 1908, is what he called the 'method of translation'. This method, used in the 1920s by R.A. Fisher and quite popular in recent times, consists of finding the coefficients of a polynomial transformation of the skew data that best fit a normal distribution under the least squares method, that is, the coefficients which minimise the sum of the square deviations between the transformed data and the normal curve.[161]

Learning from Bees and Wasps

During 1896 and 1897, Edgeworth had published several statistical works in the *Journal of the Royal Statistical Society*, all of which curiously contained illustrations of wasps and bees: 'Statistics of Unprogressive Communities', 1896, p. 258 onwards; 'Progressive Means', p. 365 onwards; 'Supplementary Notes', p. 529 onwards; and 'Applications of the Calculus of Probabilities', 1897. The illustrations were made not for the sake of Ysidro's irony but to illustrate the results of the empirical study he had conducted by observing the movements of his favourite insects in Edgeworthstown (9 September, 1896) and Oxford (throughout 1897).

Edgeworth's learning from bees and wasps did not stop here: in 1907, there appeared 'Statistical Observations on Wasps and Bees' in *Biometrika*. This article was based on his previous observations and some new ones at Edgeworthstown (27 August, 1906) and Hampstead (Hampstead Golf Links, 7 September 1906) of the wasps and bees' average time of absence from the nest by counting entrances and exits (he referred to them also as imports and exports to give them more economic appeal) either at the point of sunrise ('Method A1') or at sunset ('Method A2'). Edgeworth would insist on this subject during and after the war (see below, Section 4.9).

A Gold Medal and More Articles on Statistics

In 1907, Edgeworth was chosen unanimously as Guy Medalist (in Gold) by the Royal Statistical Society 'in recognition of your big and distinguished services to Statistical Science'.[162]

In the biennium 1908–09 he worked on a series of articles which have been highly praised by the experts. In 1908, Edgeworth published 'On the Probable Error of Frequency Constants',[163] a series of three papers, followed by an Addendum in 1909. These works, though neglected by K. Pearson and R.A. Fisher, were pioneering in dealing with the asymptotic efficiency of maximum likelihood estimates.[164]

In 1909, Edgeworth published in French in the *Bulletin de l'Institut International de Statistique*, 'Application du Calcul des Probabilités à la Statistique', a general work that was specially mentioned by J.M. Keynes:

> From 1885 onwards his [Edgeworth's] more general articles, especially his 'Methods of Statistics' . . . and his 'Application of the Calculus of Probabilities to Statistics' in the *Bulletin of the International Statistical Institute*, 1910, were of great value in keeping English students in touch with the work of the German school founded by Lexis and in sponsoring, criticising, and applauding from their first beginnings the work of the English statisticians on correlation. His constructive work, particularly in his later years, centred in highly elaborate and difficult discussions of his own 'Generalized Law of Error' [1906].[165]

3.20 WICKSELL'S ACQUAINTANCE

In 1906, the Swedish economist Knut Wicksell (1851–1926) was in London trying to gain an understanding of the work of the British economists. It is known that he gave an address to the Economic Section of the British Association which was later published in the *Economic Journal* in

1907 under the title 'The Influence of the Rate of Interest on Prices'. This paper was based on his previous work from 1898, *Interest and Prices*, and an article with a similar title from 1899, both of which were published in Swedish.

By this time, Wicksell had already mentioned Edgeworth's work in his early books; therefore, he probably sought the acquaintance of Edgeworth, who offered him the opportunity to deliver the speech and submit it to the *EJ*. During the address, Wicksell mentioned Edgeworth on one occasion when he noted that 'it was remarked by Prof. Edgeworth that if the free coinage of gold be suppressed, the Governments themselves have in their hand the regulating of general prices'.[166]

It is curious to see that in his first book published in 1893, *Value, Capital and Rent*, after a very Jevonian exposition of the theory of exchange, Wicksell referred to Edgeworth and agreed with him regarding total utility as a function of several variables, 'oxen and sheep together, as a general function of x and y'. However, he picked up Edgeworth's approach leading to the 'contract curve' not directly from *Mathematical Psychics* but from the mathematical appendix of Marshall's 1890 *Principles of Economics*. Wicksell felt that Edgeworth's utility function 'corresponds more to reality', thus espousing an opposite view to the one held by Marshall, who wrote that 'it seems less adapted to express every-day facts of economic life than that of regarding, as Jevons did, the marginal utilities of apples as functions of x simply'.[167]

Wicksell's second encounter with Edgeworth's work came via Pareto, as is shown in his 1897 and 1899 reviews of the latter's *Cours d'Économie Politique*. In order to write the reviews, Wicksell apparently read *Mathematical Psychics* and Edgeworth's two articles in Italian about barter theory.[168] Moreover, in his *Interest and Prices* of 1898, Wicksell referred to 'Edgeworth's admirable treatment' in his 'New Methods of measuring Variation in General Prices' published in 1888.[169]

Wicksell's later work, especially his *Lectures on Political Economy*, shows that he followed Edgeworth's contributions on monopoly, international trade, money and taxes – including a brief discussion of the 'tax paradox'.[170] In some way, Wicksell became Edgeworth's man at the Swedish School, in contrast with Gustav Cassel (1866–1945), who took a more Marshallian approach and never referred to him.

NOTES

1. Edgeworth (1925), II, pp. 273–312. Included in Vol. 4 of Darnell (1991).
2. Cournot (1838), p. 51.

3. Cournot (1838), Ch. VIII, Fig. 6. But Marshall would reverse the axes.
4. Jaffé (1965), Letter 286, translated from French, as are the rest of Walras's letters.
5. After Marshall, Foxwell includes a list with the addresses of J.S. Nicholson (Edinburgh), J.N. Keynes (Cambridge), L. Courtney (House of Commons), H. Sidgwick (Cambridge), R.H.I. Palgrave (Norfolk), T. Rogers and A. Toynbee (Oxford), F.Y. Edgeworth (The Savile Club, London) and H. Cunynghame, Esq. (The Temple, London). See Jaffé (1965), Letter 544.
6. Jaffé (1965), Letters 549, 573, 578, 593, 595, 618, 644, 749, 751 and 922.
7. Jaffé (1965), Letters 843, 856, 871, 872, 873, 874, 876, 881, 886, 890, 892, 900 and 903.
8. *Nature*, September 1889, pp. 434–6. The review is included in Newman (2003), pp. 554–7.
9. For an accurate exposition of this and some other discussions concerned with Walras's work, see Creedy (1986), pp. 90–91 and 105–107.
10. Bonar would point out that Edgeworth's 'remarks on Walras might not be unfairly applied to himself, and I suspect he wrote them down with his own weakness in mind', Bonar (1926), pp. 647–53. However, this does not seem to be the case, since Edgeworth's use of mathematics never implied the 'exuberance of algebraic foliage' displayed by Walras.
11. 'There is something solid that the tooth of envy cannot fret away.' *Nature*, September 1889, p. 436.
12. Jaffé (1965), Letter 910.
13. Jaffé (1965), Letters 925, 933 and 966.
14. Jaffé (1965), Letter 927 and 970.
15. Jaffé (1965), Letters 943, 944, 946, 947, 950, 951 and 952.
16. Jaffé (1965), Letters 958, 963, 964, 966 and 995.
17. 'Praise a rascal and he will punch you, punch a rascal and he will praise you.'
18. Jaffé (1965), Letters 996, 998, 999, 1000.
19. Edgeworth submitted testimonials from A. Marshall (Cambridge), H. Sidgwick (Trinity College, Cambridge), B. Jowett (Balliol, Oxford), F. Galton (Royal Statistical Society), H.S. Foxwell (University College, London), H. Wace (Principal of King's College, London) and Sir R.W. Rawson (late President of the Statistical Society, London).
20. Phelps to FYE, 23 February 1890 (EP NUFFIELD C UO, Box D 1). See also Creedy (1986), p. 10, and Whitaker (1996), I, p. 237.
21. Marshall to FYE, 10 July 1890 (Whitaker, 1996, I, p. 237).
22. Reisman (1990), p. 84.
23. Whitaker refers to a letter from Marshall to Edgeworth, 16 February 1891 (Whitaker (1996), II, p. 7, n. 9).
24. A. Marshall to H.S. Foxwell, 19 and 22 March 1890 and to Members of the Committee of Section F, 10 April 1890: Whitaker (1996), I, pp. 312, 313 and 317.
25. See Coats (1968b), pp. 349–56.
26. The Chancellor of the Exchequer, G.J. Goschen, was in the chair acting as president, with the following Council: Vice-Presidents: A.J. Balfour, H.C.E. Childers, L.H. Courtney and J. Morley; Treasurer, J.B. Martin; Council Members: Professor C. Bastable, C. Booth, J. Burnett, T. Burt, Dr Cunningham, T.H. Elliott (Honorary Secretary), T. Farrer, C. Gairdner, Prof. E.C.K. Gonner, Prof. J.E.C. Munro, Prof. J.S. Nicholson, R.H. Inglis Palgrave, L.R. Phelps, L.L. Price (Honorary Secretary), Sir R. Rawson, F. Seebohm, H. Llewellyn Smith and the ten aforementioned members. Other founding members of the British Economic Association included M. Beaufort, Lord Baden-Powell, E. Cannan, S.J. Chapman, Miss C.E. Collet, A.E. Edgeworth, H. Ellis, Mrs M. Fawcett, H. Higgs, H. Hobhouse, Mrs M.P. Marshall, T. Mozley, G.B. Shaw, L. Stephen, S. Webb. We also find some foreign names in the list of initial members: L. von Brentano, C. Menger, G. Schmoller, F.W. Taussig, L. Walras, F.A. Walker, Prof. Messedaglia and Bey Reshid, Edgeworth's Turkish friend.

27. *EJ*, 1891; LSE Archives, B.E.A.
28. Reisman (1990), p. 191 and Coats (1968b), p. 370.
29. Whitaker (1996), I, p. 287 and II, p. 7, n. 4; Skidelsky (1983), p. 63.
30. Quoted by G. Becattini in the *New Palgrave*, Eatwell et al. (1987), III, 'Pantaleoni, Maffeo'.
31. A. Marshall to J.N. Keynes, 15 January 1889: Whitaker (1996), I.
32. The exception was the review of Bonar's *Philosophy* in 1893, which was published in *Mind*.
33. Pigou (1925), p. 69.
34. Marshall to FYE: 10 June 1896, on the contents of *EJ*, no. 4: Whitaker (1996), II, p. 169.
35. EP NUFFIELD C UO Box D 10.
36. Ashley to Seligman, 13 December 1890. Quoted in Reisman (1990), pp. 195–6.
37. So, according to Reisman, 'it was a choice between alternative methodologies'. Reisman remarks on the prominent role played by Marshall in this election through Jowett, who was one of the electors and 'contacted Marshall directly for specific advice on the ranking of the five candidates'. Reisman also stresses that G.E. Goschen was on the appointing Committee 'in his capacity as Chancellor of the Exchequer, [and] he was also President of the British Economic Association. Goschen must have felt that an orthodox figure such as Edgeworth would be eminently suitable to lead dissenting Oxford back into the fold': Reisman (1990), pp. 195–6. Another Oxford expert, A. Kadish, advances the following two conjectures to explain the verdict: 'the national development in the profession and the Balliol connection. Edgeworth was a Balliol man (he matriculated in 1867) as were two of the electors – Doyle (1863) and Wallace (1864). . . . Professionally he came highly recommended by Marshall, to whom he was extremely deferential.' Just like Reisman, Kadish thinks that the key was the connection between Marshall and Jowett, Edgeworth's former tutor at Balliol: Kadish (1982), pp. 197 and 199. Whatever role these factors played it is clear that Edgeworth was the person best qualified for the position, which was in 'political economy', not, after all, in 'economic history'.
38. EP NUFFIELD C UO, Box D 3.
39. Whitaker (1996), II, p. 7 and 8, n.2.
40. Beatrice Potter to FYE, 26 February 1891 (EP NUFFIELD C UO, Box D 3/18).
41. F.Y. E. to Beatrice Potter, 24 August 1891 (EP LSE Archives).
42. Edgeworth (1925), I, pp. 3–12.
43. This quotation is contained in Malthus's biography by Bonar (1885), p. 225.
44. Edgeworth (1925), I, p. 9.
45. Whitaker (1996), II, p. 33, n. 1.
46. Price, (1926), p. 371.
47. An anecdote concerning the nudists is referred to by R. Breen and S. Mudannayake in the booklet *Oxford Oddfellows and Funny Tales* where it is explained that 'a number of them were skinny-sunbathing on the Pleasure when a group of students floated by in a punt. All but one of the startled dons covered their modesty, one placed a flannel over his head instead. When asked why he had done that, he replied haughtily, "Oh, my students know me by my face".'
48. Price (1926), pp. 374–375.
49. See Kadish (1989), p. 96.
50. Price (1946), p. 37. Quoted in Creedy (1986), p. 11 and in Reisman (1990), pp. 194 and 196.
51. See Kadish (1989), p. 96.
52. Bonar (1926), pp. 647–53.
53. See also Section 2.15. Sir Robert Giffen (1837–1910) was economist and statistician, comptroller-general of the Commercial, Labour and Statistical Department of the Board of Trade. He was the author of *Stock Exchange Securities* (1878), *Essays in Finance* (1879–86), *The Growth of Capital* (1890) and *Case against Bimetallism* (1892).

George Joaquim Goschen (1831–1907), was Chancellor of the Exchequer (1887–92) and as such in late 1890 he presided over the meeting where the British Economic Association was founded and where he was elected president. In February 1891 as well he was one of the electors who appointed Edgeworth to the Drummond Chair. Afterwards he was First Lord of the Admiralty (1895–99). He was the author of *Theory of Foreign Exchanges* (1863).

54. Harrod (1951), p. 166.
55. Bonar (1926).
56. Clara E. Collet, 'Obituary: Henry Higgs', *EJ*, 1940, pp. 546–55.
57. Reported by Coats (1968b), p. 367.
58. Included in Edgeworth (1925), II, pp. 3–62, and I, pp. 406–15 and pp. 421–42. Included also in Vol. 3 of Darnell (1991).
59. Edgeworth (1925), III, pp. 36–41.
60. See Schmidt and Weber (2008).
61. Edgeworth (1925), II, p. 5.
62. Bastable, C.F. *International Trade*, p. 10. Quoted by Edgeworth (1925), II, p. 5.
63. Edgeworth (1925), II, p. 6.
64. Edgeworth (1925), II, p. 31. See also 2.3, above. As Creedy has remarked, 'this type of emphasis came to dominate much of the later analysis in the pure theory of International Trade' Creedy (1986), p. 118.
65. Whitaker (1991), II, p. 14, n. 2.
66. *Giornale degli Economisti*, second series, No. 2, pp. 233–45 and No. 3, pp. 316–18. In Edgeworth (1925), II, pp. 313–19, translated into English as 'On the Determinateness of Economic Equilibrium', and letter from Marshall to FYE, March 1891. In Whitaker (1996), II, p. 26, n. 4.
67. Marshall to Edgeworth, 4 April 1891. In Whitaker (1996), II, pp. 27–8. A thorough exposition of this incident is contained in Creedy (2006).
68. Galton Papers 138/7, UC LONDON.
69. Mentioned above in Section 2.12. Published in December 1885 in the *Journal of the Royal Statistical Society*. See Stigler (1978), p. 300.
70. Ronald A. Fisher, the British statistician, not Irving Fisher the American mathematical economist and statistician.
71. Stigler (1978), p. 309.
72. After studying at the Universities of Heidelberg and Berlin, Carl Pearson decided to become Karl Pearson.
73. The letters referred to are filed as Pearson Papers, UC LONDON, 681.
74. Stigler (1978), p. 299.
75. *Philosophical Magazine*, August 1892, pp. 190–204 and pp. 429–38.
76. See S.M. Stigler, 'Edgeworth as a statistician', in the *New Palgrave Dictionary*, Eatwell *et al.* (1987).
77. Pearson's series of articles appeared from 1893 to 1912.
78. Letters from Edgeworth to Pearson, 25 and 30 October 1893. Pearson Papers, 681, UC LONDON.
79. Postcard dated 6 November, from Oxford.
80. 31 July (1895?).
81. 'Statistics on Unprogressive Communities' and 'Supplementary Notes on Statistics', *JRSS*, 1896, 59, pp. 358–86 and 529–39. 'The Asymmetrical Probability Curve' and 'The Compound Law of Error', *Phil. Mag*. 1896, 41, pp. 90–99 and 207–15.
82. Letter from 15 November 1895.
83. EP BODLEIAN UO, Ms. Eng. lett. c746.
84. Halsey (1976), p. 43.
85. H.J.B.'s report (EP NUFFIELD C UO, Box E 3, 9/1).
86. H.J.B. to her son (Prof. H.E. Butler), February 1926. Quoted by Creedy (1986), pp. 12–13.
87. C.V. Butler to J.M. Keynes (KP KING'S C CAMBRIDGE MA, ED/6/6/29).

88. Harold E. Butler to H.J.B., from Gordon Square (EP NUFFIELD C UO, Box D 11).
89. H.J.B.'s report. (EP NUFFIELD C UO, Box E 3, 9/1). She refers to Hare (1894).
90. Sully (1918), pp. 240–42.
91. Palgrave (1894), I, p. 473. Some of Edgeworth's Palgrave entries are included in Vol. 5 of Darnell (1991): 'Average', 'Demand Curves', 'Dupuit', 'Functions', 'Gossen', 'Index Numbers', 'Least Squares'.
92. Palgrave (1894), I.
93. Whitaker (1996), II, p. 125, n. 2. See also Reisman (1990), p. 229.
94. Pigou (1925), p. 72.
95. Regarding Marshall's complaint, see Section 3.7 above. Whitaker (1996) II, pp. 170; p. 145, 169, 176; pp. 202–3.
96. Like John Bates Clark, Irving Fisher and, after 1903, Henry Ludwell Moore, a Cournot expert who visited him often and would become one of the more intimate scientific correspondents of Walras.
97. All these letters are translated from Jaffé (1965), Letters 1015, 1080, 1081, 1051, 1549, 1215 (4), 1216, 1223, 1224.
98. *Mathematical Psychics, First and Second Reports of the Committee about Ascertaining and Measuring the Value of the Monetary Standard* and twice from his presidential address of 1889 'On the Application of Mathematics to Political Economy'.
99. EP NUFFIELD C UO, Box D 4.
100. See above, Section 3.5.
101. Letter from Pareto to Walras, 12 August 1893: Jaffé (1965), Letter 1149.
102. Letter from Edgeworth to Pareto, 4 January 1895. Pareto to Walras, 27 June 1895. Jaffé (1965), Letters 1197 and 1207.
103. Letters from Pareto to Edgeworth, 18 and 30 November 1896, see Pareto (1964–84), vol. 19, pp. 309–12 and pp. 313–16.
104. *Revue Socialiste*, June–July 1896.
105. Pareto to Pantaleoni, 9 June 1896. Jaffé (1965), Letters 1149 and 1237.
106. *Mathematical Psychics*, p. 28. A diagram on bilateral exchange is presented by Pareto in the *Cours*, p. 144, n. 1, and developed in the *Manuale*, p. 191. Bowley closed the diagram by introducing two pairs of coordinates and popularised it as the 'Edgeworth box' in his 1924 book *Mathematical Groundwork*. See Sections 3.16 and 4.12. For a discussion about the criticisms of the Edgeworth box by Tarascio, Jaffé and Weatherby see Creedy (1998), 3, pp. 267–76.
107. As pointed out by Creedy (1986) and Schmidt and Weber (2008). See Fisher (1892), Pareto (1898, 1900, 1906), Hicks and Allen (1934).
108. Pareto (1906), French ed.1909, p. 540.
109. See Edgeworth (1925), III, pp. 43–44. Edgeworth never reviewed Pareto's *Manuale* of 1906, nor its French translation with additions of 1909.
110. Pareto (1897), II, p. 229 and Pareto (1909), p. 595.
111. Wicksell (1936), p. 16; Pareto (1896), I, pp. 264ff. (vol. I); Edgeworth (1925), I, pp. 195–297.
112. Whitaker, (1996), II, pp. 43–4 and J.M. Keynes, 'Obituary: Henry Higgs', *EJ*, 1940, L, pp. 555–8.
113. J.M.Keynes, 'Obituary: Henry Higgs', *EJ*, 1940, L, p. 556.
114. C.E. Collet. 'Obituary: Henry Higgs', *EJ*, 1940, Vol. L, p. 554–5.
115. Noted by Newman (2003).
116. Edgeworth (1925), III, pp. 109–12 (Cournot's review), and II, p. 22, n. 2 (1894 article on international trade). See Creedy (1998), 20, pp. 10–18.
117. See below.
118. Edgeworth (1925), I, pp. 111–12 and 136–7. J.L.F. Bertrand, 'Théorie des Richesses: Revue de *Théories Mathématiques de la Richesse Sociale* par L. Walras et *Recherches sur les Principes Mathématiques de la Théorie des Richesses* par A. Cournot', *Journal des Savants*, 1883; Marshall (1890).

119. As stressed by Creedy (1986), p. 99.
120. Edgeworth (1925), I, pp. 136–7 and n. 1 on p. 117. Weber (2005), p. 305, points out Edgeworth's precedence in this question. And, with regards to Edgeworth's influence on Hicks, notice Hicks' declaration: 'Much of my own early work on economic theory is a continuation of his [Edgeworth's]. . . . It was in his footsteps that I was trying to walk': Hicks (1984b), p. 157.
121. As noted by Samuelson (1974). Quoted by Weber (2005).
122. Edgeworth (1925), I, pp. 139–40.
123. Edgeworth (1925), I, pp. 141–2 and pp. 132–5.
124. Edgeworth (1925), II, pp. 64–5. The article is included in Vol. 3 of Darnell (1991).
125. Quoted by Edgeworth (1925), II, pp. 68–9, n. 5. Jenkin, F., 'Incidence of Taxes', *Papers Literary and Scientific*, 1871, p. 114. Also noted by Creedy (1986), p. 113.
126. Edgeworth (1925), II, pp. 63.
127. Edgeworth (1925), I, pp. 143–71. On Edgeworth's taxation paradox and theory of taxation see Creedy (1986), Ch. 7, pp. 111–7 and Creedy (1998), 10, pp. 101–12.
128. See *EJ*, 1917, p. 407 and 1921, p. 350. Edgeworth (1925), II, p. 64.
129. Edgeworth (1925), II, p. 115. For the role of authority in Edgeworth's writings, utilitarianism and the principle of minimum aggregate sacrifice, see Creedy (1986), pp. 29–30, 114–117.
130. Edgeworth (1925), II, pp. v and 234–42.
131. This unlikely attribution was made by T.W. Hutchison and was also accepted by Peter Clarke in the *New Palgrave* entry 'Hobson, John Atkinson' (1987). See also the discussion by Newman (1987), pp. 48–9, following previous arguments by Creedy (1986), p. 22.
132. Bonar (1926), p. 651.
133. This means that Hobson could teach political economy at Oxford. He also continued to teach political economy at LSE. See Creedy (1986), p. 22.
134. Whitaker (1996), II, p. 248.
135. Whitaker (1996), II, p. 379, n. 2.
136. Whitaker (1996), II, p. 249.
137. Whitaker (1996), II, p. 250.
138. Whitaker (1996), II, p.307.
139. On some recent interpretations of the origins of the 'Edgeworth box', see Creedy (1998), 3, pp. 32–34.
140. Marshall to Foxwell, 8 May 1901: Whitaker (1996), II, p. 314.
141. Whitaker (1996), II, pp. 355–356.
142. Whitaker (1996), II, p. 356 n. 4.
143. See Kadish (1969), pp. 98–99.
144. Whitaker (1996), II, p. 382, n. 4, III, p. 3, n. 7, and pp. 24–25, n. 1.
145. Whitaker (1996), II, pp. 391–393; Pigou (1925), pp. 435–438.
146. Whitaker (1996), III, pp. 62–63, n. 4.
147. Creedy (1986), p. 100, Miller (1967) and Seligman (1962), pp. 637–638.
148. *QJE*, 18, February 1904, pp. 159–219 Whitaker (1996), III, p. 262. Reproduced also in Edgeworth (1925), I, pp. 13–60.
149. See Edgeworth (1925), I, pp. 23, 24, 51 and 55.
150. EP NUFFIELD C UO, Box D 4.
151. Throughout his active life Fisher was also director of Remington Rand (the Rand Corporation), a founder and president of the Eugenics Research Association, the Stable Money League, advocating 100 per cent deposit reserves, the Econometric Society, the American Statistical Association and a long list of other companies. James Tobin depicts Fisher as 'a congenital reformer, an inveterate crusader. He was so aggressive and persistent and so sure he was right, that many of his contemporaries regarded him as a "crank" and discounted his scientific work accordingly.' *New Palgrave* (1987), II, 'Fisher, I.' by J. Tobin.
152. Coats (1968a), pp. 184–185.

153. Groenewegen (1995), p. 382.
154. A. Marshall to J. Brentano, 18 August 1903: Whitaker (1996), III, p. 53.
155. Coats (1968a), pp. 220–21, from a letter by Foxwell to W.R. Scott, and J. Bonar, 22 November 1903.
156. A. Petridis, 'Price, L.L.', *New Palgrave Dictionary*, 1987.
157. Coats (1968a), pp. 210–13.
158. Letter from Foxwell to J. Bonar, 22 November 1903. Quoted by Coats (1968a), p. 221.
159. These five articles were included in Edgeworth (1925), Vol. I, pp. 13–60; pp. 215–26; Vol. II, pp. 320–66, pp. 367–86 and pp. 234–42.
160. Stigler, 'Edgeworth as a statistician', *New Palgrave Dictionary*, 1978, p. 98.
161. Edgeworth, 'On the Representation of Statistics by Mathematical Formulæ', *JRSS*, 1898, Part I, pp. 670–700; 'On the Probable Errors of Frequency Constants', *JRSS*, 1908, pp. 499–512. See E.S. Pearson, 'Some Reflexions on Continuity in the Development of Mathematical Statistics, 1885–1920', in Pearson and Kendall (1970), p. 343.
162. Creedy (1986), p. 10 and EP NUFFIELD C UO, Box D 4.
163. *JRSS*, 71, pp. 381–97, 499–512, 652–78.
164. As recognised by Karl Pearson's son, the historian of statistics E.S. Pearson. See Pearson and Kendall (1970), p. 415 and p. 343. See also Stigler (1978), p. 299.
165. In fact, the second article mentioned by Keynes was published in 1909 in French and translated into English in 1910 for the *EJ*: Keynes (1926), in Vol. X of Keynes (1971–1989), pp. 260–61.
166. *EJ*, XVII (1907), pp. 213–220.
167. Marshall (1890), note XII b, p. 654.
168. Edgeworth (1891), *Giornale degli Economisti*, Second Series, 2 and 3, translated into English in Edgeworth (1925), II, pp. 313–19 (Second Series 2), with the title 'On the Determinateness of Economic Equilibrium' and in Newman (2003), pp. 325–32 and 336–9 as 'Observations on the Mathematical Theory of Political Economy' (Second Series 2) and 'The Theory of Barter Again' (Second Series 3).
169. Edgeworth (1888), *JRSS*, pp. 346–388.
170. For an analysis of Wicksell's discussion of Edgeworth's tax paradox in the last edition of his *Lectures on Political Economy* (English translation of 1934), see Creedy (1998), 10, pp. 101–12.

Volunteer F. Y. Edgeworth in about 1916

Reproduced with permission from Dr David Edgeworth Butler

4. The Esquire of Edgeworthstown

The noise
Of worldly fame is but a blast of wind,
That blows from diverse points, and shifts its name,
Shifting the point it blows from. Shalt thou more
Live in the mouths of mankind, if thy flesh
Part shrivel'd from thee, than if thou hadst died
Before the coral and the pap were left;
Or e'er some thousand years have past? and that
Is, to eternity compared, a space
Briefer than is the twinkling of an eye
To the heaven's slowest orb.

(Purgatory, Canto XI, vv. 100–108).
Dante Alighieri, *The Divine Comedy*.

4.1 MARSHALL'S RETIREMENT AND OTHER ACADEMIC STORIES

The British Academy

In 1902 the British Academy for the Promotion of Historical, Philosophical and Philological Studies was established by Royal Charter. Its mission was to serve as the natural academy for the humanities and the social sciences, as the counterpart to the Royal Society, which serves the natural sciences.

After a gathering of the most prominent academies of the world in Wiesbaden in October 1899, the representatives of the Royal Society realised that there was no society in Great Britain dealing with subjects encompassed in the 'Literary' section. So under the pressure of distinguished individuals such as the former prime minister A.J. Balfour, Viscount James Bryce, Professor Sidgwick and Sir Leslie Stephen, who met on 15 December 1899, a special committee was set up by the Council of the Royal Society. The efforts of this Committee resulted in the proposal for a Royal Charter to found the British Academy and suggested a list of original fellows by subjects. The proposed list of recipients were: I History, 6 fellows; II Philology, 10, including Professor Tyrrell from Trinity College Dublin; III Oriental and Biblical Studies, 8, including Provost Salmon from Trinity College Dublin; IV Law and Politics, 6; V

Metaphysics, 7, including Sir Leslie Stephen; VI Economics 3, Professor Marshall, Professor Edgeworth and Dr Cunningham of King's, London; VII Archaeology, 6. A few more were added before the Academy was officially founded, Foxwell among them.[1]

Now that the name of Leslie Stephen has come up, we should mention that he had not lost contact with Edgeworth after the old days of the 'Sunday Tramps'. In 1902, Stephen wrote to him about a possible meeting with him at All Souls, though he (Stephen) was living 'the life of a hermit and moreover of a deaf hermit'.[2]

Section F of the British Association

Meanwhile, the members of Section F of the British Association for the Advancement of the Sciences kept celebrating their annual assemblies. In August 1904 the meeting of the Section was held in Cambridge. In the end, Marshall threw a house party which Edgeworth attended, though he left before a final photograph was taken.[3] One anecdote from that gathering of the British Association is reported by Bowley:

> In 1904, a party of economists was bicycling out of Cambridge, and, with some danger to the traffic, Edgeworth began to discuss the method of translation or some similar topic; Professor Cannan drew up alongside and said, 'Put on the pace, Bowley, he can't talk mathematics at more than 12 miles an hour.'[4]

Marshall's Retirement and Succession

There are no direct letters between Marshall and Edgeworth on academic matters from 1902, the date of Marshall's letter from South Tyrol, to April 1909, when he sent Edgeworth three long and sympathetic letters. In these letters Marshall discussed the economics of cereals and alluded to his concept of elasticity: 'I am even more perplexed by what you say about elasticity of demand. . . . I object to the phrase negative elasticity, because I think it tempts people to carry analytical mathematics beyond their proper scope.' He also referred to one of the mottos espoused in his *Principles*: 'One in the many and the many in one' ('i.e. the relations of details to fundamentals, a matter on which the experience of age is some atonement for its stupidity').[5]

There are three reasons explaining Marshall's new proximity to Edgeworth, and all had one root cause: his retirement from academia in 1908 when he was 66 years old. The first reason was the proposal by Edgeworth, Price and some other members of the Economic Society to finance a portrait to be painted of Marshall. After circulating a printed appeal through the Society, there were 117 contributors: 'I hope your consent will not be refused to so many friends and admirers.' Marshall accepted: 'My face is not worthy to be painted. But as you will it, I may not say nay.'[6] The portrait was painted by William Rothenstein.

The second reason for Marshall's new flow of correspondence with Edgeworth in 1909 was that now he had more free time. And the third reason was the election of Marshall's successor in the Chair of Political Economy at the University of Cambridge.

Edgeworth lived the entire election process that took place in 1908 intensely since he was one of the electors. The rest of the appointed electors were: A.J. Balfour, the former prime minister; Lord Courtney; J.N. Keynes; J.S. Nicholson; R.H. Inglis Palgrave; V.H. Stanton; and W.R. Sorley. The candidates were Foxwell, then 58 years old, a former student and old friend of Marshall at St John's College and professor at both Cambridge and London Universities; Ashley, then 48 years old, who in 1901 had prevailed over Foxwell to land the new Chair of Commerce at the University of Birmingham in spite of Marshall's and Edgeworth's support of Foxwell; Cannan, then 46 years old, a professor at the London School of Economics; and Pigou, then thirty years old, who against all odds was chosen. Pigou had the full backing of Marshall. Through the entries of the diary of J.N. Keynes, we can trace the events:

> May 29. Edgeworth (who is staying with Marshall) arrived at 9:30 to discuss tomorrow's election. We [Keynes, Nicholson and Palgrave who were staying at Keynes's house] gathered that he had been sent by Marshall.
>
> May 30. Balfour did not come. . . . Pigou was elected. I am extremely sorry for Foxwell. The whole thing has worried me very much. We dined with the Marshalls.

J.N. Keynes does not detail how the electors voted, but Foxwell was convinced that if Balfour had been there the outcome would have been different. Among all the electors, Foxwell always grouped Edgeworth, along with Courtney, Sorley and Stanton, as Pigou's backers.[7] And he was quite probably right. The fact is that Edgeworth, independently of Marshall's pressures, had always praised Pigou as a scholar: in his review of Pigou's book, *The Riddle of the Tariff*, Edgeworth wrote that 'the power with which he [Pigou] wields the organon of economic theory is of the highest promise'.[8]

One could object that Edgeworth's opinion was biased, since Pigou was clearly on the free-trade side and had bravely defended the *Manifesto*. He also quoted Pigou in praising terms in his 'Appreciations of Mathematical Theories'.[9] In fact, Pigou's style and method were very appealing to Edgeworth since they were philosophically based and wrapped up in mathematics. On the other hand, Edgeworth's friendship with Foxwell during

Jevons's times at Hampstead had faded into the past. So, Edgeworth did not go against his inner convictions when he voted in favour of Pigou.

Foxwell thought that Marshall's opposition to him was a direct consequence of the free-trade controversy of 1903:

> This is Marshall's rather savage revenge for a letter I wrote in *The Times* a propos of the ridiculous professoral manifesto. . . . A member of Council writes me that 'this election will condemn the Cambridge school of economics to the same level of impotence to which Edgeworth has reduced the Oxford school.'[10]

The upshot of the election was that Edgeworth lost an old friend.

4.2 HOMAGE TO AND DEATH OF WALRAS AND THE INCIDENT WITH MOORE

In 1898, Pareto came into a great deal of money by inheritance and eventually left his chair at Lausanne in 1907, when he was only 59 years old.[11] Edgeworth, who since 1896 had had strong differences with Pareto after his criticism of Pareto's distribution law, was most likely not aware of this. He was also probably ignorant of the tense personal relationship between Pareto and Walras when in 1909 he wrote to Pareto in French, on the occasion of Walras's 75th birthday, that he was 'very sorry' that he could not 'attend the ceremony through which your University [Pareto had left Lausanne two years before!] will honour the great pioneer of Mathematics applied to Political Economy'.[12]

Moreover, Edgeworth included an article in the Festschrift issue of the *Giornale degli Economisti* in honour of Walras.[13] Edgeworth's lengthy, 60-page article entitled 'Apprezzamenti di Teorie Matematiche' was the Italian translation of an article published in the *Economic Journal* in 1907–1908 (see above). In spite of his multiple citations, the homage article never alluded to Walras. This omission did not pass unnoticed by the French professor.

Walras died just six months later, on 5 January 1910.[14] The differences between him and the English neoclassical school were never reconciled during his lifetime.

These differences persisted with Walras's American admirer, H.L. Moore (1869–1958), who took as his life's mission the task of empirically proving the findings of the Marginal School, starting with Walras's general equilibrium.

Before earning a PhD in Economics from Johns Hopkins in 1896, Moore spent some time in Europe, where he attended lectures in mathematical statistics by Karl Pearson in London and went to study at the

University of Vienna. He became familiar with the work of von Thünen and Cournot, and he forged contacts with Walras, Bortkiewicz, Pareto, Marshall and Edgeworth.[15]

Moore sent copies of his 1911 book *Laws of Wages: An Essay in Statistical Economics* to both Edgeworth and Marshall, who confessed to Edgeworth in rather frank terms:

> Moore is a nightmare to me. [. . .All] these pictures and decimals and learned terms are intended to prepare the way for a future generation of workers, who can put all the faith . . . into a group of mathematical machines and turn the handles. But neither our statistics nor our mathematics is ready for this work. We are like the first assailants of a fortified position; our corpses will fill the trenches so Moore can get on . . . He seems to me to have only proved there was *some sort* of causal connection in cases in which no one would doubt there was one.[16]

Marshall's words about the uses and abuses of mathematics and statistics by future generations of economists were premonitory.

Edgeworth reviewed Moore's book in the *EJ* dated March 1912, adopting the strategy of concentrating all his differences with the author on a specific chapter:

> From this general description of the work we must except one part, the chapter which purports to verify the hypothesis that differences in wages correspond to differences in 'efficiency' (or 'general sagacity and energy'). . . . The thinness of his conclusion is disguised by adventitious involutions . . . having overlaid a simple matter with useless and cumbrous technicalities. . . . Not only has he employed a steam-engine to crack a nut; but the nut is blind.[17]

Moore replied respectfully but unconvincingly through the *EJ*, whereupon Edgeworth counterattacked with his entire battery of ironies:

> It was not suggested – as Professor Moore seems to suppose – that the real relation between capacity and wages was a simple affair; it was only his method of ascertaining the relation that was regarded [in the review] as simple – when divested of disguise. Because a simple, or 'simpliste' method of treating a subject is adopted, it does not follow that the subject is simple.[18]

Marshall, for his part, was much more extreme, and wrote a letter to Moore on 5 June 1912 attaching a part of his aforementioned note to Edgeworth, which contained all the criticisms we have detailed. Moore replied sarcastically that he would 'treasure your letter and the outspoken criticism you sent to Professor Edgeworth. I doubt whether I shall ever again receive so frank and so generous treatment from a fellow-worker.'[19]

There is no proof of any further contact between Edgeworth and Moore.

4.3 1906–11: THIRD STAGE OF THE *ECONOMIC JOURNAL*

In 1905, Henry Higgs left his duties as both Secretary and Assistant Editor of the *Economic Journal*, obliged by 'the pressure of higher official duties', these duties being Higgs's appointment as the Prime Minister's Private Secretary. This opened a new period with Edgeworth as the sole editor again, which did nothing to stop his own flow of reviews, notes and articles, which are listed below in Appendix L.

Edgeworth's Reviews, Articles and Notes

Edgeworth could not always avoid sharp comments, such as in his review of Amery's *The Fundamental Fallacies of Free Trade*, when he stated that the 'language seems better adapted to edify the faithful of the compatriots' club than to win new converts from the educated world outside'. And in his review of Colson's *Cours d'Économie Politique*, Edgeworth indulged in irony by referring to his Latin classics: '*Disjecti membra poetæ* cannot represent an epic.'[20]

As we have stressed before, his articles and notes during this period of the *EJ* were primarily surveys. On the other hand, they became more and more alike in length, and it is hard to know the criterion under which a work was designated as a note or an article.

The note of 1907, 'Correspondence of Ricardo with Maria Edgeworth', shows that this correspondence is, on the whole, quite premonitory, especially on Maria's side, of what might and indeed did happen a quarter of a century later, the potato famine. It includes Maria's letter to Ricardo dated 9 July 1822 on Irish potatoes and beef exports; a letter from Ricardo to Maria dated 13 December 1822 on the potato question; a third letter from Maria to Ricardo continuing the potato discussion; and Ricardo's reply on 11 January 1823. The last letter shown is from Ricardo to Maria dated 26 May 1823, and it contains the surprising paragraph that takes issue with the state of affairs in Ireland, 'Your restless nation. . .', which was reproduced in Section 1.4. To obtain these documents, Edgeworth took advantage of Aunt Harriet's last letter of June 1888, in which she informed him that Ricardo's third son Mortimer 'is also dead but left sons, one of whom has his uncle Osman's property of Bromesberrow'.[21] This note in the *EJ* enabled Piero Sraffa to find some previously unknown letters from Ricardo more than thirty years later.

In 1907, Edgeworth wrote an Obituary note in the *EJ* on the death of his official tutor at Trinity College Dublin, Professor John Kells Ingram (1823–1907). Ingram occupied the Chair of English Literature and the

Chair of Greek. He was made a fellow in 1846, became Vice-Provost in 1898 and issued *Poetical Sonnets* in 1900. Here is part of Edgeworth's obituary note:

> Almost all the Muses might weep for the many gifted Ingram. The efficacy of literary form to enhance the value of knowledge is conspicuous in Ingram's economic writings . . . for instance in his *History of Political Economy*. . . . Even those who resent this vein of scepticism [against excessive Ricardian abstraction] derive profit and pleasure from the narrative. . . . Ingram delighted in the exercise of pure reasoning. Like many among the mighty ones of old times at T.C.D. – like Graves and Salmon and Rowan Hamilton – he combined pursuits elsewhere almost irreconcilable, literature and the higher mathematics. Salmon in his *Higher Plane Curves* makes honourable mention of Ingram's contribution to that branch of Mathematics. Hamilton used to submit his mathematical thoughts to Ingram [who is reported to say] 'Nothing has ever been to me such a source of intellectual pleasure as pure geometry'. [However,] a sweeping condemnation of that method is contained in his remarks on Cournot and Jevons . . . The paradox of Ingram's hostility to the mathematical, and more generally the deductive, method in economics is explained by his devotion to the philosophy of Comte [who] denounced abstract reasoning about human affairs as mere metaphysics.

More Obituary Notes

In 1910 there is an obituary for Walras (1834–1910) written by Vilfredo Pareto: 'Walras did for Political Economy that which Lagrange effected for rational mechanics and his fame is bound to grow with every advance of science.' Lagrange is clearly the standard of comparison. Edgeworth had made a similar comparison between William Rowan Hamilton and Lagrange in a 1889 review in *The Academy*.[22]

In 1910 as well, there appears an unsigned obituary of Sir Robert Giffen (1837–1910), which was probably written by the editor: 'Common sense, rather than abstract reasoning, may be assigned as the special excellence of Giffen's economic work . . . Giffen deserves to be honoured with the masters of Statistical Science.' With regard to other authors contributing to the *EJ* who may be of particular interest to us, we may mention that after 1906 there appear several reviews of books on socialist topics by C. Violet Butler, daughter of Edgeworth's cousin Harriet Jessie Butler.

4.4 PRIVATE LIFE, 1909–12

Edgeworth's private life was quite devoid of memorable events. In his fifties, sixties and seventies, routine dominated his way of living: Oxford during the academic terms, Hampstead in vacations, some mountaineering on the Continent or visits to Ireland during the long vacation; jogging, walking and bathing in the early mornings, sometimes with Price in Oxford and with Bonar, who was also usually his mountaineering companion, on Hampstead Heath.

The only change was technological, since with the advent of the bicycle, Francis became an expert cyclist. He had also tried the pleasures of motor cars as early as 1906, when he was invited by a friend, William R. Amon, who wrote to him in August 1906 about motoring trips to the Cotswold Hills and Malvern Hills.[23] But the bicycle had the advantage over the motor car in that it involved physical exercise.

Though the bicycle had its risks, he did not have any trouble with this new way of riding, but curiously enough, he had a bad fall when riding a horse at the end of September 1907. In January 1908 Bonar received a typewritten letter from Edgeworth:

> As to the machine, its use is one of the uses of adversity. I contracted the habit while unable to write autographically during a great part of last term owing to the misfortune of having broken my right arm. The second day of the term, riding too recklessly through an unknown country I encountered a huge trench, in jumping which I parted company with my steed and fell in such wise as to break my arm in two places.[24]

According to Bonar, the broken arm 'seemed to cause no breach of good spirits. Edgeworth used the typewriter till he was well again.'[25]

With regard to his family, Edgeworth remained close to his nieces Mia, Fay, Laura and Rosa Sanderson after his sister's death at Winchfield in 1893. He also became more involved with his cousin Harriet and her family after the death of her husband Arthur Gray Butler in 1909. As she described many years later, Francis 'changed and *steadied* and mellowed greatly as years went on'.[26]

But the most important occurrence for Francis, a sad one, was the death of his brother Antonio Eroles Edgeworth in 1911. Francis visited Antonio Eroles and his wife at Edgeworthstown at least during late August and early September of 1884, 1896 and 1906, when he took observations of the movement of wasps and bees. Eroles visited him in Oxford from time to time, as reported by their cousin Harriet Jessie Butler. Since Eroles had no descendents, he was succeeded at the Edgeworthstown's estates by his only surviving brother, so Francis inherited Edgeworthstown in spite of being the sixth son of a ninth son of his grandfather Richard Lovell Edgeworth.[27] However, the estate of Edgeworthstown was not as productive from a landlord's standpoint as it once was, and was mostly the source of administrative headaches.

After informing Marshall that the death of his brother would oblige him to leave the editorship of the *Economic Journal*, Francis had to travel to Edgeworthstown as the new landlord. His first decision as such, under pressure from the Tenant's League, was to reduce all rents by 20 per cent, as reported in a cutting from *Freeman's Journal* (Dublin), 22 July [1911]:

> Edgeworthstown Branch, Friday. A large meeting under the auspices of the local branch of the Town Tenants' League was held in the Courthouse here on Wednesday evening last, Mr. T. Fenion, Co.C., President of the Town Tenants' League, in the chair. The chairman said he was glad to announce that he had that morning received a communication that their landlord, Mr. Edgeworth, had acceded to their demands (applause). They had held out for a reduction of 20 per cent and at last it had been granted to them (applause). . . . This success should increase the loyalty of the people of Edgeworthstown to the Town Tenant's League.[28]

This cutting was included in a letter from Edgeworthstown to Marshall in July 1911.

A second decision that Francis took, and that Harriet Jessie Butler detailed in her often mentioned report, was that 'after Eroles's death, his widow [Mme Françoise Delcher Edgeworth] continued to live at Edgeworthstown and keep up the house for Francis'.[29] We do not know when she died or left the manor house, but we have information that after the war, Edgeworth's niece Rosa married and the couple settled at Edgeworthstown with Francis's full consent and delight.[30]

We also have a description of Francis's family contacts with his cousin Harriet and her children after 1911, from Ruth Florence Butler, her second daughter:

> He was very kind to all of us and proud I think of Harold Edgeworth Butler and Christina Violet Butler. He rode with us occasionally (badly; he had a bad fall once, – not with us). He skated (well, I think) and tried to teach us. He showed me the right way to wear my Master of Arts' hood. . . . He also sculled well and had a special boating party on the Upper River every summer, generally with Dr Gowdy to help. He always invited Harriet Jessie Butler and one of us to the All Souls' lunch and gave us tickets . . . two in the semicircle when Harold had to read . . . – a glorious occasion for Butlers and Edgeworths alike. He used to have his nieces to Oxford and to entertain them. . . .
>
> Francis Ysidro Edgeworth was fond of bicycling till quite late in life and went on a cycling tour in the West of Scotland with Harold Edgeworth Butler in the early 1900s – and also I think in Ireland – Connemara.[31]

This report from Ruth Butler contributes to the description of the sporting activities that Edgeworth practised until a few days before his death. These included, throughout his life, jogging, tramping, walking, mountaineering, swimming, sculling, horseback riding, cycling, ice skating and, last but not least during his later years, playing golf. His leisure activities also included concerts with his friends and, when motion pictures appeared, he also liked to go to the cinema with them.

Harriet's four children were now grown up. In 1911, Harold was 33 years old and was in the process of becoming a Professor of Latin at University College London; Olive was 31 and would become Warden of Lady Margaret Hall Settlement, Lambeth; Ruth was 30 and on her way to being appointed Fellow at St Anne's College in Oxford, and in 1905 she produced an article on industries for the Victoria County History; and Violet was just 27 and working as a social policy worker while preparing her most important book, *Oxford Survey*, published in 1912. From 1903 to 1905, Violet had been reading modern history at the Society for Home-Students, where she earned a first class. There she was taught some economics by Mrs H.A.L. Fisher, and in 1906–07 she took an economics diploma that included some tutorials on socialism from Sidney Ball. After 1907, she occasionally reviewed books with a social content for the *Economic Journal* and eventually Violet Butler would become a highly esteemed sociologist.[32]

4.5 SULLY AGAIN

As mentioned above, in 1897 James Sully quitted Hampstead and abruptly disappeared from Edgeworth's life for a period of more than fifteen years. In fact, he did not leave London. He went to live on the west side of London's metropolitan area, at number 10 Park Hill, Ealing.[33] Meantime, in January 1898 Sully opened an experimental psychology laboratory at University College, London, and in 1901 – as we have seen through Sophie Bryant in Section 2.22 – he and a group of nine other colleagues, Bryant included, founded the British Psychological Society. Sully held the Grote Chair from 1892 until his retirement in 1903.

Then in 1913 Sully wrote to Edgeworth, who replied to him. These letters have probably been lost, since in the Edgeworth personal files at Nuffield College there are only two subsequent letters from Sully in which he tries to regain his past friendship with Edgeworth. The first one is written from Frutzen, Berne Canton, Switzerland, on 9 August 1913, in which Sully acknowledges that 'your letter of July 13 brought me much of that purer and quieter sort of pleasure which makes one, even when on the boundaries of querulous old age, feel sure that happiness is still within his reach'. He also remarks on 'the combination of harmony and good sense' when Francis touches some 'deep themes' and stresses how important Edgeworth was for him during his Hampstead years:

> I knew from the days when you first took me in hand at the Savile that you were to be my rejuvenator and I owe you more than I can tell for the kind persistence with which you kept me up to the youthful mark. If I could tell you what that uprooting at Hampstead was to me in 1897 you would not doubt that you had been an invaluable stimulus and support to me.

Sully afterwards refers dramatically to this 'uprooting at Hampstead' in 1897. 'Since then I seem to have been another man, to have retired more in upon myself, to have stiffened in all the intellectual and moral joints. When the harsh fate overtook my poor wife, my gloomy destiny seemed to be completed.' But he does not tell Edgeworth either in this letter or in the second one, the exact cause of his leaving Hampstead. Sully only writes that the leaving 'was no result of choices of weighing pro's and con's' but 'a family reason which I cannot explain'. And he confirms in the second letter, which was written in Worthing and dated 15 December 1914, 'what a pang – a long slowly healing wound – it was to leave it in 1897' and that 'life has never been the same to me since; it seems to have shrunk into something pitifully poor in comparison with what it had been'. In spite of this sense of defeat, the conclusions of both letters are optimistic, as Sully still feels confident that old friendships can be revived:

> I cling to the remains of old friendships – such as with you, Boulding, C. Read and others – with a sort of savage greed. I have even developed new ones, among others one with A.F. Sharid whom I think you know with whom I stay sometimes when I am in town. And I feel that these rare meetings with friends, and communication with them by pen and paper keep me going.

In 1913 he finally dares to propose a meeting with Edgeworth:

> I should dearly like to meet you at Hampstead so as to see you again as in the old days. I should expect you with your more eager impulse to exertion to cut the talk even shorter than you used to do. In truth, it is a delightful thought to me that you keep so much of your old youthful energy and alacrity. I will write and let you know when I am likely to be in London.[34]

At the end of 1914, Sully, who was working on his 'reminiscences', insisted to Edgeworth that he 'would much like to have a talk about the war, though this is a subject about which we are all agreed'.[35]

Sully's reminiscences were published in 1918 under the title *My Life and Friends*. We have used these memoirs to get a picture of Edgeworth and Hampstead in the 1870s. After the publication of *My Life and Friends*, we have no further news about Sully, who died in November 1923.

4.6 1912–18: FOURTH STAGE OF THE *ECONOMIC JOURNAL*

In the letter Edgeworth wrote to Marshall from Edgeworthstown on 25 July 1911, he suggests:

> I am glad to think that J.M. Keynes may be willing to become a candidate for the [*EJ*] editorship. . . . I have been receiving £50 a year the last two or three years; but previously £60. I gave up £10 the better to endow a colleague [H. B. Lees-Smith had assisted F.Y.E. since 1908] (who gets I think the magnificent sum of £25 – more or less, between £20 and £30). The Council would be well advised in returning to the figure of £60 for the Editor without cutting down the Assistant.[36]

On 17 October 1911, the council, chaired by Marshall, accepted Edgeworth's resignation and appointed John Maynard Keynes to take over the editorship of the *Economic Journal* while Edgeworth became Chairman of the Editorial Board. According to the minutes of the Society there were two names: Professor W.J. Ashley of the University of Birmingham, previously a professor at Harvard and Toronto, and Keynes. As Ashley was 'quite unable to find time to undertake the work', Keynes was unanimously elected editor of the *EJ* 'starting on 1 January 1912 or an earlier date to be agreed with Edgeworth'.[37]

J. M. Keynes as Sole Editor

On the same day, Edgeworth sent a letter to Keynes in which he 'gladly hailed' him as editor of the *Economic Journal* and 'congratulate myself on being associated with you as Chairman of the Editorial Board.'[38] On the other hand, Harrod, Keynes's first biographer, describes the state in which Edgeworth left the task:

> On taking over, he [J.M. Keynes] discovered heavy arrears. Edgeworth often found it difficult to make up his mind. . . . Keynes maintained the tradition of having some contributions from persons outside the academic field and of combining realistic with more theoretical studies. Indeed, Edgeworth, oddly enough, had been very inhospitable to purely theoretical work from any pen other than his own, and Keynes . . . judged it necessary to redress the balance in favour of academic theorists.[39]

Regardless of this, Edgeworth's work as editor could not have been as reproachable as Harrod depicts it, since in October 1915, Keynes asked Edgeworth to assist in the editing, because he was becoming increasingly involved with the Treasury. As the Minute Books of the Royal Economic

Society from October 1915 confirm, it was accepted that Edgeworth would help Keynes by performing some tasks as effective joint editor of the *Journal*.[40]

Thus, we do not know if the idea of announcing the correspondent in each academically relevant country on the first page of the journal starting in 1915 came from Keynes or Edgeworth, but we can see that from that moment on Edgeworth had a special preference for reviewing the books written by the correspondents.[41]

In 1919, when Keynes was busy at the Versailles Peace Conference, Edgeworth officially became Joint Editor of the *EJ* with Maynard. This joint editorship lasted until Edgeworth's death in 1926. In fact, 17 October 1911, when Maynard Keynes officially joined the *EJ*, also marks the beginning of a great friendship, as Harrod has recognised:

> Despite his early rage at Edgeworth's obtuseness on the subject of index numbers, he [J.M.K.] had come to appreciate the qualities of that great man. The appreciation was reciprocal. Edgeworth could not say too much in praise of Keynes. . . . I once told him that Keynes was staying in Christ Church. 'Ah,' he said, 'you have the pure milk of economics with you . . .' and, flinging his arms above his head, he proceeded with an inarticulate eulogy.[42]

Edgeworth's confidence in Maynard Keynes as an economist moved him to invite Keynes to stay at All Souls in May 1913 in order to 'communicate some sound doctrine to these same youth and maidens'.[43]

Thus, there is a seven-year period from 1912 to 1918 when J.M. Keynes was the sole editor of the *Economic Journal*. In spite of his less committed position, Edgeworth's written contribution to the journal did not diminish in either quantity or quality compared to earlier stages. Through Edgeworth's reviews from these years – detailed in Appendix M at the end of the book – we can infer his interest in the economic and non-economic aspects and effects of war, as 10 out of his 17 reviews of this period were on this topic.

Edgeworth's Reviews, Articles and Notes

As usual, the reviews were written in Edgeworth's ornamental style. In the 1913 review of Pigou's book *Wealth and Welfare*, one of the longest written by Edgeworth, his final comments are:

> How far do our author's theories belong to the category of practical wisdom, or to that higher kind of science which the philosopher distinguishes as grand and wonderful and difficult, but not useful for human purposes? Mathematical economics are certainly useful to some extent; but does the further elaboration which that study has received in his treatise imply a correspondingly large

> contribution to the Art of Political Economy? The analogy of mathematical physics does not help us to answer this question; the calculus of utility and probability is something so peculiar and unique. 'Ai posteri l'ardua sentenza.'[44]

In the reviews dating from 1918, Edgeworth, sensing the end of the war, regains his full sharpness, especially in his reviews of books not dealing with warfare. On Anderson's book: 'Our readers may be advised to suspend their judgment until they hear the other side.' On Arias's book: 'One may sympathise with that cautious student of whom it is told that he would not assent to the axioms of Euclid until he knew what use was to be made of the admission.' On Moulton's and Phillips' books: 'We hesitate which of the two versions to prefer.' On Smith-Gordon and Staples's view on Ireland:

> Notwithstanding some gloomy signs, the authors take a cheerful view of the future. At least, their predictions are of a kind sometimes employed by the Hebrew seers when they prophesied blessings conditional on good conduct . . . We are indeed in the presence of a poet, one of those by whom 'the world is wrought into sympathy' with new motives and ideals. The mere economist must recognise a creative force which transcends logical analysis.

And referring to the future, Edgeworth quotes in this review the words of another poet: the earth will be 'something other than the wildest modern guess of you and me'.[45]

Edgeworth's list of articles during this period from 1912 to 1918 is also significant but shorter than in other periods, partly because of his work on probabilities and statistical inference, five articles in total, and partly because he did not publish the four lectures that he gave on the subject of the economics of war in the *Economic Journal*. The list includes papers on index numbers and on mathematical economics, and a four-part work about the microeconomic theory of railway rates.[46]

Mathematical Economics: New Light on Utility and Welfare

In his customary survey of the period, published in March and June 1915, entitled 'Recent Contributions to Mathematical Economics', Edgeworth draws special attention to the developments in the area of 'utility', still resounding with the echoes of Pareto's ophelimity's index. He centres his chronicle around the content of six books – by A. Osorio, E. Antonelli,[47] E.W. Zawadski, J.S. Nicholson and M. Fanno. Pareto's *Manuale d'Economia Politica* of 1906 or its 1909 translation into French with some additions are not included. In fact, Edgeworth never reviewed this book and uses here Osorio's *Théorie Mathématique de l'Éxchange*, to relate Pareto's advances to the work of some authors who had preceded him, such as the French mathematician Poincaré, who wrote in a letter to Walras:

> Can satisfaction be measured? I may say that one satisfaction is greater than another, because I prefer one to the other; but I cannot say that one is two or three times greater than another . . . Satisfaction is then a magnitude, but not a measurable magnitude [. . . and] therefore not amenable to mathematical theory? By no means . . . you may define satisfaction by an arbitrary function, provided that the function continually increases along with the satisfaction it represents.

Then, Edgeworth adds that 'Poincaré's ruling is in accordance with the view' of the mathematician A.E. Love 'that the capacity of numbers to express the results of counting and measuring may be regarded as a secondary property derived from the more fundamental one of expressing order'. And Edgeworth refers again to A. Voigt's work, already mentioned in the 1894, 1900 and 1907 *EJ* where Voigt had anticipated Poincaré's dictum:

> Perhaps is better to say with Professor A. Voigt that no unit is required: quantities like utility are to be measured only by *ordinal* numbers. In confirmation of this conception Professor Voigt refers to the view among mathematicians [Professor Love] which sees in ordinal number rather than in cardinal the primary conception of number.[48]

After these references, Edgeworth clearly indicates that Pareto's ophelimity index theory had its precedents:

> Professor Pareto is therefore in very good company when, scrupling to designate utility as function (say *u*) of quantities of commodities (say *x, y*), he contemplates a family of successive *indifference-curves* . . . in the plane *x, y* . . . such that the advance from any one indifference-locus to the next in succession affords an *index*, rather than a measure, of the advance in satisfaction, or as Professor Pareto prefers to say, *ophelimity*.[49]

Despite Edgeworth's recognition of the pertinence of ordinal utility for the theory of demand, he reminds the *EJ* reader of the need in some specific areas of establishing interpersonal comparisons of utility or welfare:

> Jevons's suggestion that the theory of utility is limited to the motions of a single mind, that 'no common denominator of feeling seems to be possible' appears to us untenable. The contrary is postulated throughout large tracts of economic science; for instance, the theory of taxation and that of industrial conciliation. Even a more fundamental part of political economy, the theory of value and distribution, involving the equation of net advantages in different occupations, suggests at least, if it does not require the comparison between, the welfare of different persons.[50]

Edgeworth used comparative devices in his works on monopoly, taxation, international trade, distribution and bargaining, like the introduction of a constant marginal utility of money as the basis of some specific measurements – a device also applied by authors working on the consumer's surplus, such as Dupuit, Marshall and Pigou.[51]

'Railway Rates'

Published in 1911, 1912 and 1913 as 'Contributions to the Theory of Railway Rates', in 1925 Edgeworth reissued this work as 'Railway Rates'. This is a survey that examines specific theoretical aspects of the application of the theory of monopoly.

Edgeworth mainly investigates the case in which 'a monopolist dealing with a whole class of mutually competitive customers *discriminates* between different classes of customers'. He analyses discrimination 'on the lines of Dupuit' and his *rente des acheteurs*.[52] He reconsiders 'the vexed question of the relation between *through fares* lowered by competition and fares to intermediate localities which are subject to monopoly'. The discrepancies that Edgeworth finds between 'expert-practice and abstract theory' can be reduced by 'taking into account the monopolist's concern for interest in the distant future', for 'altruistic motives' and for the presence of some type of competition.[53]

With the work of Dupuit, Edgeworth mentions and discusses the works on this topic by Johnson and Huebner, Cohn, Hadley and Ackworth, Ripley, Schipfer, Marriott, Taussig, M. Kirkman, Meyer, Newcomb and J.M. Clark.[54]

4.7 THE WAR YEARS

What was Edgeworth's reaction to the war? We have some evidence that he was not overwhelmed by physical panic during the bombings and that he tried profoundly to analyse the phenomenon of war. James Bonar reports that 'on one occasion, during an air raid, he [Edgeworth] discussed philosophy with a famous living professor of it, by Jack Straw's Castle, where the anti-aircraft gun was doing its best to defeat the enemy and deafen West London'.[55]

In his correspondence with J. Maynard Keynes, Edgeworth alludes to the war. In a letter dated 27 December, no year (most likely 1914) he first refers implicitly to Keynes's 'War and the Financial System' and 'The Prospects of Money', two articles recently published in the *EJ*,[56] by saying flatteringly: 'I am disposed to think that writing on Currency, like Poetry, does not admit of mediocrity.' He then cites a theory by Ricardo about economic cycles that is reminiscent of Smith's theory of public deficit: 'Economic malaise is attributable to the changes in the direction of consumption (and accordingly of production) between peace and war and back again.'

Political Economy and War

In 1915, Edgeworth gave a lecture in Oxford, 'On the Relations of Political Economy to War'.[57] This lecture is most important for apprehending Edgeworth's insight into the war. He starts by quoting Tacitus and Cicero: 'Money is the sinews of war', and he goes on:

> Political Economy tends in time of peace to develop the sinews of war, not only by prescribing useful legislation, but also by warning against noxious kinds of Governmental interference. . . . I go on to consider not the art of Political Economy as rendering the means of waging war more effective – but the science of Economics as rendering the ends for which war is waged less desirable.[58]

To diminish the desirability of war's ends, Edgeworth alludes to a new responsibility for economists: that of pointing out the misconceptions about political economy which favour conflict and affect 'firstly the efficient causes of war (means), secondly its final causes (motives)'. For instance, 'the erroneous notion, from which even Voltaire was not free, that in trade one country cannot gain without another losing'.[59]

Another misconception is 'the belief that the stimulus given by war to industry is likely to be of permanent benefit to work-people'. A third error is that war is favourable to capitalisation, about which Edgeworth remarks that 'against the possibility that peace may prove adverse to capitalization is to be set the certainty that war is destructive of capital'.

Then Edgeworth comes to his own terrain by turning to the widely heralded 'likeness between war and competition'. But 'the economic analogue of war . . . is, however, not perfect competition but the case of an industrial dispute', the case of the bilateral monopoly. The indeterminacy of the final result in bilateral exchange implies a negotiation between the barterers that leads to either conciliation when the negotiators are 'representatives of the parties themselves', or arbitration when diplomatic representatives of a third party or authority are brought to issue a resolution that will be enforced for both parties. When conciliation is not possible because war has been declared, Edgeworth argues in favour of arbitration, since a range of arbitration necessarily exists.

He finally proposes that 'the hundreds of millions now spent by the Europeans in massacring each other or preparing for that eventuality might be directed to the investigations of Nature's secrets'.[60] And he adds:

> For may we not expect that when the brains and means now wasted on war are directed to scientific discovery, immense new powers over Nature will be obtained? Personally, . . . I dare to hope that access would be won to the stupendous store of energy which lies hid in the world of atoms. Innumerable particles rushing at rates of many thousand miles per second will be harnessed to the service of man – inexhaustible sources of energy, immeasurably surpassing Niagara and Victoria Falls.[61]

Edgeworth was, therefore, one of the first persons – after the physicist Frederick Soddy[62] – to envisage atomic energy. Edgeworth fantasised about its peaceful uses and conjectured the deterrence of war through them. Unfortunately, the uses of this energy in warfare invalidated such an optimistic conjecture.

'Economists on War'

In fact, during this war period, Edgeworth was particularly aware of the literature published by economists about the war and did not limit his reviews in the *EJ* only to the economic aspects of the books dealing with this subject. The works reviewed are numerous and from authors on both sides of the matter. As an illustration of Edgeworth's perception of war, we shall focus on the 1915 article 'Economists on War', which contains the reviews of several books: *Händler und Helden* by W. Sombart (German), *The Neutrality of the United States in Relation to the British and German Empires* by J.S. Nicholson (Scottish), *An Economic Interpretation of the War* by E.R.A. Seligman (American) and *Les Causes et les Conséquences de la Guerre* by Y. Guyot (French).[63]

In 'Economists on War', Edgeworth selects 'the judgments pronounced by eminent economists of different nationalities' and announces that they 'are calculated to be mutually corrective', although his own opinion is unambiguously added when required.

With regard to Werner Sombart's book, Edgeworth accuses him of the 'sin of over-generalisation' by presenting the war as mainly due to the opposition between the character and mentality of Germany and England, the 'incomparable superiority of the German heroic spirit' over the 'English mercantile character'. According to Edgeworth, Sombart depicts an 'overlaid caricature, distorted by objectionable features'.

To avoid the 'danger of being carried away by Professor Sombart's turgid eloquence', Edgeworth prescribes 'as an antidote Professor Nicholson's incisive wit'. Edgeworth admires Nicholson's powers of satire and agrees with him about 'the motives which have led us British into the present war', which may be boiled down to 'the Greatest Happiness Principle in relation to Liberty', since 'with the mass of British people the supreme test is liberty'.

On the other hand, Professor Seligman's opinion, as summarised by Edgeworth, is that 'the present war is mainly due to the opposition between the material interests of Germany and England'. Nations simply follow 'the same law which is found in all life from the very beginning of the individual cell – the law of expansion or of self-preservation'. According to Seligman, there are three stages in this expansion. The first, common protection, through which the USA has 'built up an enormous industrial power'. This is not the case for England, which after having 'built up her industry, . . . now found it her interest to go over from a system of protection to one of free trade' in the second stage of development. The third stage is reached when 'the emphasis is transferred from the export of goods to the export of capital' and imperialistic policies are set. Germany had reached this third imperialistic stage at the beginning of the twentieth century, and the war between England and Germany emerges as the final stage in a competition involving not only the export of commodities but also the export of capital. Seligman prophetically announces that 'we are likely to see during the next few generations wars on an even greater scale than the present one'.

With Guyot's book, Edgeworth stresses the inadequacy of overly harsh post-war policies. However, according to Guyot, the political dissolution of the German Empire and accompanying measures 'to satisfy the aspirations of numerous groups' are required, like the evacuation of Belgium and Alsace-Lorraine, the division of the German colonies among the Allies, the imposition of an indemnity without a spirit of revenge, the taking of the German railways and the Prussian mines, the imposition on Germany of free trade and 'some other trifles of that sort'.

Guyot's post-war measures devised in 1915 became real in 1919, and Edgeworth's position against them may be viewed as a precursor of the position manifested by J.M. Keynes in the book *The Economic Consequences of Peace*. Edgeworth concludes the joint review with a final pessimistic sentence: 'Alas! none of our authorities lead us to expect a speedy end of the present war, much less of war in general.'[64]

Other Reviews and Lectures on War

Pigou's 1916 book, *The Economy and Finance of the War*, was extensively reviewed by Edgeworth in the *Economic Journal*. Edgeworth remarks that Pigou accepts his favourite doctrine of the minimum aggregate sacrifice 'as the criterion of proper taxation' and defends the principle that 'the aggregate amount of sacrifice involved in providing any given revenue would be at a minimum if the whole of the funds required were lopped off the few largest incomes enjoyed in the country'. Edgeworth affirms that 'the essence of the reasoning commends itself to practical good sense', and he maintains that, though 'we are indeed far from ignoring the danger of applying such doctrines to ordinary practice, . . . we think that Professor Pigou has guarded against that danger by prescribing only for the exceptional cases of an unprecedented war'. As in 'Economists on War', Edgeworth does not address his comments to magnify economic topics; on the contrary, he stresses Pigou's statement:

> Compared with what this war has cost and is costing in values outside the economic sphere – the shattering of human promise, the accumulated degradation in the thought and feeling of many who have remained at home – compared with these things the economic cost is, to my mind, trivial and insignificant.[65]

Following these broader lines, in the review of Gill's 1917 book (see note 63) on the causes of modern warfare, Edgeworth comments: 'The pacifist doctrine has the peculiarity that it is only true if universally accepted.' At any rate, as he hinted at in his 1915 Oxford lecture, Edgeworth agrees with Gill that 'if the world were set free from the remnants of mercantilist ideas there would be comparatively little ground of dispute'.[66]

In 1917, too, Edgeworth approvingly quotes Lehfeldt's declaration: 'The war will probably leave behind it a lesson that the State both can and ought to do far more than the nineteenth century thought right.' They were both anticipating J.M. Keynes's thoughts from the 1930s.[67]

As seen above, Edgeworth did not concentrate his activities on the subject of war in just a handful of reviews, he also tried to express his convictions by delivering lectures in Oxford. Besides the aforementioned 'On the Relations of Political Economy to War' (1915); he delivered 'The Cost of War and Ways of Reducing It Suggested by Economic Theory' (1915); 'Currency and Finance in Time of War' (1917); and 'A Levy on Capital for the Discharge of Debt' (1919). These four lectures were published afterwards under the title *Economics of War: Four Lectures*.[68]

The Platoon

Edgeworth, who was 70 years old in February 1915, wanted to get more involved in work related to the war. He volunteered for No. 4 Platoon. At least, this surprising move is what may be inferred from a typewritten letter dated 30 June 1915 from a certain F.J. Wylie, who thanks his companions for:

> the splendid glasses that I am now using. They will remind me of the many hours we have spent together, which I at least have unaffectedly enjoyed. And they will be a reminder to me always of how forgiving a Platoon ('No. 4' at any rate) can be. I shall value the glasses coming from my friends in 'Godley's Army'.[69]

In fact, this move was so furtive that none of Edgeworth's biographers had detected it.[70]

4.8 EDGEWORTH'S LAST DRUMMOND YEARS

In 1915, a committee was appointed to reorganise the studies in economics and other social sciences at the University of Oxford. Its ranks included Edgeworth and Price. The committee put forward a recommendation to establish a new degree; however, the recommendation was not pursued because of the war. The question was raised again after the war, and in 1920 economics was placed among the 'Modern Greats' in a new honours school in philosophy, politics and economics. Price did not like this solution since it did not give economics enough space.[71] In his obituary profile of Edgeworth, Price attributes his acceptance to 'a loyalty, with which I sympathized, towards the ancient school of Literæ Humaniores that had been our common training-ground as undergraduates . . . and partly also from instinctive cultivated liability to bend in turn to views, pressed successively with force, however divergent they might be'. Moreover, Price complains, Edgeworth did not try to exert any personal influence on the boards where he was *ex officio* member, such as the Board of the Faculty of Modern History, whose meetings he rarely attended.[72]

In 1922, Edgeworth – along with Pigou, Flux, Chapman and J. Maynard Keynes – participated in the committee which drew up an 'Address to Dr Marshall' to be published in the September 1922 issue of the *EJ* on the occasion of Marshall's eightieth birthday. Foxwell did not sign it because he was still resentful towards Marshall: 'I thought it exaggerated and fulsome; but mainly because he played me such an ingeniously dirty trick just before the election to the Professorship here [in Cambridge, May 1908, when Pigou succeeded Marshall] that . . . I broke off all communications with him.'[73]

Also in 1922, Edgeworth retired from the Oxford chair at the age of 77, a ripe old age for retirement, though his retirement from academia was not complete since he was appointed emeritus professor and remained coeditor of the *Economic Journal*.[74]

Recent writers who have examined Edgeworth's performance in the Drummond Chair in Oxford from an institutional standpoint have not been at all enthusiastic. This is not surprising since they maintain a historicist point of view. They reproach Edgeworth for avoiding university politics, for relying upon the mathematical method and for being unaware of the capacity of his students.[75]

Finally, we have one student's opinion about Edgeworth's responsiveness as a teacher. The student has not been chosen at random; her full name is Christina Violet Butler, the daughter of his cousin Harriet, who studied social sciences at Oxford:

> He really was extremely lucid and businesslike especially in dealing with written work as a teacher, though his mind worked in such a way that people I know, rather thought that he was not; and the undergraduate who was taking economics was rather puzzled by his lectures. The illustrations ranged from rapid quotations from Homer, Shakespeare or many 18th c. poets to well assorted contemporary bluebooks, and the *Economic Journal* (and other quarterlies) for the past twenty years.[76]

4.9 MORE CONTRIBUTIONS TO STATISTICS; PEARSON AGAIN

Edgeworth's work on statistics continued through the 1910s and 1920s. In 1911, a new edition of the *Encyclopædia Britannica* appeared, in which the entry 'Probability and Expectation', written by Edgeworth, just as in the previous edition, had been modified.

During the long vacation of 1912, Edgeworth participated actively with Bowley in the section on 'Statistical, Economic and Actuarial Mathematics' of the Fifth International Congress of Mathematics, held in Cambridge from 22 to 28 August.[77]

In 1912, too, Edgeworth was elected President of the Royal Statistical Society, a post he was to hold from the end of 1912 to 1914.[78] In his presidential address, 'On the Use of the Theory of Probabilities in Statistics Relating to Society', delivered on 17 December 1912, he surveys the use of probabilities in physics to substantiate the following proposition:

> The use of Probabilities in Statistics relating to human affairs may properly be introduced by the use of the theory in physics. For as there is not one sort of arithmetic for social and another for physical phenomena, so the principle of Probabilities is essentially the same in these two regions.

This presidency gave him an additional impetus to publish works in the society's journal. From 1913 to 1917, Edgeworth contributed to this magazine with his presidential address and two four-part surveys, 'On the

Use of Analytical Geometry to Represent Certain Kinds of Statistics' and 'On the Mathematical Representation of Statistical Data'.

The period 1921–25, which in fact witnessed Edgeworth's last years of statistical contributions, was particularly prolific and interesting. His publications included a two-part survey of 'Molecular Statistics', two articles on index numbers, 'Mr. Correa Walsh on the Calculation of Index Numbers' and 'The Element of Probability in Index Numbers', and two articles on pure statistical techniques.[79]

Edgeworth also published several reviews in the *Journal of the Royal Statistical Society*,[80] as well as several contributions to statistics in other journals, such as the *Philosophical Magazine*, including 'A Variant Proof of the Distribution of Velocities in a Molecular Chaos' (1913) and 'An Astronomer on the Law of Error' (1918).[81] In *Metron*, he published 'Entomological Statistics' (1920), a work about his latest observations of wasps and bees. Bonar furnishes additional details on this research:

> When I rejoined him [Edgeworth] after the war, he was inquiring into wasps and their movements in and out of a mound at the Golf ground. The fruits of this investigation went, I believe, to Professor Karl Pearson, the record is printed in Gini's *Metron*, July 1920 ['Entomological Statistics']. 'Lubbock attacked the problem directly by marking individual insects. I have employed the logic of Statistics to obtain an inferential result' [wrote Edgeworth]. One of his favourite passages in Dante referred to insects:
> 'Non v'accorget voi, che noi siam' vermi
> Nati a formar l'angelica farfalla,
> Che vola alla giustizia sensa schermi?' [82]

Pearson Again

Towards 1925 there was a minimal rekindling of the friendship between Pearson and Edgeworth after nearly thirty years of indifference prompted by Pearson's resentment. In fact, after 1896 the correspondence between the two was merely bureaucratic: only five letters are saved from 1900 (an exchange of articles), one from 1906, one from 1912 and eleven from 1915. This last bunch was about Isserli's doctoral thesis, for which Edgeworth served on the panel.

As time went on, Pearson's unjustified resentment grew, as is shown in 'Notes on the History of Correlation', a paper that on 14 June 1920 he read to the Society of Biometricians and Mathematical Statisticians, which was closely connected to the Department of Applied Statistics of University College, London, run by Pearson himself:

> Edgeworth replaces Galton's 'Index of Correlation' and Weldon's 'Galton's Function' by the term 'coefficient of correlation'... I should sum up Edgeworth's

> work of 1892 by saying that he left the problem of multiple correlation at least in a very incomplete state. He probably knew what he was seeking himself, but he did not give the requisite attention to the wording or printing of his memoir to make it clear to others, and accordingly in looking back at the matter now I am very doubtful whether in 1895 I ought to have called the problem of multiple correlation, 'Edgeworth's Problem'. . . I think I am justified in saying this for I have not to my recollection come across any treatment of multiple correlation which starts from Edgeworth's paper or uses his notation.[83]

In June 1925, Edgeworth attended the dinner that the Society of Biometricians held in honour of Professor Westergaard, author of *Scope and Methods of Statistics* (1917), a book that he had reviewed. Edgeworth met Pearson on this occasion, and not long afterwards, we find the last letter from Edgeworth dated 28 November 1925 claiming that 'you were so kind the other day as to give me a sort of general invitation to your biometrical discussions'. In fact, he accepted Karl's invitation to attend the meetings of the Society of Biometricians, and in December 1925 he did so. About his talk and his contributions to biometrics, Karl Pearson made the following 'nice' comment at the Galton dinner in February 1926, a few days after Edgeworth's death:

> Only last December he came and spoke as he had always spoken . . . and his criticism failed as it had always failed, because he spoke not the language of the people [but was . . .] *Magister Obscurantissimus* in his dialectic. Besides we owe him something, . . . and, if any of you at any time wonder where the κ in Biometrika comes from, I will frankly confess that I stole it from Edgeworth.[84]

So, despite Karl Pearson's initial acknowledgement of Edgeworth's influence in his early work on correlation, he went on downgrading his findings in successive recollections and came finally to reduce Edgeworth's contribution to statistics to a humble Greek letter.[85]

4.10 FAMILY NEWS, 1916–22

During the war and in the years after it, Edgeworth kept in touch with his nieces, especially with Rosa, who married and went to live at Edgeworthstown with her husband. However, his closest family contact during his final years was with his cousin Harriet Jessie Butler and her son and daughters.

We have several letters from Francis to Harriet. In the first letter, written from All Souls College on 2 July, probably in 1916, he informed her that 'the operation of last Tuesday was most successful'. But in the second later dating from 6 July, he admitted that 'success has not continued to crown my operations'.[86]

Contrary to what one might expect, these operations did not refer to Edgeworth's health and were not surgical in nature. In fact, he was informing his cousin about his observations on the length of bumblebees' voyages.

'In Francis' last years', Harriet Jessie Butler informed Maynard Keynes, 'his memory and agility of mind were already at that time remarkable' and she stressed 'how well he still remembered [in 1925] the poetry he had learnt in his youth, and complete books of Milton, Pope, Virgil, and Homer would readily come to his memory'.[87]

Some interesting letters from Francis to Harriet were written between 1917 and 1919. In them, Edgeworth commented on a number of literary matters, like certain sentences used in the first half of the eighteenth century by R. Blair Campbell, such as the expression 'short and far between'. He alluded to the sentence 'Good breeding sends the arrow to the heart.' In another letter, Francis asked his cousin: 'Have you ever considered the chronology of Castle Rackrent?' He referred to dates concerning Sir Condy and Lady Rackrent that do not fit. The tone of these letters is reminiscent of those between Francis and the first Harriet Butler, their aunt. In these letters, written from All Souls, Hampstead and the Royal St George Yacht Club, Kingstown, we learn also about his nieces, his Essex cousin from the Abbé branch of Edgeworths and Harriet's children:

July 1917?: 'I have a letter from Rosa full of romance.'

August 1917?: 'No more news from Rosa. It is not known when Kenneth [Essex Edgeworth, from the Abbé branch] will return from the front for the ceremony.' 'Rosa seems exceedingly flourishing'.

July 1919: 'Kenneth [Essex] Edgeworth is engaged to be married to Mrs. Eve, the widow of an Indian official. I have put off going to Ireland for a few days and am entertaining Mia during her short respite from her work at Reading. She seems very well and to like her work. I trust that Ruth has quite recovered. I had already congratulated Harold and now I congratulate you and the whole family on the approaching auspicious event [the birth of Harriet's grandchild].'[88]

On the other hand, Edgeworth's letters to Violet Butler, Harriet's youngest daughter, were more academic in tone. In June 1920, he discussed with her 'On the Banking Principle vs. the Currency Principle'. In April 1922, Edgeworth commented on several educational books and mentioned Charles Gide. But he also referred to his leisure activities and to his niece Rosa and her husband:

I have just returned from a short excursion in the South Downs where I think there is some of the best walking that is to be found in England. Rosa and Cuthbert are expected about the 4 of May. They will go down in a day or two to Devonshire where I will join them for a few days at Lynmouth.[89]

4.11 KEYNES, CORREA WALSH AND EDGEWORTH ON PROBABILITY AND INDEX NUMBERS

In 1922 J.M. Keynes published his *Treatise on Probability*, in which he quotes Edgeworth and admits that he does not always differ utterly from Edgeworth's point of view on specific probability:

> There are very few writers on probability who have explicitly admitted that probabilities, though in some sense quantitative, may be incapable of numerical comparison. Edgeworth admitted that 'there may well be important quantitative, although not numerical estimates of probabilities.' . . .
>
> He writes 'The Calculus of Probabilities is concerned with the estimation of degrees of probability; not every species of estimate, but that which is founded on a particular standard. That standard is the phenomenon of statistical uniformity . . .'. This use of terms is legitimate, [. . .but] it leaves aside the most important questions. The calculus of probabilities, thus interpreted, is no guide by itself as to which opinion we ought to follow, and is not a measure of the weight we should attach to conflicting arguments . . .
>
> I think with Edgeworth that the hypothesis of the equal distribution of ignorance is, within the limits of practical life, justified by our experience of statistical ratios, which *à priori* are unknown, i.e. such ratios do not tend to cluster markedly round any particular value. . . . The ultimate basis of the theory of statistics is thus not mathematical but observational.[90]

These allusions moved Edgeworth to reaffirm his position on the subject of probability in response to J.M. Keynes. Edgeworth did this, not only in the aforementioned review of Keynes's book in the *Journal of the Royal Statistical Society*, but in the 1922 article published in *Mind*, 'The Philosophy of Chance', which, in fact, is the most detailed review written on Keynes's dissertation and at the same time may be considered to be Edgeworth's final testament on this crucial topic.[91] In the conclusions he states:

> (1) That probability in general presents gradations has not been disproved. . . .
>
> (2) New light is thrown by Mr. Keynes on the *a priori* probabilities, sometimes ascribed to the Principle of Sufficient Reason, which play a considerable part both in social science and in Physics. He strengthens the defence of these propositions against Dr. Venn's polemic. At the same time Mr. Keynes exhibits and guards against the errors and exaggerations to which the Principle of Sufficient Reason has often led. . . . .

(3) Dealing with the Logic of Statistics, Mr. Keynes reaffirms effectively the important truth that the Methods of Induction are largely dependent on preexistent knowledge. . . . There are water-tight compartments in Nature. . . . The character of natural law seems closely connected with that *independence* which plays a great part in Probabilities. Independence is the prime condition for the fulfillment of that 'law of error' (or deviation of magnitudes from their average) which Quetelet and Galton celebrated. . . .

(4) The use of mathematical Expectation as a guide to conduct is not barred by the difficulty of obtaining precise data.[92]

Despite the differences in their positions, Edgeworth tried to restrain his criticisms, wrapping them in very soft sentences. For instance, he did not like Keynes's attitude of gross denunciations of most of his predecessors, and he pointed to this lack of respect but finished with a final word of justification:

> I cannot . . . acquiesce in the frequent disparaging remarks and occasional sweeping denunciations which the author bestows on his eminent predecessors. Laplace, Poisson, Quetelet, Cournot, Mill, Jevons, and many others, even his favourite author Lexis in one passage at least, all according to him have gone astray. I find it, indeed, necessary to defend myself against the imputation of attaching too much importance to the opinions of one who attaches so little importance to the opinions of the leading authorities on the subject. The imputation would no doubt be serious with respect to some sciences. But it is a peculiarity of our study, one which it shares with economics, that you can retain respect for one who speaks disrespectfully of high authorities.[93]

To illustrate that, he finally compared Keynes with John Rae, in the sense that 'Rae does not mince his expressions of dissent . . . referring to the Father of Political Economy. And yet Rae is praised in the highest terms by J.S. Mill. Mill's words may be transferred, with the alteration of a name only, Laplace being substituted for Adam Smith, to the case before us.' Mill's words were:

> The author unites much knowledge, an original vein of thought, a considerable turn for philosophical generalities . . . The principal fault of the book is the position of antagonism in which, with the controversial spirit apt to be found in those who have new thoughts upon old subjects, he has placed himself towards Adam Smith. I call this a fault (though I think many of the criticisms just and some of them far-seeing. . .).[94]

Index Numbers

Three years later, in 1925, Edgeworth entered the arena to defend the probability view of index numbers which had been his position from the beginning, when he partly relied upon Jevons on this subject. In that year he wrote 'The Element of Probability in Index-Numbers' to stress this probabilistic nature, in contrast to the position maintained by Irving Fisher, Correa Walsh and Allyn Young in 1923 and 1924 that the problem of index numbers does not fall within 'the field of the theory of errors'.[95]

Edgeworth always maintained that there were as many kinds of averages and index numbers as purposes. He had written in 1888, in 'New Methods of Measuring Variation in General Prices':

> There are as many kinds of average as there are purposes; and we may almost say in the matter of prices as many purposes as writers. Hence much vain controversy which has been applied to metaphysics, one party makes a good stroke at billiards, and thinks he has scored off another who is playing chess.[96]

Moreover, from 1881 onward, when *Mathematical Psychics* was published, Edgeworth upheld the theory that prices were by nature stochastic variables, as he had stated explicitly in his short article dating from 1884, in which he tried to reveal the causes for 'the fact of price'.[97] In his 1925 article, Edgeworth maintained this probabilistic point of view, considered the different applications and shapes of index numbers and emphasised their essentially stochastic nature, as opposed to Correa Walsh's opinion, and he concluded:

> The uncertainty which the nature of things has attached to estimates of utility and probability is intolerable to Mr. Correa Walsh. He will not acquiesce in what Mill calls 'the necessary indefiniteness of the idea of general exchange value.' He believes in an objective general exchange value. Not content to walk with the rest of the world in the twilight of Probability, he cherishes a light visible only to himself:
> 'The dark lant-horn of the spirit
> Which none see by but those who bear it.'[98]

Edgeworth took this quotation, one of his favourites, from Samuel Butler, but he neglected to mention the source.

4.12 1919–25: FIFTH STAGE OF THE *ECONOMIC JOURNAL*

With Keynes getting more and more involved in the negotiations for the peace treaty at Versailles, Edgeworth was busier with the editing of the *Economic Journal* and, as we have seen above, was once again named official editor, this time in conjunction with Maynard, who did not quit the post. Keynes wrote in 1926 the following about this collaboration:

> His practical gifts as an editor were quite other than might have been expected from his reputation as an unpractical, unbusinesslike person, remote from affairs, living on abstractions in the clouds, illuminating the obscure by the more obscure. As one who was associated with him in the conduct of the Journal for fifteen years [1911–1926], I can report that this picture was the opposite of the truth. He was punctual, businesslike, and dependable in the conduct of all routine matters.

And Keynes followed with a description about their work as joint editors:

> He exercised his editorial powers with great strictness to secure brevity from the contributors (he invented what he termed a law of diminishing returns in the remuneration of the articles, by which the rate falls after ten pages have been exceeded and sinks to zero after twenty pages), and invariably cast his influence in favour of matter having topical interest and against tedious expositions of methodology and the like. . . . He established and was always anxious to maintain the international sympathies and affiliations of the *Journal*. I am sure that there was no economist in England better read than he in foreign literature. He added to this what must have been the widest personal acquaintance in the world with economists of all nations. Edgeworth was the most hospitable of men . . . He had a strong feeling for the solidarity of economic science throughout the world.

After all this praise, Keynes concluded: 'All his eccentricity and artistic strangeness found its outlet in his own writings. All his practical good sense and daily shrewdness was devoted to the *Economic Journal*.'[99]

In fact, Keynes and Edgeworth worked as an excellent team. Since the flow of proposed material that arrived at the *Economic Journal* was excessive, they had to choose carefully. The unfavourable opinion of one of the co-editors was enough to reject a paper. Keynes was not always right in his rejections. Bertil Ohlin, Nobel Prizewinner in Economics in 1977, pointed out:

> At Cassel's suggestion, I sent a paper [in 1922] containing a brief version of my thesis [about international trade theory] to Professor Edgeworth who was then co-editor with Keynes for the *Economic Journal*. It presented equation systems as a basis for an analysis of the causes and effects of international trade. At that time, equations were not so popular as diagrams. Anyhow, Edgeworth sent my paper to Keynes and asked for his opinion. Keynes wrote on a piece of paper which followed the manuscript via Edgeworth back to me: 'This amounts to nothing and should be refused. J.M. Keynes.' I still retain this little note as a valuable document.[100]

Ohlin was not resentful, and in 1929 he discussed the German reparation problem with Keynes both in the *Economic Journal* and privately.[101] Likewise, he included a favourable comment – 'admirable paper' – on

Edgeworth's long article 'The Pure Theory of International Trade'[102] in his 1933 book, *Interregional and International Trade*.

Edgeworth's Articles

Edgeworth's contributions to the *Economic Journal* during the last seven years of his life were not as numerous and fundamental as his works on statistics already discussed. In fact, he published only six articles, as listed below in Appendix N.

We can see that the six articles cover three topics, two of them classical for Edgeworth, taxes and index numbers, and a new one, women's wages. With regard to the methods of graduating taxes, Edgeworth complained about the neglect of the appointed ad hoc commission: 'The grounds on which the Commission rejects the use of graduation formulæ are to be examined here . . . The Commission seems not to have done justice to the use of mathematical formulæ for the purpose of interpolation.'[103]

As for the pair of papers on index numbers, it is interesting to note his anti-dogmatic proposal in the 1925 article 'The Plurality of Index Numbers', in line with the content of his aforementioned 1923 paper, 'Index Numbers according to Mr. Walsh':

> I propose to reconsider an old question, whether the formulæ which purport to indicate change in the value of money are all of one type or admit of diversities adapted to different purposes. . . . I plead only for toleration of some additional tenets. It would be too much to ask economists, what Cromwell asked theologians, to think it possible that they might be mistaken. Each maker of index-numbers is free to retain his conviction that his own plan is the very best. I only ask him to think it possible that others may not be entirely mistaken.

'Equal Pay to Men and Women'

He had never before published anything on the third topic, women's wages, if we exclude his review in 1893, about thirty years earlier, of Smart's book *Women's Wages*. In 1922, Edgeworth was elected President of Section F of the British Association for the Advancement of Science for the second time. His presidential address, 'Equal Pay to Men and Women for Equal Work', was published in the *EJ* and was followed in 1923 by a second article on the subject, 'Women's wages in Relation to Economic Welfare'. In both of them, Edgeworth was quite direct: 'Equal pay to men and women for equal jobs is unfair to men, because women do less quantity of work.' He was not for discrimination in favour of women in work in order to pay them equal wages with men. However, he did accept a voluntary scheme of compensation:

> To sum up; equal pay for equal work, in the sense of free competition between the sexes, has been advocated with some reservations and adjustments. Desperate disordered competition, tending to the degradation of labour is supposed to be excluded. There are suggested compensations to families for the loss sustained by the male breadwinner through the increased competition of women. Among such compensations*, the endowment on a large scale by the State is not included. The advantages weighed are economic in a strict sense. The balance may be affected when welfare or well-being in a wider sense is taken into account.

What type of compensations does Edgeworth suggest? Here we have his answer in a footnote:

> *Compensations: a simpler plan is suggested. It is open to any association of men – a trade's union, for example – to resolve that each member of the association should contribute a quota of his earnings towards the formation of a fund which is to be distributed among the wives of the members in accordance with the size of their families. . . . Bachelors and childless husbands should be persuaded to support a fund which they may hope one day themselves to benefit from as future fathers of families.

One of the newspapers that appeared the next day, which Edgeworth kept in his files, headlined the conclusions of his presidential address in the following harsh way: '"Equal pay to men and women, unfair to men", says Professor.'[104]

In the second article, 'Women's Wages in Relation to Economic Welfare', Edgeworth waxed ironically philosophical:

> Milton, when galled by the yoke of an ill-assorted marriage, . . . proposed, as a remedy for marital troubles, freedom of divorce . . . 'I doubt not but with one gentle stroking to wipe away ten thousand tears out of the life of man.' But the economist, remembering how often the appearance of easy remedies for human ills, in his sphere at last has proved deceptive, will not expect much from a stroke, gentle or violent, intended to revolutionise established institutions which have worked well for the production of wealth and economic welfare. The only reforms of such institutions which the economist can approve are tentative and gradual.

Edgeworth's Reviews

If there was a downward trend in the articles he produced, the flow of reviews followed its usual pace, as shown in Appendix N. With his long and abundant experience of penning reviews, Edgeworth had become the most outstanding reviewer of his time. In this final period, most of the reviews are interesting either for his display of erudition or for the florid compliments and ironic criticisms that he no longer takes the trouble to rein in. As illustrations, we have his review of F. Bernis's *La Hacienda Española: Los Impuestos*, which Edgeworth praises as 'an important contribution to the art of Political Economy', mainly because Bernis was the *EJ*'s Spanish correspondent. However, afterwards he reprimands him diplomatically with a striking display of erudition:

> We have reason to accept his facts, having compared samples thereof with the observations and statistics which his eminent countryman, Professor Florez [Flores, *sic*] de Lemus, presented in the Spanish Supplement to the *Times* (29 June 1914) . . . We cannot agree with the principle, which he seems to entertain, that protective duties should form the rule, their excess only being objectionable (pp. 194, 202) . . . The author might perhaps be likened to a distinguished predecessor, Ustarez [Uztáriz, *sic*], whose judgment on a matter of taxation (the effects of the Alcabala) obtained the approval of A. Smith. But it would be unjust to rank him with a mercantilist, however eminent. Rather our author, recalls Florez Estrada . . . a valuable contribution to the economic literature of his country.

Edgeworth's most enthusiastic review was the one dedicated to Professor Cassel's *Treatise*. Recognising the importance of the book and its author, Edgeworth wrote a six-page paper in his most peculiar laudatory style:

> In the view of that soaring genius [W.R. Hamilton], astronomy and metaphysics generally require the exercise of a faculty akin to the artistic imagination. There is produced 'an imitation, not a copy, of Nature. It is a creation of the mind so framed as to resemble in an immense number of particulars what we know of the external universe'. . . . Professor Cassel appears to us to have remarkably well defined the province and limits of abstract general reasoning in economics. 'We proceed like the astronomers who first determine the motion of a planet as if it were not influenced by the other planets . . . afterwards take into consideration the disturbances caused by the other planets . . . In order to be able to pursue this method one must be assured that the motion first described represents the essential portion of the phenomenon'.

And, certainly, Edgeworth's most enthusiastic praise was reserved for Marshall's *Money, Credit and Commerce*:

> Leading by easy routes to the most difficult parts of economic science, this work is destined to be a powerful aid to scientific education. . . . If much of it might have been written in the 'eighties of last century, much of it will be read in the 'eighties of this century. It is, as far as what relates to human affairs can be, *in specie eternitatis*.

In the complimentary review of his friend and disciple Bowley's *The Mathematical Groundwork of Economics*, Edgeworth saluted the work that would popularise his diagrams of barters in *Mathematical Psychics* through the four-coordinates diagram known as the 'Edgeworth box':

A long-felt want is satisfied by this clear, concise and correct statement of the leading propositions and methods which mathematics contributes to Political Economy. . . . By steps that are neither violently abrupt nor tediously circuitous he reaches the heights from which the mutual dependence of all economic quantities can best be contemplated. At those heights, too, are observed some curiosities of theory, like Alpine flowers, found only at great altitudes.

In his review of J.S. Nicholson's *The Revival of Marxism* and A. Loria's *Karl Marx*, Edgeworth's position with regard to Marx's work becomes explicit:

We have much sympathy with those who hold that the theories of Marx are beneath the notice of a scientific writer. However, the refutation of prevailing fallacies has always been recognised as part of the economist's province. . . . Professor Sombart, along with profuse eulogy of Marx, makes the admission [in *Der Moderne Kapitalismus*, 1902] that he contributed nothing to the technique of the science. The importance of Marx's theories is, indeed, as Professor Nicholson shows, wholly emotional.

However, Edgeworth's position, though a far cry from Marxism, was not strictly conservative. His conception of the importance of the bargaining process in non-competitive situations, settled in 1881, is the underpinning of his encouraging review of J. Maurice Clark's *Studies in Economics of Overhead Costs*:

J.M. Clark describes as prevalent in industries with overhead costs [or fixed costs] a situation of semi-monopoly. . . . The departure from the rule of pure competition is important, in view of modern theories of distributive justice and modern conditions of industry. While Professor J. B. Clark [the author's father] teaches that the ideal remuneration of labour is its marginal product, Professor J. M. Clark finds that wages are commonly below that marginal product.

Likewise, the topic of his 1922 presidential address led him to sympathise with feminist positions such as those expounded by Mrs H.A.L. Fisher in *The Economic Position of the Married Woman*:

Mrs. Fisher argues well: 'Economically it is a sheer waste to turn a skilled teacher or designer or writer or singer into an unskilled housekeeper. . . . If she gives up her work and takes to domestic life she is equally keeping some one else out of a job. Some one less efficient than she does the work which some other woman could do better.' [Nevertheless, F.Y.E. adds:] Experiment as well as reasoning must be interpreted with caution.

Edgeworth's final stage at the *EJ* is the richest in ironic and devastating comments, which Maynard enjoyed greatly. Here we just offer a handful of selected morsels.

On Andréadès's *The Economic Institutions of Ancient Greece*:

> The mastodon and the ichthyosaurus will not again appear on earth; but the prejudice against interest, the postulate that prices should be 'just' and other ancient or mediæval dogmas seem continually to revive. . . . He not only arranges economic fossils and explains their significance: he also researches and finds new specimens.

On Amoroso's *Lezioni di Economica Matematica*:

> But while differing from our author on some cardinal points, we recognise that he has evinced great ability in defence of positions which we regard as indefensible. If they were defensible, '*Si . . . dextrâ / Defendi possent, etiam hâc defensa fuissent*'.[105]

On Rathenau's *In Days to Come*:

> Altogether we are disposed to class Rathenau with Carlyle and Ruskin as a preacher rather than a teacher; an authority about ends rather than means. He lifts our eyes to distant and sublime heights; he directs our feet by paths which may prove unsafe or impracticable.

On Walsh's *Index Numbers*:

> It is as if the prescriptions given by medical science for different cases were applied indiscriminately by one not conversant with the symbols of pharmacy.

On Pigou's *The Economics of Welfare*, second edition, Edgeworth complained about the rearrangements and rewritings with regard to the first edition (1920), which included much of *Wealth and Welfare:*

> The continual change in the order of topics breaks the links of memory, tangles the chains of reasoning, and is fatal to artistic form. The shifting of pages and paragraphs acts like the scattering of the leaves on which the oracles of the Virgilian Sybil were inscribed. The disorder repelled those who came seeking inspired counsel: '*Inconsulti abeunt, sedemque oder Sibyllæ*.'[106]

Obituary Notes

With Keynes's approval, in 1923 Edgeworth accepted a final obituary note by Pantaleoni and Loria dedicated to Pareto, who had just died in Céligny, Switzerland. They wrote: 'His ardent genius found scope in applying to Political Economy that mathematical method in which Cournot, Walras, Jevons, Edgeworth had gathered splendid laurels.'

Pareto's obituary was Pantaleoni's last contribution as he died of a heart attack on 29 October 1924 after concluding an address. Barone had also died in 1924, on 14 May. The *EJ* published obituaries for both of them, written by Loria.

However, the longest obituary from 1924 in the *EJ* was the one written by Maynard Keynes and devoted to Alfred Marshall, who died on 13 July. We shall shortly return to Marshall's death.

4.13 EDGEWORTH IN HIS RETIREMENT YEARS

In his reminiscences of the period from 1922 to 1926, Robert Graves (1929) alludes to one of Edgeworth's most celebrated peculiarities, that of avoiding 'conversational English persistently using words and phrases that one expects to meet only in books', T.E. Lawrence ('Lawrence of Arabia'), who was by then a Fellow of All Souls, 'returned from a visit to London, and Edgeworth met him at the gate. "Was it very caliginous in the metropolis?" "Somewhat caliginous but not altogether inspissated"', Lawrence replied gravely.'

We also have several written descriptions about Edgeworth in his seventies and beyond. Keynes says:

> His health and vigour of body were exceptional. He was still a climber in the mountains, bather in the cold waters of the morning at Parson's Pleasure, unwearying pedestrian in the meadows of Oxfordshire, after he had passed his seventieth year. He was always at work, reading, correcting proofs, 'verifying references', working out on odd bits of paper long arithmetical examples of abstruse theorems which he loved to do, . . . writing letters, building up his lofty constructions with beautiful bricks but too little mortar and unclear structural design. Towards the end of his life it was not easy to carry through with him a consecutive argument *viva voce* – he had a certain dissatisfied restlessness of body and attention which increased with age and was not good to see. But on paper his intellectual powers even after his eightieth year were entirely unabated; and he died, as he would have wished, in harness.[107]

He felt full of energy even when he turned eighty. There is a letter from Edgeworth to Maynard Keynes where he refuses a certain proposal from Keynes related to his age, probably to mention his eightieth birthday in the *Economic Journal*:

> It is friendly of you to think of alluding to my case in the *Journal*. But I would decidedly prefer that there should be no allusion. I do not like to be differentiated by attention called to *Anno Domini* from many with whom I feel an equality in respect of energy.[108]

Some of Keynes's assertions were also held by David Hutchinson Macgregor, former disciple of Marshall and professor at Leeds who became Edgeworth's successor in the Drummond Chair, lived at All Souls College as well and was his good friend. Macgregor wrote to Keynes, immediately after Edgeworth's death, that 'his physical vigour was astonishing', and he described him as 'a curious mixture of English doggedness, Irish humour and Spanish sprightliness and courtesy. He was the merriest and the kindest of men.' Macgregor also mentioned that 'with his college friend W.P. Ker [Edgeworth was] a faithful attendant often of All Souls Sunday Tramps'. It is also reported that, in his seventies, Edgeworth often played golf on the course at Cowley.[109]

William Paton Ker, ten years younger than Edgeworth, was an old friend who was mentioned back in 1876 as a participant in Leslie Stephen's Sunday Tramps. Born and educated in Glasgow and Balliol, he was Professor of English at Cardiff (1883) and London (1889), and Professor of Poetry at Oxford (1920), where he lived at All Souls College and renewed his friendship with Edgeworth. Described as a 'talker, lecturer, and writer of prodigious learning and vitality, he wrote *Epic & Romance* (1897), *The Dark Ages* (1904), and *The Art of Poetry* (1923).'[110] Ker, just like Edgeworth, was an excellent mountaineer. He especially loved the vistas of sweeping landscapes. In 1923, when he was 68 years old, he once again visited one of his favourite sites, the Italian Alps. This time he chose to walk up the slopes of the Pizzo Bianco in Macugnaga. While he was ascending, he paused and exclaimed, 'I thought this was the most beautiful spot in the world, and now I know it.' These were his last words. He fell dead on the spot of a heart attack.

Meanwhile, Alfred Marshall was slowly fading at his home at Cambridge due to a chronic digestive ailment.[111] He still assembled some old lectures and produced the book *Money, Credit and Commerce*, which Edgeworth received in February 1923 at Hampstead and, as we have seen, reviewed for the *EJ*. In his letter to Marshall acknowledging receipt of the book, Edgeworth told Marshall that he had just arrived back at Hampstead after a month of absence.[112] Marshall died on 13 July 1924. As a final token of friendship, Edgeworth contributed an article, 'Reminiscences', to the book *Memorials of Alfred Marshall*, edited by Pigou in 1925.

Even in his last years, when Edgeworth was nearly 80, he had the ability to attract new young friends, as was the case with Roy F. Harrod, who was a 24-year-old Fellow at Christ Church, Oxford when in April and May 1924 he exchanged letters with Edgeworth on the subjects of economic dynamics and international trade. Edgeworth wrote to him:

> I on the contrary hold that the essential characteristic of international trade is the inequality of the salary to effort and sacrifice. Thus only do we escape from the Ricardian rule according to which value is proportional to quantity of labour (effort & sacrifice).[113]

Harrod had studied money and international trade in Cambridge as a postgraduate under Maynard Keynes during 1922 and 1923. After Edgeworth's death, Harrod wrote to Keynes that lately 'he had grown very fond of him' and:

> he was in superb form at our Political Economy Club at a meeting last term. Joseph had read a paper trying to show that economists used the word value in no intelligible sense. This roused Edgeworth to defend the notions of total and marginal utility at considerable length. . . . He ended with a rhetorical laudation of the Jevonsian economist, '*vicisti, Galilei*'.

Harrod also referred to 'his last set of lectures, given about two years ago. They were most exciting. For a time he tumbled about, inaudible and confused, paving the air with his hands, as though making some great effort, and then he suddenly emerged into a fascinating passage of perfect lucidity.'[114] Later on, in his 1951 biography of J.M. Keynes, Harrod would describe the emotions that Edgeworth displayed when lecturing:

> In old age his face was largely concealed in his beard, and his sunken eyes were not very expressive; but at a critical moment one could gauge that his feelings were overcoming him. One such was when after many hours of lecturing and after many passages of digression with quotations from the classics and analogies from physics, he at last made the supply curve intersect the demand curve on the blackboard. One knew it was a great moment. He wagged his beard and muttered inaudible things into it. He seemed to be in a kind of ecstasy.[115]

Notwithstanding Keynes's aforementioned opinion of 'a certain dissatisfied relentlessness of body and attention which increased with age', Harrod would also stress that in his last years Edgeworth's attention was undiminished and that 'right up to the end he was able to make detailed comments'.[116]

4.14 THE FIRST REVIVAL OF EDGEWORTH'S WORKS

Papers Relating to Political Economy

In 1923, the Royal Economic Society decided to assemble and publish Edgeworth's articles on economics. He worked to check and comment on the entire edition, which was published in three volumes in 1925 under the title *Papers Relating to Political Economy*. The third volume included a selection of his reviews in the *EJ*. Since the selection was made by Edgeworth himself, it is interesting to see the books he considered most important in those 35 years of reviewing. Yet one gets the impression that

Edgeworth's main criterion of choice concerned not only the quality of the text but also his close acquaintance with the author.[117] Some of the selected reviews were from books in foreign languages: German (books by von Bortkiewicz, Pareto, Pierson, Dietzel, Cassel), Italian (three books by Graziani, plus Preziosi and Arias), French (Gide, de Foville, Colson) and Greek (two books by Andréadès).

He spent a lot of time choosing the articles and reviews, preparing and even updating the material and writing brief introductions to each article. But he enjoyed this task, as Keynes comments:

> The publication of his Economic Papers was a great satisfaction to Edgeworth. His modest and self-effacing ways would always have prevented him from undertaking such an enterprise on his own initiative. But as soon as others were prepared to take the responsibility, the business of selection and preparation for the press was a congenial task. Moreover, the publication proved a great success in every way, and was reviewed in learned journals throughout the world with expressions of esteem . . . I think that Edgeworth was genuinely surprised at the extent of his international reputation, and it gave him as much pleasure as surprise.[118]

When *Papers Relating to Political Economy* appeared, Edgeworth felt so proud that he dared to ask for a copy of the three volumes in March 1925 for his cousin Harriet Jessie Butler and expressed the 'trust that the Statistical Society, following the handsome precedent set by the Economic Society will in due time republish the numerous papers which I have contributed to their Journal during a period of forty years'.[119] Taking into account the influence of Karl Pearson on the Statistical Society, it is no surprise that Edgeworth's statistical papers were not reissued.

As a direct consequence of the publication of *Papers*, Schumpeter wrote an article on Edgeworth's work in October 1925: 'Edgeworth und die neuere Wirtschaftstheorie' for *Weltwirtschaftliches Archiv*. Gustavo del Vecchio, too, wrote about him: 'F.Y. Edgeworth e l'Economia Matematica', *Giornale degli Economisti* e *Rivista di Statistica*, September 1925.

The Revised Edition of Palgrave's Dictionary

During 1924 and 1925, Edgeworth wrote four new entries for the revised edition of *Palgrave's Dictionary of Political Economy* that appeared in 1925 and 1926. He contributed with the following long articles:

1. 'Economics, Teaching of', in which Edgeworth affirms that in those times Oxford students had to work from Walker's *Political Economy* and Smith's *Wealth of Nations*.

2. 'Pantaleoni, M. (1857–1924)', in which Edgeworth writes: 'The attack on the socialist Jaurès, who is accused of traitorous correspondence with his country's enemies, reveals, or at least suggests, the presence of an element not favourable to economic science – some heat of political passion.'
3. 'Pareto, V. (1848–1923)' in which Edgeworth summarises the differences between the so-called Lausanne School (Walras-Pareto) and the English School (Jevons-Marshall-Edgeworth): 'Some of the conceptions which Pareto adopted from Walras have not escaped criticism. He perhaps attached too much importance to the exhibition of simultaneous equations involving numerous unknown quantities; in contrast to the simple *curves* of Marshall. . . . The doctrine that the entrepreneur makes neither gain nor loss is a hard saying to some [an implicit reference to Marshall and Marshallian disciples like Pigou] . . . Upon the whole it appears that Pareto has very justly distinguished the functions of abstract reasoning and concrete knowledge and very successfully cultivated both branches of economic science'.
4. 'Pareto's Law', in which Edgeworth writes: 'While the logical foundation of Pareto's law is open to controversy, there is no doubt about the validity and importance of the law.'[120]

4.15 EDGEWORTHSTOWN ON THE HORIZON

During his final years, Francis's interest in Edgeworthstown rose, and he was not deterred by the Irish situation. In fact, he followed the sequence of political events closely: the 1920 Government of Ireland Act divided the island into two self-governing areas, Ulster and the rest. This act later resulted in the Anglo-Irish Treaty of December 1921, which established the new dominions of Northern Ireland and the Irish Free State in the British Empire community. Then, from June 1922 until May 1923, there was the Civil War between the Republicans of de Valera and Brugha and the Free State Army of Griffith and Collins, who died in an ambush in August 1922. He witnessed the constitution of the Free State in December 1922 and the Cosgrave ministries, the end of the armed resistance of the Republicans and their victory in the elections of August 1923 that allowed Cosgrave's party to continue in government because de Valera's Republicans refused to sit in the new Dáil (Parliament).

All these moves towards the complete independence of the land of his Anglo-Irish ancestors did not change Edgeworth's plans for spending his 'happy old age' in Edgeworthstown. He was glad that his niece Rosa Sanderson and her husband were living there, probably encouraged by Francis himself for fear that the manor house might be raided and destroyed by forces of any faction if found uninhabited. In fact, out of the nearly 2000 owned by landlords, seventy large houses were destroyed, mainly in the west and south of the country. In the Edgeworthstown environs, only minor incidents were registered with armed men: at Pakenham Hall, the largest house nearby with almost one hundred rooms, they took three bicycles and a pair of scissors.[121]

One month after his eightieth birthday, in March 1925, Edgeworth informed his cousin Harriet that he was planning to travel to Ireland with his niece Laura.[122] And in the notes that Harriet wrote to Maynard Keynes in April 1926, as information for Edgeworth's obituary, she explained:

> He [Francis] had taken great interest lately in gathering up all the old family traditions and in endeavouring to restore Edgeworthstown house to something of its former tradition, under the care of a married niece and her husband [Rosa and Cuthbert Montagu]. Circumstances had kept him away from it for many years, but his vacation visits to the 'property' had of late been a great joy to him, especially during the last two or three years and he declared himself to be 'looking forward to a happy old age' in the home of his forefathers.[123]

4.16 EDGEWORTH'S DEATH

We do not know whether he went to Edgeworthstown during the spring of 1925, as he had planned, or whether he delayed the journey until summer. But in June 1925 Edgeworth was at the aforementioned dinner offered to Professor Westergaard by the Society of Biometricians where, according to Karl Pearson, he showed that 'aged eighty he was still a boy in his pleasures and spirits' by dancing a 'gambolade, *pas de seule*, . . . down the cloisters . . . in front of the procession returning from the refectory to the laboratory after the dinner'.[124]

As we have also seen, in December 1925 Edgeworth attended the regular meeting of this Society of Biometricians, where he participated in the customary academic tradition of offering academic critiques of the contents of the session.

We do not know exactly where Edgeworth spent his last Christmas, but on the following Sunday he visited his niece Fay – Frances Harriet Sanderson – at Burnham Abbey, Maidenhead. On the first day of 1926, Francis, writing from Hampstead, wished his cousin Harriet a 'Merry happy New Year' and offered her a photographic portrait of himself as a present.[125]

That January, Edgeworth wrote – with excellent handwriting that belies the hypothesis of his death from Parkinson's disease – to Ruth Butler, Harriet's daughter:

> You might be able to put me in the way of obtaining some tribute to the memory of Vinogradoff [Russian historian, Professor of Jurisprudence in Oxford 1903–1925], some sort of obituary notice of himself and his work such as ought to appear in the *Economic Journal*. [Ruth had attended Vinogradoff's lectures at Oxford.][126]

Edgeworth was working on the next issue of the *Economic Journal* as well as on the correction of the proof sheets of his article for the *Journal of the Royal Statistical Society*, 'Mr. Rhodes' Curve and the Method of Adjustment'. We do not know if he felt ill before or after his eighty-first birthday on 8 February, since Price's obituary for the *Journal of the Royal Statistical Society*, is not very precise:

> During the illness which proved fatal he was engaged in the correction of the proof sheets of the last of this long series of learned papers, that which appeared in our issue dated January of the current year.[127]

Four days after his birthday, on Friday 12 February, Edgeworth wrote two letters. In both the handwriting is awful, showing a vast difference from the aforementioned letter to Ruth. The first was written from All Souls, is addressed to Maynard Keynes, refers mostly to the *Economic Journal* and informs him that he is going to Acland Home due to his 'attack of illness'. We transcribe everything we could decipher:

> Dear Keynes,
> I think that I am not quite certain that I sent you almost in . . . [Meshatel?] and receiving it a letter from Doley explaining the positions of his signature.
> I not will communicate with . . . [?] and . . . [?] a *modus vivendi*. My attack of illness getting worse I have to retire to the Acland Home but continue . . . without having time to . . . which cannot now be got at.
> I have declined Hilton Young's [a Member of Parliament].
> Yours sincerely.[128].

The second was written at the Acland Home and sent to his cousin Harriet: 'Mrs. Butler; 14 Norhan Garden, Oxford.' Despite the alarming handwriting, it is perfectly readable:

> My dear Harrie, If you would come here at 4 p.m. tomorrow, I could give you tea in my somewhat palatial room. Dr. Brooks will I believe have told you that my complaint is not infectious.[129]

However, the not infectious 'complaint' was to prove fatal. In the morning of next day, Keynes received a telegram from Macgregor informing him that 'EDGEWORTH DIED THIS MORNING.'[130] Francis's cousin

Harriet Jessie Butler and her daughter Violet went to see his body. Harriet Jessie wrote the next day:

> He [Francis] had become very cousinly & affectionate of late, tho' one could never go very near! Violet and I went last night to the Chapel at Acland Home – where he lay on a bier with a sort of stateliness and *keenness* of air, only so *still*. It was a very fine face in its entire repose. Certainly he was one of the cleverest men I have known and the very quickest in mind, I think.[131]

One of the Sanderson nieces, Fay, wrote from Maidenhead to Harriet:

> He was like a father to us. He was here the Sunday after Christmas Day and so particularly well and happy – we had such a happy afternoon together. He only complained of the cold in the early mornings and how difficult it was for him to get warmed up. . . . I am very glad the service is to be at All Souls and buried in Oxford where he was so happy. I have not heard from Rosa. . . . Poor Rosa she will be dreadfully grieved – it made her so happy that he so enjoyed being at Edgeworthstown. . . . I send a little bunch of snowdrops and violets from the garden . . . to be put in the coffin. Perhaps Rosa will not arrive too late to see him.[132]

According to the order of service, Rosa and her husband arrived from Edgeworthstown in time for the funeral, which took place at All Souls College on Wednesday, 17 February at 2.15 pm. The burial was at Holywell. The details of the funeral, taken from the order of service, were as follows: 'I am the Resurrection and the Life, saith the Lord'; Psalm XC, 'Lord, thou hast been our refuge.' The last words recited over the grave at Holywell cemetery were: 'The grace of our Lord Jesus Christ, and the love of God, and the fellowship of the Holy Ghost, be with us evermore. Amen.' The Bishop of Gloucester and the Rev. A. H. Johnson, Fellow and Chaplain conducted the funeral. The chief mourners were Mrs Butler (Harriet Jessie), Professor H. E. Butler (Harold), Miss R.F. (Ruth) and Miss C.V. (Violet) Butler, Mr and Mrs C. Montagu (niece Rosa B. Sanderson, and husband Cuthbert Montagu), Mrs Moss and Lieutenant Colonel K.E. Edgeworth (Kenneth Essex Edgeworth, a cousin from the Abbé side).[133]

Obituary Notes and Articles

On 16 February, *The Times* published an obituary note, probably based on Price's information:

> Francis Ysidro Edgeworth was an insatiable reader, but his love of walking, mountaineering, golf and boating with his strict and regular habits, maintained to the last his wonderful vitality. Every summer, even at the age of 80, he used to bathe at Parson's Pleasure before breakfast, and he would often be seen riding his

> bicycle in the country round Oxford or playing on the course at Cowley. A lifelong friend has never known him to be out of temper or speak an ill-word to others.

The Times then mentioned that Edgeworth had been President of the Economic Section of the British Association in 1889; President of the Royal Statistical Society, 1912–1914; Fellow of the British Academy and of King's College, London; and member of the Athenæum, the Savile and the Alpine Clubs; and that he succeeded to Edgeworthstown in 1911. It also mentioned that the Chair of King's College, London, that he occupied in July 1888, had been vacant from the time of Jones and Senior.[134]

Maynard Keynes kept in his files the obituary note by J.C. Stamp in *The Press* from 16 February 1926, which concluded: 'An athlete and a boy to the last, he carried his youth with him for 81 years.'[135] Keynes asked Edgeworth's relatives and friends for information about Francis in order to write an obituary for the *Economic Journal*. He received a long report from Harriet Jessie Butler, as well as notes from her daughter Violet which we have already reproduced. Most of the family aspects of Keynes's article on Edgeworth come from this source. We do not know whether Harriet was prompted by Keynes's demand for biographical data or whether she had already decided together with her son Harold to edit *The Black Book*, written by her Irish great-grandfather Richard Edgeworth, followed by the *Memoirs of Richard Lovell Edgeworth Esq., Begun by Himself and Concluded by His Daughter Maria Edgeworth*, which had been published in 1820.[136] They added an epilogue on Harriet Butler (Harriet's and Francis's aunt) and on Francis Beaufort Edgeworth (Francis's father) with details of his December 1831 marriage to Rosa Florentina Eroles, and about his acquaintance with Carlyle in 1836. The book, published in 1927 under the title *The Black Book of Edgeworthstown & Other Edgeworth Memories 1585–1817*, is dedicated 'to Ysidro Francis Edgeworth'.

Meanwhile, Maynard Keynes received some more private information on Karl Pearson's opinion of Edgeworth from his friend the statistician G.U. Yule. Like Edgeworth and Yule, Keynes could not stand Pearson's egocentric personality. Yule wrote: 'Greenwood and I attended last Monday a terrible dinner in Pearson's lab at Univ. College when Pearson made some of his characteristically tactless and tasteless remarks about both Bateson and Edgeworth.'[137] Yule also stated that Greenwood asked him 'to support Isserli's suggestion that we should inscribe upon dear Edgeworth's tomb, "Here lies a mathematical statistician who was also a gentleman."'[138]

In spite of Karl Pearson's 'tactless and tasteless remarks', the Royal Statistical Society engaged L.L. Price to write an obituary of its President from 1912–1914 and Guy Medalist (in Gold, 1907), which was published in the next issue of the Journal of the Society and offered the lovable and realistic portrait of Edgeworth that we have referred to profusely.[139] The Society also acknowledged Edgeworth as 'a colleague so greatly valued and regarded with so much affection'.[140]

Edgeworth's memoir written by Maynard Keynes appeared in the March 1926 *Economic Journal* and strove to offer a touching portrait of the man, while at the same time, following the common appreciation of the day, it fell rather short in praising most of his academic works. This article became in fact the official profile of Edgeworth among economists. We have used and sometimes criticised the information contained in it, but we have to acknowledge our respect for Keynes as a biographer when he tries to portray Edgeworth as a person:

> On anyone who knew Edgeworth he must have made a strong individual impression as a person. But it is scarcely possible to portray him to those who did not. He was kind, affectionate, modest, self-depreciatory, humorous, with a sharp and candid eye for human nature; he was also reserved, angular, complicated, proud, and touchy, elaborately polite, courteous to the point of artificiality, absolutely unbending and unyielding in himself to the pressure of the outside world.[141]

At the end of 1926, also in the *Economic Journal*, James Bonar published a note called 'Memories of F. Y. Edgeworth', another of the texts which distils affection and that characterised Edgeworth's personality after 1890.[142]

Independently of these friends' memoirs, attention towards Edgeworth's works in political economy and statistics has varied over the years and their relevance has switched from fading away into oblivion to being exalted by present scholars. But the story of this academic projection goes beyond the overtly biographical character of this book.

Edgeworthstown

Francis Edgeworth was succeeded in the Edgeworthstown estates by his niece Maria E. Sanderson (Mia), eldest daughter of his sister Maria (Mary) Sanderson; she eventually transferred it to her sister Rosa, who was living there with her husband, Mr Cuthbert F. Montagu. In 1935 Rosa Montagu sold the estate together with 200 acres of land for £3000.

It was purchased by Mr Bernard Noonan, originally born near Edgeworthstown, who made a fortune in New York dealing in real estate. He is better known as the builder of Noonan Plaza, a giant apartment house at 105 West 168th Street, New York, one of the most modern and artistically ornamental buildings of its type in America. He partially renovated the manor house in Edgeworthstown and installed several modern conveniences. In a report written by Harriet Jessie Butler around 1938 she informs us:

[Francis] had been a good owner as far as he could, but he was unable to reside there and his death was sudden. His niece Maria Sanderson succeeded him, and her sister Rosa (Mrs Montagu) lived there for a while, but the house was too expensive for them, and the house was sold [in 1935] by the trustees to an enthusiastic Edgeworthstown man, named Noonan, who returned from America on purpose to buy it. He does not live there regularly, I think, but I have little knowledge of the place since Rosa left it. I visited it once in Francis's time, . . . It was much changed since the old days. [143]

NOTES

1. British Academy files. Edgeworth's election is cited by Newman (1987), p. 40.
2. EP NUFFIELD C UO, Box D 4. Leslie Stephen would die in February 1904. For his work as the editor of the first twenty-six volumes of the *Dictionary of National Biography* and his books *The Playground of Europe* (1871), *Hours in a Library* (1874–79) and *History of English Thought in the Eighteen Century* (1876); the biographies of *Johnson* (1878), *Pope* (1880) and *Swift* (1882); *Science of Ethics* (1882), *Life of Henry Fawcett* (1885), *An Agnostic's Apology* (1893) and *Studies of a Biographer* (1898–1902) among others, Stephen was named Knight Commander of Bath in 1902. See Annan (1984).
3. Whitaker (1996), III, p. 94.
4. Bowley (1934), p. 119.
5. Marshall to FYE, 21, 22 and 27 April 1909: Whitaker (1996), III, pp. 220–24; Pigou (1925), pp. 438–42.
6. FYE. to Marshall, 9 February 1907 and Marshall to FYE, 11 February 1907; Whitaker (1996), III, pp. 152–3.
7. Letter from Foxwell to Clara E. Collet, quoted by Coats (1972).
8. *EJ*, Vol. 14, (1904), p. 65. Quoted by Coase (1972).
9. *EJ*, Vol. 17 (1907), pp. 221–6.
10. Coats (1972), p. 494.
11. Roncaglia (2001), p. 372.
12. Jaffé (1965), Letter 1762. See also Section 3.14.
13. June 1909, Series 2, Vol 20, No. 6, 'Omaggio a Leone Walras'.
14. Jaffé (1965), Letter 1768, note 3.
15. Mirowski (1990), p. 589.
16. Marshall to FYE, 16 January 1912; see Whitaker (1996), III, pp. 290–291. Quoted also in Mirowski (1990), p. 594 and Mirowski (1994), p. 430, but erroneously presented in both works as a letter from Edgeworth to Marshall.
17. *EJ*, March 1912. Reprinted in Mirowski (1994), pp. 392–396.
18. *EJ*, June 1912. Reprinted in Mirowski (1994), p. 400.
19. Moore to Marshall, 6 June 1912: Whitaker (1996), III, p. 290.
20. Horace, *Satires*, Book I, satire IV, v. 61.
21. See Section 2.19 above.
22. Review of R.P. Graves's *Life of Sir William Rowan Hamilton*, 26 October 1889, *The Academy*.
23. Letter from W.R. Amon to FYE, 21 August 1906 (EP NUFFIELD C UO, Box D 4).
24. Bonar sent this letter dated 18 January 1908 to J.M. Keynes on 30 October 1926 (KP KING'S C CAMBRIDGE, EJ/6/6).
25. Bonar (1926), p. 648.
26. H.J.B.'s report (EP NUFFIELD C UO, Box E 3, 9/1).
27. J.M. Keynes wrongly reports that he was the 'fifth son of a sixth son': Keynes (1926), p. 251.
28. Kept in the Keynes Papers (KP KING'S C CAMBRIDGE MA, L/M).

29. EP NUFFIELD C UO, Box E 3/9.
30. Harold E. Butler to H.J.B., from Gordon Square (EP NUFFIELD C UO, Box D 11).
31. Reported in February 1970 (EP NUFFIELD C UO, Box E 3/10).
32. In 1976 her disciples produced a book of essays, *Traditions of Social Policy: Essays in Honour of Violet Butler*, where she declared that she had 'received much help from F. Y. Edgeworth: he behaved rather like an uncle . . . He was very kind to me.' This homage book also stressed that Edgeworth 'was interested in the question of unequal wages between men and women, a subject which Miss Butler was soon to study in some detail'. We shall further examine this topic in Section 4.12. See Halsey (1976), pp. 43 and 45.
33. See, for instance Sully's return addresses on his letters in the Browning Papers kept at King's College, Cambridge, Modern Archives.
34. EP NUFFIELD C UO, D 4/10.
35. EP NUFFIELD C UO, D 4/11.
36. Whitaker (1996), III, pp. 285–286. This letter contradicts Mirowski's theory that it was a Cambridge plot against Edgeworth to name J.M. Keynes as the *EJ* editor.
37. Royal Economic Society's Minute Books, Vol. II, 17 October 1911: Reported by Moggridge (1992), p. 208.
38. KP KING'S C CAMBRIDGE MA, EJ/6/6/2nd part.
39. Harrod (1951), p. 185.
40. Royal Economic Society's Minute Books, Vol. II, 1 October 1915: Reported by Moggridge (1992), p. 253.
41. In 1919, for instance, Edgeworth reviewed the book *La Hacienda Española*: *Los Impuestos* by Professor F. Bernis (University of Salamanca), who is mentioned every year as correspondent in Spain.
42. Harrod (1951), p. 438.
43. Letter from F.Y.E. to J.M. Keynes, 20 May 1913 (KP KING'S C CAMBRIDGE MA, EJ/6/6/2nd part).
44. 'To the posterity the arduous sentence' (Alessandro Manzoni referring to Napoleon).
45. The poet, not mentioned by Edgeworth, is Lord Tennyson (1809–1892).
46. Reproduced in Edgeworth (1925), I, pp. 172–91 and 429–91, II, pp. 344–68 and pp. 384–405.
47. Not to be confounded with G.B. Antonelli, the Italian author of the first work on utility's integrability, *Sulla Teoria Matematica della Economia Politica* (1886), prior to I. Fisher (1892). Étienne Antonelli was French and in p. 66 of the book referred to – *Principes d'Économie Pure* (1914) – he includes the letter from Poincaré to Walras mentioned below.
48. Edgeworth (1925), II, pp. 322–3. See Schmidt and Weber (2008).
49. Edgeworth (1925), II, p. 473.
50. Edgeworth (1925), II, p. 475.
51. A.J. Dupuit, 'De la Mesure del'Utilité des Travaux Publics', 'De l'Influence des Péages sur l'Utilité', 'De l'Utilité et de sa Mesure', *Annales des Ponts et Chaussées*, 1844, 1849 and 1853; Marshall (1890); Pigou, 'Monopoly and Consumer's Surplus', *EJ*, 1904, p. 388, *Wealth and Welfare* (1912). See Creedy (1986), Chs 6 and 7.
52. Dupuit, op.cit.
53. Edgeworth (1925), I, p. 172.
54. Edgeworth (1925), I, pp. 172–91.
55. Bonar (1926), pp. 647–8.
56. *EJ*, August and November 1914, pp. 460–96 and pp. 610–34.
57. Edgeworth's lecture is included in Newman (2003), pp. 340–60.
58. Newman (2003), p. 340.
59. 'Patrie', in Voltaire's *Dictionnaire Philosophique*.
60. Newman (2003), p. 359.
61. Newman (2003), p. 360.
62. Edgeworth quotes his work *Interpretation of Radium*.
63. In a separate review from the same year, 1915, Francis dealt with *L'Economia del*

Mondo, Prima, Durante e Dopo la Guerra Europea by M. Alberti (Italian). In 1916 he reviewed *The Economy and the Finance of the War* by A.C. Pigou (British) and *La Germania all Conquista dell'Italia* by G. Preziosi (Italian). And in 1917 he covered *Some German Economic Writings about the War* with contributions by several authors (German), *After War Problems* with contributions by several authors, including A. Marshall (British); *National Power and Prosperity: A Study of the Causes of Modern Warfare* by C. Gill (British); *Economics in the Light of War* by R.A. Lehfeldt (South African); *War Finance Primer* by the National Bank of Commerce, New York (American) and *A Constructive Criticism of the U. S. War Tax Bill* by E.R.A. Seligman (American). See Appendox M.

64. Edgeworth (1925), III, pp. 194–201.
65. Edgeworth (1925), III, pp. 189–94.
66. Edgeworth (1925), III, pp. 212–14.
67. Edgeworth (1925), III, pp. 214–15.
68. Oxford University and Clarendon Press, 1919. The first lecture is included in Newman (2003), pp. 340–60.
69. EP NUFFIELD C UO, Box D 4/12. See Edgeworth's photograph at the entrance of Chapter 4 in a military uniform. On his chest is pinned a cloth cross. Did he enlist as Voluntary Aid for the Red Cross? He also wears a dark arm-band, which may well be red.
70. In Newman (2003), the posthumous editors – Margot Levy and Donald Winch – label the photograph at the begining of this chapter: 'F.Y. Edgeworth in an unidentified uniform'.
71. Petridis (1987), 'Price, L.L.'.
72. Price (1926).
73. Whitaker (1996), III, p. 382–3, n. 3.
74. Creedy (1986), p. 10 and Newman (1987), p. 40.
75. Kadish, writing in 1982, affirms that 'he was the last man to bring about too much needed change in the institutional status of Economics in Oxford. His concept of the nature and method of Economics was essentially different from the concepts of the young Oxford economists and his aversion to university politics was detrimental to any efforts to change the academic status of the young economists' careers. . . . Once he started lecturing he soon outreached the capacity of his listeners', Kadish, (1982), pp. 201–7. G. M. Koot writes in 1987: 'Edgeworth's tenure prevented the flowering of promising developments in historical economics at Oxford. Moreover, his unassertive personality, his work in obscure mathematical economics, and his deference to Marshall also prevented the creation of a theoretical school of economics at Oxford': G.M. Koot, *English Historical Economics, 1870–1926*, Cambridge University Press, 1987, p. 92, quoted in Reisman (1990), p. 197. In contrast to them, Creedy writes: 'I would be critical of Edgeworth if he had made such a bad judgment as to neglect his work for "politicking" in Oxford.'
76. KP KING'S C CAMBRIDGE MA, ED/6/6/29.
77. Whitaker (1996), III, p. 296.
78. Creedy (1986), p. 10 and Newman (1987), p. 40.
79. The articles on pure statistical techniques were 'Untried Methods of Representing Frequency' (1924) and 'Mr. Rhodes' Curve and the Method of Adjustment' (1926, posthumously published). Other articles published during this last period in the *Journal of the Royal Statistical Society* were 'Psychical Research and Statistical Method' (1919) and 'Statistics of Examinations' (1923).
80. Such as those of W. Zawadski's *Les Mathématiques Appliquées à l'Économie Politique* (1914); H. Westergaard's *Scope and Methods of Statistics* (1917); A.L. Bowley's *Change in Distribution of National Income* (1920); J.M. Keynes's *A Treatise on Probability* (1922) and G.U. Yule's *A Mathematical Theory of Evolution based on the Conclusions of Dr. J.C. Willis* (1925).
81. Moreover, in the last period 1921–26, Edgeworth published 'The Philosophy of Chance' (1922, repeating the title of his 1884 paper) in *Mind*, (see below Section

4.11) and 'On the Use of Medians for Reducing Observations Relating to Several Quantities' (1923) in the *Philosophical Magazine*.

82. Verses from *La Divina Commedia*, Purg. X, 124–6, quoted by Bonar (1926), p. 649. [Don't you know that we are worms, born to become the angelic butterfly that flies towards justice without shield?]
83. Quoted in Pearson and Kendall (1970), pp. 202 and 204.
84. Kendall (1968), pp. 274–5.
85. This is not the opinion of Egon S. Pearson, son and biographer of Karl, who in 1970 describes a close relationship between K. Pearson and Edgeworth noting that 'they were both influenced by Francis Galton, they were certainly in personal contact and they were working at the same time on asymmetrical frequency curves and multiple correlation. They were, however, both of them clearly originators with very distinct lines of approach and any attempt to assess priorities would be meaningless': Pearson and Kendall, (1970), pp. 339–40. See also Stigler (1978), p. 299. Stigler writes: 'He [K. Pearson] was, I think, overtly defensive and insufficiently generous'.
86. EP NUFFIELD C UO, Box D 1.
87. Keynes (1926), in Vol. X of Keynes (1971–1989), p. 254.
88. Letters from F.Y.E. to H.J.B. 27 July (1917?), 19 and 20 August (1917?), 16 October 1917, 9 July 1919 (EP NUFFIELD C UO, Box D 1).
89. EP NUFFIELD C UO, Box D 1.
90. Keynes, (1921), in Vol. VIII of Keynes (1971–1989), pp. 30 n., 105 n. and 413. The quotations from Edgeworth, taken from 'Philosophy of Chance', *Mind*, 1884, pp. 225 and 223, were mentioned at Section 2.12 above.
91. Also mentioned in Section 4.9, n. 81, above.
92. Mirowski (1994), pp. 344–8.
93. Mirowski (1994), p. 347.
94. Edgeworth, 'The Philosophy of Chance', in Mirowski (1994), pp. 347 and 348.
95. A. Young, 'Fisher's "The Making of Index Numbers"', *Quarterly Journal of Economics*, (1924), pp. 342–64. See also Correa Walsh, 'Professor Edgeworth's Views on Index-Numbers', *QJE*, (1924), pp. 500–519; and I. Fisher, 'Professor Young on Index Numbers', *QJE* (1923), pp. 742–55; See also Mirowski (1994), pp. 351–69.
96. F.Y. Edgeworth, 'New Methods of Measuring Variation in General Prices', in Mirowski (1994), p. 211. Included also in Vol. 4 of Darnell (1991).
97. 'The Rationale of Exchange', examined above in Section 2.12.
98. Mirowski (1994), p. 364.
99. Keynes (1926), in Vol. X of Keynes (1971–1989), pp. 264–5.
100. B. Ohlin, 'Autobiography', http://nobelprize.org/nobel_prizes/economics/laureates/1977/online-autobio.html (accessed 12 October 2009).
101. B. Ohlin, 'The Reparation Problem: A discussion', *EJ*, 1929.
102. This is the title of Edgeworth's article updating in Edgeworth (1925) his own 'Theory of International Values' dating from 1894.
103. *EJ*, 1920, reprinted in Edgeworth (1925), II, pp. 261 and 270.
104. EP NUFFIELD C UO, Box C 4.
105. Virgil, *Æneid*, II, pp. 291–292. 'If [the towers of Troy] could be defended by a right hand, they would have been defended by this right hand', that is, Hector's. Interpretation by P. Newman: see Newman (2003), p. 376.
106. 'Uncounselled, men depart and loathe the Sibyl's seat', Virgil, *Æneid*, III, verse 452.
107. Keynes (1926), in Vol. X of Keynes (1971–1989), pp. 265–266.
108. F.Y.E. to J.M. Keynes, 13 February 1925 (exactly one year before F.Y.E's death) (KP KING'S C CAMBRIDGE MA, ED/6/6).
109. (KP KING'S C CAMBRIDGE MA, EJ/6/6/29). See below, the obituary in *The Times*, 16 February 1926.
110. *Chambers's Biographical Dictionary*.
111. Keynes (1924), in Vol. X of Keynes (1971–89), p. 231.
112. Whitaker (1996), III, p. 390.

113. Edgeworth to Harrod, 29 May 29 1924 (Harrod's correspondence).
114. R. Harrod to J.M. Keynes, 25 March 1926 (KP KING'S C CAMBRIDGE MA, EJ/6/6/53).
115. Harrod (1951), p. 438.
116. Harrod, retirement speech to the Sub-Faculty of Economics, 1967, quoted in Young and Lee, *Oxford Economics and Oxford Economists*, 1993, pp. viii–ix.
117. We find the reviews of the books by J.N. Keynes (*The Scope and Method of Political Economy*), Marshall (*Principles*, 2nd ed. and *Taxation after the War*), Sidgwick, Böhm-Bawerk (*The Positive Theory of Capital*), Fisher (*Mathematical Investigations in the Theory of Values and Prices*; *Infinitesimal Calculus* and *Cournot's Mathematical Principles*), Bastable (*Public Finance*), the Webbs (*History of Trade Unionism*), Price, Nicholson, Bonar and Hollander, J.B. Clark, Darwin, H.G. Wells (*Anticipation of the Reaction of Mechanical and Scientific Progress upon Human Life and Thought*, 1902), Pigou (*Wealth and Welfare*), Bowley, Cunynghame, Carver, Taussig, Mitchell, Ashley, J.M. Clark, Mrs Fawcett (*Women in Economic Life*, 1918), Cannan, Loria, and more.
118. Keynes (1926), in Vol. X of Keynes (1971–1989), p. 263.
119. F.Y.E. to H.J.B., 18 March 1925 from The Atheneum, Pall Mall, SW1 (EP NUFFIELD C UO, Box D 1).
120. Palgrave (1925–26), III, pp. 710, 711, 713.
121. Pakenham (2000), pp. 10 and 175.
122. EP NUFFIELD C UO, Box D 1.
123. Notes from H.J.B. to J.M. Keynes, 23 April 1926 (KP KING'S C CAMBRIDGE MA, ED/6/6).
124. Kendall (1968), pp. 274–5.
125. EP NUFFIELD C UO, Box D 1.
126. EP NUFFIELD C UO, Box D 1.
127. Price (1926), pp. 371–7.
128. KP KING'S C CAMBRIDGE MA, EJ/6/6.
129. EP NUFFIELD C UO, Box D 1.
130. KP KING'S C CAMBRIDGE MA, EJ/6/6.
131. Quoted by Creedy (1986), pp. 12–13.
132. EP NUFFIELD C UO, Box D 11.
133. KP KING'S C CAMBRIDGE MA, EJ/6/6.
134. EP NUFFIELD C UO, Box E 3.
135. KP KING'S C CAMBRIDGE MA, EJ/6/6.
136. Printed for R. Hunter, London. Second edition 1821. Third edition Richard Bentley, London, 1856.
137. William Bateson (1861–1926), Professor of Biology, had also died. He had maintained a Mendelian viewpoint against Pearson and the biometricians.
138. G.U. Yule to J.M. Keynes, 26 February 1926 (KP KING'S C CAMBRIDGE MA, EJ/6/6/35).
139. Price (1926).
140. EP NUFFIELD C UO, Box D 11.
141. Keynes (1926), in Vol. X of Keynes (1971–1989), p. 265.
142. Bonar (1926).
143. EP NUFFIELD C UO, Box E 3/9. In 1939 the house together with fifty acres of land was donated by Mr Noonan to the Sisters of Mercy, a Roman Catholic congregation. Certain clauses of the donation stipulated that the manor house was permitted to be used by the army during the war and the Sisters could not take possession of it until 1947, by which time the house, despite Mr Noonan's renovations, was in an advanced state of disrepair. The house was then converted into what is now, Our Lady's Nursing Home, a residence for the elderly.

Last picture of Francis Ysidro Edgeworth, in 1925

Reproduced with permission from Dr David Edgeworth Butler

Appendices

APPENDIX A

Questionnaire of the Central Relief Committee completed on 1 January 1847 by Maria and Frances Anne Edgeworth and their neighbour Mr Parnell:

Parish of Edgeworthstown, *Co.* Longford. *Extent:* Around 5 [sq.] miles, appr. 5000 inhabitants of which 3000 stand in need of public relief. *Employed in the ordinary manner:* appr. 100, wage 8 to 10d per day. *Employed in public works:* appr. 400. *Earnings not sufficient to preserve their families from want?* Not at the present price of food. *Are there manufactures?* No. No fishing. *Farm's size:* Small. Tillage, not grazing. *Are the small farmers cultivating their ground?* They are not, nor can they as all their labour is going to procure provisions. *Any stock of potatoes stored?* No. *Is it expected to plant a considerable quantity of potatoes for next season?* No. *Persons incapable of labour?* More than 500 (widows, children and old people). *In what Poor Law Union does the place lie?* Longford. *How far is the Poor House?* 7 miles and it is full. *How much were the last poor rates?* 7½ and 5d. *Private subscriptions towards relief?* £186, plus £92 contributed by the Government. The Irish Relief Committee have granted £20 for a soup shop. *How these funds have been applied?* In selling products at a reasonable rate to the poor since Sept. last. *Absentee proprietors and their contributions:* Charles Sneyd Edgeworth subscribed £25, Mr. Tuite subscribed £19. Mr. C.S. Edgeworth has a resident agent. Moreover there is one resident proprietor with a few large farmers. They have subscribed to the Relief Fund. No other relief associations. All the families relieved have been visited before. There is a great deal of sickness [consumption] and some fever [cholera]. *Nearest storage place:* Longford, 7 miles. *Soup shop established?* Yes, it distributes 30 gallons a week, at 1 pence per quart. *An increased supply of soup [. . .]?* Yes, without increasing prices. *Reference:* Mr. John K. Pomele[?] , Secr. of Relief Committee. Signed: Maria Edgeworth, Frances Anne Edgeworth and Mr. Parnell.

Additional observation: The want of shoes is great and affects health and the poorer labour in draining work. Addressed to Dr. Harvey.[1]

APPENDIX B

Ysidro Francis Edgeworth's academic record in the registers of Trinity College, Dublin:

1 July 1861. Entrance exams results: English Composition 7; Latin Composition 10; Greek Verse 10; Greek Prose 10; Latin Verse 10; Latin Prose 10; History 5: Arithmetic and Algebra 7. Average 8.6 = No. 1.

Trinity Term of 1862. Junior Freshman class exams: Arithmetic and Algebra 9, Greek 10, Latin 10. Composition 8. Average 9.1 = First.

Michaelmas Term of 1862. Honours exams: Mathematics 0.405 (3rd place), Classics 0.620 (3rd place). Junior Freshman class exams: Arithmetic and Algebra 6, Greek 9, Latin 10, Composition 10. Average 9 = First.

Hilary Term of 1863. Senior Freshman class exams: Greek 10, Latin 10, Composition 10, Logic 7, Mathematics 5. Average 8.4 = First.

Michaelmas Term of 1863. Honours exams: Classics 0.795 (1st place).

Hilary Term of 1864. Honours exams: Classics 0.745 (1st place). Junior Sophister class exams: Logic 5.5; History 4, Greek 9.5; Latin 9, Composition 6. Average 6.8.

Trinity Term of 1864. Honours exams: Classics 0.705 (3rd place).

Michaelmas Term of 1864. Honours exams: Classics 0.765 (2nd place). Senior Sophister class exams: Latin 10, Greek 8.5, Composition 6, Physics 5, Mathematics 4, Logic 2, Astronomy 1. Average 5.2.

Hilary Term of 1865. Honours exams: Ethics 0.544 (4th place). Senior Sophister class exams: Greek 10, Ethics 6.5, Latin 6, Composition 6, Mathematics 4, Astronomy 1.5. Average 5.6.

Michaelmas Term of 1865. Candidate Bachelor exams: Physics 10, Greek 10, Composition 10, Latin 9, Ethics 8, Mathematics 6, Astronomy 5.5. Average 8.3. = First and No. 1.[2]

APPENDIX C

Memory of Frances Anne Edgeworth distributed at her funeral:

Frances Anne was born July 28, 1769 at Flower Hill, near Navan (Co. Meath), where her grandfather Daniel Cornelius Beaufort was rector. He had come from Holland where his brother Louis, author of *Incertitude de l'Histoire Romaine*, continued to reside.

Daniel Cornelius with son Daniel Augustus [1739–1821], also a clergyman, who married in 1767 Mary Waller, daughter of William Waller Esq.

of Allenstown (Co. Meath) lived for some years with them and their young family. in Wales, first at Penyan and then at Piercefield, the beautiful seat of Mr. Morris. The two eldest children, thoroughly well instructed, were Fanny [Frances Anne] and William.

After the death of grandfather at 89 years old, Fanny was with father and mother in London, where she had the best masters for painting. Daniel Augustus returned to Ireland with all the family where he was appointed at the Vicarage of Collon (Co. Louth) and rector of Navan where he succeeded his father. The walls of the sitting rooms at the Vicarage were hung with her paintings. Her father possessed much knowledge in every subject. They resided at Flower Hill, near Navan and became intimate with Mrs. Ruxton and with Mr. Edgeworth, often at Black Castle. On 31st May, 1798 Miss Beaufort was married to Richard Lovell Edgeworth. They went from church St Anne's in Dublin to Edgeworthstown with the children. She had 6 children of her own, but she never biased or lessened her affection for the others.

In 1803 she accompanied Mr. Edgeworth, Miss Maria Edgeworth and Miss Charlotte Edgeworth to Paris. In 1813 visited London with her husband and Maria Edgeworth, getting acquainted with interesting people.

In 1817, at Mr. Edgeworth's death, his son and successor, Lovell, requested his stepmother to remain with her children at Edgeworthstown and there the Miss Sneyds and Maria Edgeworth continued to reside as long as they lived. Maria Edgeworth died in 1849 but Mrs. Edgeworth's life was prolonged sixteen more years. She inherited from her mother a serene mind. She was conversing with her usual animation on the 1st of Feb., when about 5.00 o'clock in the evening she lost suddenly her power of speech. But she preserved her consciousness and continued by her looks showing that she distinguished the countenances of her children and grandchildren, till the 10th when she expired without pain or struggle at 8.00 o'clock in the evening at 95 years old.[3]

APPENDIX D

Will of Frances Anne Edgeworth:

I nominate my son Michael Pakenham Edgeworth, my grandson David Reid Edgeworth and William Webb of Castle Nugent [old agent of both Sneyd and Pakenham], Esquires, Trustees and Executors of this my will. The proceeds to arise are to be applied and disposed of by my said trustees and executors to and for the use and benefit of my three younger grandchildren in equal shares and proportions, namely David Reid Edgeworth, Richard Lestock Edgeworth and Ysidro Francis Edgeworth and the survivor or survivors of them, but it is my will that in case any or either of my said three grandsons shall, by death or failure of issue of their elder brother Antonio happen to succeed to and become owner of Edgeworthstown house with the settled estates, such child so succeeding and being amply provided for shall from thenceforth give up and relinquish all right under this my will. . . . I give and bequeath the whole of the stock and capital now held by me in the Imperial Continental Gas Comp. to my granddaughter Maria. Edgeworthstown furniture for Michael Pakenham Edgeworth, Harriet and

Lucy. Books for my children and grandchildren in the way they select. Silver to Michael Pakenham Edgeworth. To sister Harriet Beaufort the engraving of Sir Francis Beaufort. This Sept. 6, 1864.[4]

APPENDIX E

'Memory of David Reid Edgeworth', published in the *Annual Report of the Institution of Civil Engineers*, 17 December 1872:

David Reid Edgeworth, fourth son of Mr. Francis Beaufort Edgeworth, was born at Edgeworthstown the 14th of April 1842. Educated at home with his three brothers till his 14th year, he entered the Royal Navy as midshipman in Summer 1855, going out first at the entrance examination. After one and a half year partly spent in HMS *Firebrand* at the Crimea, he left the Navy disliking the monotony of the life. He was then articled to Mr. G.W. Hemans (Vice-President of the *Institution of Civil Engineers* and on the termination of his pupilage in 1860 he was employed by Mr. Hemans on various lines of railway in Ireland, being successively: Assistant Engineer for one year on the Longford and Sligo railway, Resident Engineer, for two years on the Streamstown and Clara railway, for one year on the Athenry and Ennis railway and District Engineer for one and a half year on the Enniskillen, Bundoran and Sligo railway. Latterly became interested in the reclamation and cultivation of lands on the coast of Wexford. He died on the 14th of October 1871, from bronchial inflammation induced by exertions during bad weather while inspecting reclamation works.

David Reid Edgeworth had been elected a Member of the Institution on the 4th of Feb. 1868.[5]

APPENDIX F

Statements by the executors of David Reid Edgeworth and Richard Lestock Edgeworth's estates:

1. Executors' Statement. 'In Re David Reid Edgeworth, Esq. Deceased 14 October 1871.' This was a financial document about the principal of the estate of David Reid Edgeworth at his death: Total assets £1,256/5/1[6] and total liabilities, paid by executors, £1,650/2/6, among them 'Funeral expenses' £39.

In addition a sum of £500 was due by the Deceased to Mr. Francis Edgeworth, being share of a charge of £3,000 on Antonio Eroles Edgeworth's Estate [Edgeworthstown] to which Deceased and Mr. Francis Edgeworth were jointly entitled and which he joined Deceased in mortgaging to raise £2,000. Only outstanding assets: Lands of Curry held for 999 years from 1754 subject to a head rent of £35/9/10 and a mortgage of £1,000. This would be the only asset available to meet the above mentioned sum due to Y.F. Edgeworth and the sum of £545/18/6 paid by A. Eroles Edgeworth [to the executors, which included the £35 that David Reid debited to Richard Lestock mentioned in the following.]

2. Executors' Statement 'In Re Richard L. [Lestock] Edgeworth. Died 8th May 1869' (prepared at the same time as David's in October 1871). This was a financial document about the principal of the estate of Richard Lestock Edgeworth at his death: Total assets £362/3/5 including £35 of 'Cash due by David Reid Edgeworth'. Total liabilities £148/8/6, among them 'Funeral expenses paid by the British Consul at Naples £85'.[7]

APPENDIX G

Memory of Michael Pakenham Edgeworth:

Deceased July 30, 1881. [Born in May 2, 1812]. In 1823 he entered Charterhouse, in 1827 he went to Edinburgh where he began to learn Oriental languages and Studied Botany.

After a very distinguished career at Haileybury [Nr. 1 of his class at the East Indies' College where, at that time, T.R. Malthus was Professor of Political Economy], he went to India in the Civil Service. He was stationed at Scharunpore and at Ambala where his administration received the high approval of his superiors and the grateful appreciation of the Natives. He was present at the interview between Lord Auckland and Runjut Singh with whom he had by a chance a remarkable *tête à tête*. In 1842 came home on leave and in [Feb.] 1846 he married Christina MacPherson, daughter of Dr. MacPherson H.C. of King's College, Aberdeen, returned to India where he was stationed at Banda till 1850, when he was selected one of the five Commissioners for the settlement of the Punjab, first at Mooltan and afterwards at Julhundhur, leaving that country owing to a sunstroke in 1855. To his deep regret he was never able to return. Resided after 1856 in Anerly, but in 1877, on the marriage of his only daughter, he left Anerly and took a house in London. He died suddenly in the Island of Eigg on the 30th July 1881. He published many papers in the *Asiatic Journal of Calcutta*, a grammar of Cashmiri, also in the [. . .Sirdan?] Society and lastly a book on Pollen in 1878.[8]

APPENDIX H

Francis Ysidro Edgeworth's report to the Electors of Professorship of Political Economy at the University of Oxford, the Drummond Chair, vacant due to the death of Thorold Rogers on 12 October 1890.[9]

I am a graduate of Oxford which I entered in 1867, bringing with me several terms kept at Trinity College, Dublin, being permitted to count them as terms kept in Oxford.

I had won at T.C.D. many of the distinctions which may be obtained for proficiency in Classics in the earlier part of the course, for instance a Scholarship, the first prizes in Greek Prose and Verse Composition and the first place in the first class in two examinations for Honours in Classics. In Oxford I took a First Class at the final Examination in Literæ Humaniores, 1869. After Oxford, I studied Mathematics for some years. I am the author of several papers on mathematical subjects: relating to the Calculus of Probabilities and the abstract theory of Statistics published in the *London Philosophical Magazine*, *Transactions* of the Cambridge Philosophical Society and the *Journal of the Statistical Society*.

I have endeavoured to apply mathematical conceptions to Political Economy in *Mathematical Psychics*, 1881, . . . to apply abstract reasoning to the measuring of the variations in the value of money (Economic Section of the British Association, 1887, 1888, 1889, as Secretary of the Committee). I have also acted as Secretary for a Committee appointed by the British Association for enquiring and reporting on the amount of the precious metals in use as money. In 1889 I was President of the Economic Section of the British Association. Presidential Address: 'On the Application of Mathematics to Political Economy'.

I have not confined my studies to the abstract side of Economic Science. I have studied and reviewed many works of branches of Political Economy in *Academy*, *Nature*, *Journal of Education*, *Giornale degli Economisti* (edited by M. Pantaleoni) and *Revue d'Économie Politique* (edited by C. Gide).

I have had much practice in teaching. I lectured for two years at Mr. Walter Wren's institution, on Logic and Mental and Moral Science. I have been employed as a Lecturer on Logic and the cognate subjects prescribed for the B.A. and M.A. Examinations of the London University for ten years at King's College, London. During these years I have, from time to time, given a course of elementary lectures on Political Economy at the Lady's Department of King's College. In 1888 I was appointed to a Professorship of Political Economy at King's College: a chair which had remained vacant since it had been resigned by Senior. In the present year [1890] I was elected to the Tooke Professorship of Economic Science and Statistics, on its being resigned by the late Prof. Thorold Rogers. . . .

I am a member of Balliol College, M.A. of University of Oxford, D.C.L. of University of Durham; Vice-President of the Royal Statistical Society, Examiner in Political Economy for the College of Preceptors; member of the organising Committee, as being an Ex-President of the Economic Section of the British Association; member of the International Statistical Institute; Secretary of the British Economic Association and editor of its *Journal*.

APPENDIX I

List of the reviews, articles and notes published by F.Y. Edgeworth in the *Economic Journal* during the period from 1891 to 1895.

Reviews

1891

March: First issue of the *Economic Journal*. The first volume from 1891 includes reviews by F.Y.E. of J. Neville Keynes *The Scope and Method of Political Economy*; Alfred Marshall *Principles of Economics* (2nd ed.); and Henry Sidgwick *The Elements of Politics*.

1892

F.Y.E. reviews Margaret Benson *Capital, Labour and Trade and the Outlook*; Eugene Böhm-Bawerk *The Positive Theory of Capital*; Richard Cantillon's *Essai sur le Commerce*; Luigi Cossa *Introduzione allo Studio dell'Economia Politica*; Dr C. Dusing *Das Geschlechtverhältniss der Geburten in Preussen*; J. Jacobs *Studies in Jewish Statistics*; John Biddulph Martin *The Grasshopper in Lombard St.*; William Smart *An Introduction to the Theory of Value on the Lines of Menger, Wieser and Böhm-Bawerk*; and W.F. Willcox *The Divorce Problem*.

1893

Edgeworth reviews Irving Fisher *Mathematical Investigations in the Theory of Value and Prices*; Fred E. Haynes *The Reciprocity Treaty with Canada of 1854*; Prof. T. Hertska *Freiland*; J.B. Martin *The Currency of the United States*; Sir Rawson Rawson *Analysis of the Maritime Trade of the U.K.*; W. Smart's *Women's Wages*; H.M. Thompson *The Theory of Wages*; and Dr Walsh *Bimetallism and Monometallism*.

1894

F.Y.E. reviews Lord Brassey *Papers and Addresses*; the contents of the *Bulletin de l'Institut Internacional de Statistique*; J.R. Commons *The Distribution of Wealth*; A. Duckworth *Comparison of Populations and Rates of Mortality in New South Wales and Victoria*; T.G. Spyers *The Labour Question*; Sydney and Beatrice Webb *The History of Trade Unionism*; and F. von Wieser *Natural Value*.

1895

F.Y.E. reviews *Economic Classics*, edited by Prof. W.J. Ashley; Edward Atkinson *The Use and Abuse of Legal Tender Acts*; G. Jamieson, T.H.

Box and D.O. Coal *The Silver Question*; Simon J. Maclean *The Tariff History of Canada*; Alfred Marshall *Principles of Economics*; George V. Mayr *Statistik und Gesellschafstlehre*; and Prof. F.S. Nitti *La Misura delle Variazioni di Valore della Moneta.*

Articles and Notes

1891
'An Introductory Lecture on Political Economy' (delivered at the University of Oxford, 23 October).

1892
Note: 'Recent Attempts to Evaluate the Amount of Coin Circulating in a Country'.

1894
'The Theory of International Values' (69 pages in three parts); notes on 'Prof. J.S. Nicholson on "Consumer Rent"'; 'Recent Writings on Index Numbers' (Included in Edgeworth (1925), I, pp. 344–50 under the title 'Variorum Notes on Index-Numbers'); 'The Measurement of Utility by Money'; 'Prof. Böhm-Bawerk on the Ultimate Standard of Value'.

1895
Notes: 'Mr Pierson on Scarcity of Gold' (Edgeworth (1925), I, pp. 351–5); 'Thoughts on Monetary Reform' (18 pages, Included in Edgeworth (1925), I, pp. 421–42 as 'Questions connected with Bimetallism); 'The Stationary State in Japan'.

APPENDIX J

Entries by F.Y. Edgeworth in *Palgrave's Dictionary of Political Economy*:

1. On political economy and economics: 'Absentee', 'Agents of Production','Bailey, Samuel, on Value', 'Barter and Exchange', 'Bastiat as a Theorist', 'Bounties', 'By-products, Theory of Value of', 'Competition and Regulation', 'Curves', 'Debasement of Coin', 'Deferred Payments', 'Demand Curves', 'Depreciation of Monetary Standard', 'Difficulty of Attainment', 'Distance in Time as an Element of Value', 'Doctrinaire', 'Doses of Capital', 'Efficiency of Money', 'Elasticity', 'Exchange, Value in', 'Facts', 'Fallacies', 'Fixed Incomes', 'Forced Currency', 'Functions', 'Higgling of the Market', 'Ideal Money', 'Income', 'Inconvertible

Currency', 'Index Numbers', 'Indifference, Law of', 'Indirect Utility', 'Intrinsic Value', 'Joint Production', 'Luxury', 'Margin (in Economics)', 'Mathematical Method in Political Economy', 'Maximum Satisfaction', 'Monopoly', 'Multiplication of Services', 'Negative Quantities', 'Numerical Determination of the Laws of Utility', 'Over-production', 'Pleasure and Pain', 'Total Utility', 'Unit of Value', 'Utility', 'Wealth'.

2. On statistics and actuarial science: 'Aleatory', 'Average', 'Birth Rate', 'Census', 'Death Rate', 'Duration of Life as an Element of Well-being', 'Error, Law of', 'Luck', 'Marriage Rate', 'Means, Method of', 'Probability and Calculus of Probabilities', 'Risk'.
3. Biographies: 'Cournot, A.A.', 'De Moivre, A.', 'De Quincey, T.', 'Dupont de Nemours, P.S.', 'Dupuit, A.J.E.', 'Gossen, H.H.', 'Hagen, K.H.', 'Helferich, J.', 'Jenkin, H.C.F.', 'Jennings, R.', 'Jones, R.', 'Mill, J.', 'Mill, J.S.', 'Rae, J.'.
4. Short biographical references about economists, philosophers, engineers, statisticians and mathematicians: 'Aickin, J', 'Attwood, T.', 'Baxter, R.D.', 'Beldam', 'Berkeley, G.', 'Blake W.', 'Brassey, T.', 'Brindley, J.', 'Brougham, H.', Buckle, T.', 'Buquoy, G.F.' 'Buridan, J.', 'Burke, E.', Burlamaqui, J.J.', 'Camerarius, J.', 'Campanella, T.', 'Cary, J.', 'Cayley, E.', 'Child, J.', 'Clarkson, T.', 'Clayton, D.', 'Colquhoun, P.', 'Conduitt, J.', 'Corbet, T.', 'Cotton, R.B.', 'Cowell, J.W.', 'Cradocke, F.', 'Craig J.', 'Croumpe, S.', 'Culpeper, T.', 'Deparcieux, A.', 'Dieterici, K.F.W.', 'Eden, F.M.', 'Eliot, F.P.', 'Ellis, W.', 'Fullarton, John', 'Hearn, W.E.', 'Peacock, G.', 'Playfair, W.', 'Porphyry'.

APPENDIX K

List of the reviews, articles and notes published by F.Y. Edgeworth in the *Economic Journal* during the period from 1896 to 1905.

Reviews

1896

F.Y.E. reviews J.S. Nicholson, *Strikes and Social Problems*; N.G. Pierson, *Leerboek der Staathuisondkunde*, Vol. I; and L.L. Price, *Money and its Relation to Prices*.

1897

F.Y.E. reviews C.F. Bastable *The Theory of International Trade*; A. Graziani, *Istituzioni di Scienza delle Finanze*; Six Oxford Men's *Essays on Liberalism*; and N.G. Pierson *Leerboek der Staathuisondkunde*, Vol. II.

1898

F.Y.E. reviews M. de Cérenville *Les Impôts en Suisse*; a joint review for the English translation of A. Cournot, *Researches into the Mathematical Principles of the Theory of Wealth*; and I. Fisher *A Brief Introduction to the Infinitesimal Calculus Designed Especially to Aid in Reading Mathematical Economics and Statistics*.

1899

F.Y.E. reviews J. Davidson *The Bargain Theory of Wages*; A German Coal Miner *How the English Workmen Lives*; C.W. Macfarlane, *Value and Distribution*; Clemente Vidaurre, *Economía Política* (3rd edn); F. Virgilii and C. Garibaldi, *Introduzione alla Economia Mathematica*; and A. Ferrin Weber, *The Growth of Cities in the XIX-th Century*.

1900

F.Y.E. reviews C.F. Bastable, *The Theory of International Trade* (3rd edn) and *La Théorie du Commerce International* (2nd edn); J. Bonar and J.H. Hollander, *Letters of David Ricardo 1811–1823*; J.B. Clark, *The Distribution of Wealth*; W. Smart, *Taxation of Land Values and the Single Tax*; J.H. Hollander, ed., *Studies in State Taxation*.

1901

No reviews written by F.Y.E. were published.

1902

L.G. Chiozza, *British Trade and the Zollverein Issue*; H. Cox, *The United Kingdom and its Trade*; C. Gide, *La Coopération*; D. Schloss, *Les Modes de Remunération du Travail* (translation); S. and B. Webb, *The History of Trade Unionism* and *Industrial Democracy*; H.G. Wells, *Anticipations of the Reaction of Mechanical and Scientific Progress upon Human Life and Thought*.

1903

F.Y.E. reviews C.F. Bastable, *Public Finance*; J.W. Root, *The Trade Relations of the British Empire*; and W.J. Ashley, *The Tariff Problem*.

1904

H. Dietzel, *Vergeltunzölle*; A. Graziani, *Istituzioni di Economia Politica*; J.S. Nicholson, *Elements of Political Economy*; A.C. Pigou, *The Riddle of the Tariff*; Sir Horace Plunkett, *Ireland in the New Century*; and R. Dalla Volta, *Sulla Repercussione e la Incidenza dei Dazi Doganali*.

1905
F.Y.E. reviews T.N. Carver, *The Theory of Distribution*; H. Cunynghame, *A Geometrical Political Economy*; J. S. Nicholson, *Rates and Taxes as Affecting Agriculture*; and F.W. Taussig, American Economic Association's Presidential Address *The Present Position of the Doctrine of Free Trade.*

Articles and Notes

1896
'Further considerations on Index-Numbers', *EJ*, 6 (March 1896) pp. 132–42 as a reply to N. G. Pierson's articles on the subject in *EJ* (included in Edgeworth (1925), I, pp. 356–68, under the title 'A Defence of Index Numbers').

1897
'Prof. Seligman on the Theory of Monopoly' (included in Edgeworth (1925), I, pp. 111–42 and 143–71); 'The Pure Theory of Taxation' (also in Edgeworth (1925), II, pp. 63–125).

1898
Note: 'Prof. Graziani on the Mathematical Theory of Monopoly'.

1900
Note: 'Defence of Mr. Harrison's Calculation of the Rupee Circulation' (5 pages), and article 'The Incidence of Urban Rates' (65 pages).

1901
Notes: 'Mr. Walsh on the Measurement of General Exchange Value' (pp. 404–16) and 'Disputed Point in the Theory of International Trade', (pp. 582–95).

APPENDIX L

List of reviews, articles and notes by Edgeworth at the *EJ* during the period from 1906 to 1911:

Reviews

1906
L.S. Amery *The Fundamental Fallacies of Free Trade* and A.S. and E.M. Sidgwick *Henry Sidgwick: A Memoir* ('All that we learn about the personality of Sidgwick confirms our deference to his practical wisdom.')

1907
A. de Foville's *La Monnaie* and Y. Gouyot's *La Science Économique* ('He might even be compared with Bastiat').

1908
W.C. Mitchell, *Gold Prices and Wages under the Greenback Standard.*

1909
H. von Dietzel *Theoretische Sozialökonomic*; W. S. Jevons *Investigations on Currency and Finance* (short review for this new edition); R. Rea *Free Trade in Being*; H. Whithers *The Meaning of Money* ('Extreme lucidity accompanied with modest doubt').

1910
J.M. Clark [son of J.B. Clark] *Standards of Reasonableness in Local Freight Discriminations* ('Mr Clark, inheriting a name distinguished in economic literature, has added to it new lustre'); C. Colson, *Cours d'Économie Politique Professé à l'École Nationale des Ponts et Chaussés*; and J.S. Mill *Principles of Political Economy* (new edn).

1911
M.B. Hammond, *Railway Rate Theories of the Interstate Commerce Commission.*

Articles and Notes

1906
Article: 'Recent Schemes for Rating Urban Rates' *EJ*, 1906 (Edgeworth (1925), II, pp. 215–33, as 'Further Considerations on Urban Rates').

1907
Article: 'Appreciations of Mathematical Theories' (I and II) *EJ*, 1907, which continues in *EJ* 1908 (Edgeworth (1925), II, pp. 320–39 and 340–67, as 'Variorum Theories on Consumer's Surplus, Rent, Duopoly, Entrepreneurs' Remuneration' and 'Mr. Bickerdike's Theory of Incipient Taxes & Custom Duties').

Note: 'Correspondence of Ricardo with Maria Edgeworth' in the possession of Mrs F. Ricardo, Bromesberrow Place, Ledbury.

1908
Article: 'Appreciation of Mathematical Theories' (continuation III and IV – see 1907).

1910
Note: 'Applications of Probabilities to Economics' (I, pp. 284–304; II, pp. 441–65) (Edgeworth (1925), II, pp. 387–428).

1911
Note: 'Contributions to the Theory of Railway Rates' (I, pp. 346–70; II, pp. 551–71).

Note on Bickerdike's 'F. Y. E. on Application of Probabilities to Monopoly and Differential Prices', pp. 143–8.

APPENDIX M

Reviews, articles and notes by Edgeworth at the *EJ* during the period from 1912 to 1918.

Reviews

1912
F.Y.E. writes an adverse review – already mentioned – of H.L. Moore's *Law of Wages* in *EJ*, 22 (March, 1912), pp. 66–71.

1913
A.C. Pigou *Wealth and Welfare*.

1915
M. Alberti *L'Economia del Mondo, Prima, Durante e Dopo la Guerra Europea*.

1916
A.C. Pigou *The Economy and the Finance of the War*; G. Preziosi (a friend of Pantaleoni) *La Germania alla Conquista dell'Italia*.

1917
Various authors, *Some German Economic Writings about the War*; various authors, including A. Marshall: *After War Problems*; C. Gill *National Power and Prosperity: A Study of the Causes of Modern Warfare*; R.A. Lehfeldt (South Africa) *Economics in the Light of War*; National Bank of Commerce (New York) *War Finance Primer*; E.R.A. Seligman *A Constructive Criticism of the U. S. War Tax Bill.*

1918

B.M. Anderson (Harvard) *The Value of Money*; G. Arias *Principii di Economia Commerciale*; H.G. Moulton (Chicago) *Principles of Money and Banking*; C.A. Phillips (ed.) *Readings in Money and Banking*; L. Smith-Gordon and L.C. Staples *Rural Reconstruction in Ireland*; A. Loria *The Economic Causes of War*.

Articles and Notes

Articles, 1912 and 1913

'Railways Rates' (III, 1912, pp. 198–218 and IV, 1913, pp. 206–26).

1915

'Economists on War' (In fact, a review of several books by W. Sombart, J.S. Nicholson, E.R.A. Seligman and Y. Guyot); 'Recent Contributions to Mathematical Economics', I and II (works by A. Osorio (Portugal), Antonelli, Zawadski, J.S. Nicholson, W.E. Johnson and M. Fanno).

1918

'The Doctrine of Index Numbers According to Prof. W. Mitchell' (pp. 176–197) where 'chained index numbers' are mentioned (p. 181).

APPENDIX N

Reviews, articles and notes by Edgeworth for the *EJ* during the period from 1919 to 1925.

Reviews

1919

F. Bernis *La Hacienda Española: Los Impuestos*; E. Cannan *Money: Its Connection with Rising and Falling Prices*; A. Andréadès *The Economic Institutions of Ancient Greece*; R.A. Lehfeldt *Gold Prices and the Witwatersrand*; Interim Report of the European Commission of the National Industry Conference.

1920

Prof. Cassel *Treatise*; S. and B. Webb *The History of Unionism*, revised edition; J.S. Nicholson *The Revival of Marxism*; and A. Loria *Karl Marx*.

1921

A. Hoare *The National Needs of Britain*; G. Gough *Wealth and Work*.

1922

L. Amoroso *Lezioni di Economia Matematica*; A.C. Pigou *The Political Economy of War*; W. McDougall *National Welfare & National Decay*; R.H. Brand *War and National Finance*; G. Subercaseaux (University of Chile) *Le Papier-Monnaie*; and *El Sistema Monetario y la Organización Bancaria de Chile*.

1923

W. Rathenau *In Days to Come*; A. Marshall *Money, Credit and Commerce*; seven members of the Labour Party *The Labour Party's Aim: A Criticism & a Restatement*; F. Bernis *Consequencias* [*sic*] *Económicas de la Guerra*; C.M. Walsh *Index Numbers*.

1924

A.L. Bowley *The Mathematical Groundwork of Economics*; Mrs H.A.L. Fisher *The Economic Position of the Married Woman*; A.C. Pigou *The Economics of Welfare*, 2nd edn; J. Maurice Clark *Studies in Economics of Overhead Costs*.

Articles and Notes

1919

'Methods of Graduating Taxes on Income and Capital' (pp. 138–153).

1920

Note: 'Mathematical Formulæ and the Royal Commission on Income Tax'.

1922

Presidential address of Section F of the British Association: 'Equal Pay to Men and Women for Equal Work'.

1923

'Index Numbers according to Mr. Walsh' and 'Women's Wages in Relation to Economic Welfare'.

1925

'The Plurality of Index Numbers' (pp. 379–88).[10]

NOTES

1. EP NL IRELAND, Ms. 989.
2. TRINITY CD registers V/23/6, V/24/4, V/30/21 to 24, V/31/3.
3. Memory of Frances Anne Edgeworth. (EPBLUO, Ms. Eng. misc. c899.
4. EP BODLEIAN L UO, Ms. Eng. misc. c 899.
5. EP BODLEIAN L UO, Ms. Eng. misc. c 899.
6. That is, £1256, 5 shillings and 1 penny.
7. EP BODLEIAN L UO, Ms. Eng. misc. c 899.
8. EP BODLEIAN L UO, Ms. Eng. lett. c730.
9. EP NUFFIELD C NO, BOX D10.
10. Included as 'Graduation of Taxes' and as 'Formulæ for Graduating Taxation' in Edgeworth (1925), II, pp. 243–59 and pp. 260–72.

Bibliography

LETTERS AND DOCUMENTS

Crampton Papers at the Library of Trinity College Dublin, Manuscripts' section. (CP TRINITY CLD, Ms. 4178).

Edgeworth Papers at the Bodleian Library, Oxford (EP BODLEIAN UO, Ms.Eng.lett. c698-746, d529, Ms. Eng. misc. c899, Ms.Eng.poet. e188-189).

Edgeworth Papers at the British Library of Political and Economic Science of the London School of Economics (EP LSE, Lett. 6-1889, 8-1891).

Edgeworth Papers at the National Library of Ireland, Dublin, Manuscripts section (EP NL IRELAND, Ms. 989, 8145, 11 132-6 and 22 822).

Edgeworth Papers at Nuffield College, Oxford (EP NUFFIELD C UO, Boxes A to E).

Edgeworth Papers and *John Maynard Keynes Papers* at the Library of King's College Cambridge, Modern Archives (KP KING'S C CAMBRIDGE, Ms. ED/6/6, EJ/6/6).

Galton Papers and *Pearson Papers* in the Library Services Special Collections of University College London (GP UC LONDON, Ms. 138.7 and PP UC LONDON Ms. 681).

Oriental and India Office Collections at the British Library, London (Bombay, N/3/37/ Folio 121).

Parish Register Books of Pla de Sant Tirs, (I) 1740–1802, (II) 1803–1860. Kept at the Arxiu Diocesà de La Seu d'Urgell.

Trinity College Registers (V/23/6; V/24/4; V/30/17 to 24; V/31/3).

University of Barcelona's Historical Files: Studies in 'Lleis'; 'Expediente de D. Isidro Eroles y Eroles', 1842.

BOOKS AND ARTICLES

Alcalá Galiano, A. (1878/1955), *Recuerdos de un anciano*, in *Obras Escogidas de D. Antonio Alcalá Galiano*, Biblioteca de Autores Españoles, (2 Vols).

Allen, W.R. (1992) 'Irving Fisher, F.D.R., and the Great Depression', in M. Blaug (ed.), *Pioneers in Economics 41*, Aldershot, UK and Brookfield, USA, Edward Elgar Publishing.

Annan, N. (1984), *Leslie Stephen: The Godless Victorian*, New York, Random House.
Auspitz, R. and R. Lieben (1889), *Untersuchungen über die Theorie des Preises*, Leipzig, Duncher and Humblot.
Azcárate, P. de (1960), *Wellington y España*, Madrid, Espasa-Calpe.
Baccini, A. (2007), 'Edgeworth on the foundations of ethics and probability', *European Journal of the History of Economic Thought*, 14:1, 79–96.
Balaguer, V. (1863), *Historia de Cataluñya y de la corona de Aragón*, Vol. 11, Barcelona, S. Manero.
Barbé, L. (2002), *Retrat de Familia Sobre Fons de Trèvols*, Barcelona, Ed. 62. There is a 2002 Spanish version, Barcelona, Muchnik ed.
Barbé, L. (2004a), 'Els orígens alturgellens d'un economista famós: F.Y. Edgeworth', Les Faldades de *Lo Banyut*, No. 1, spring.
Barbé, L. (2004b), 'Francis Ysidro Edgeworth's Catalan grandfather', *European Journal of the History of Economic Thought*, 11:2, 295–307.
Barratt, A. (1869), *Physical Ethics*, London, Williams and Norgate.
Bergson, A. (1938), 'A Reformulation of Certain Aspects of Welfare Economics', *Quarterly Journal of Economics*, February, 310–34.
Black, R.D.C. (1977), *Papers and Correspondence of W.S. Jevons*, Macmillan in association with the Royal Economic Society (5 volumes).
Blaug, M. (1962), *Economic Theory in Retrospect*, London, Heinemann.
Bonar, J. (1885), *Malthus and his Work*, London, Macmillan.
Bonar, J. (1926), 'Memories of F.Y. Edgeworth', *Economic Journal*, 36, 647–53.
Bowley, A.L. (1934), 'Francis Ysidro Edgeworth', *Econometrica*, 2, 113–24.
Bridel, P. (ed.) (2001), *The Foundations of Price Theory*, (6 volumes set), London, Pickering and Chatto. Contains in Vol. 4 F.Y. Edgeworth's 'Economical Calculus' and 'On Professor Jevons' Formulæ of Exchange' from *Mathematical Psychics* (1881), and 'The Pure Theory of Monopoly' (*EJ*, 1894).
Burke, B. (1886), *A Genealogic and Heraldic Dictionary: History of the Landed Gentry of Great Britain and Ireland*, 7th edn, 2 Vols, London, Burke's Peerage.
Butler, H.J. and H.E. Butler (1927), *The Black Book of Edgeworthstown and Other Edgeworth Memories 1585–1817*, London, Faber and Gwyer.
Butler, M. (1972), *Maria Edgeworth: A Literary Biography*, Oxford, Oxford University Press.
Butler, M. and C. Colvin (1971), 'A Revised Date of Birth for Maria Edgeworth', *Notes and Queries*, September, p. 339.

Camp, F. (1945), 'Joaquín Ibáñez-Cuevas, Barón de Eroles', *Boletín de la Real Academia de Buenas Letras de Barcelona*, XVIII, 251–9.

Carlyle, T. (1851), *The Life of John Sterling*, Boston, Phillips and Sampson; and London, Chapman and Hall, 1852.

Carter, T.T. (1884), *Harriet Monsell: A Memoir*, New York, E.P. Dutton.

Casey, E.O.B. (1877), *Illustrious Irishwomen*, London, 2 vols.

Castells, I. (1989), *La Utopía Insurreccional del Liberalismo: Torrijos y las Conspiraciones Liberales de la Década ominosa*, Barcelona, Ed. Crítica.

Clarke, D. (1965), *The Ingenious Mr Edgeworth*, London, Oldbourne Press.

Clarke, I.C. (1949), *Maria Edgeworth: Her Family and Friends*, London, Hutchinson and Co.

Coase, R.H. (1972), 'The appointment of Pigou as Marshall's successor', *Journal of Law and Economics*, 15, 473–85.

Coats, A.W. (1964), 'The role of authority in the development of economics', *Journal of Law and Economics*, 7, 85–106. Also included in his *The Sociology and Professionalization of Economics*, London, Routledge and Kegan Paul, 1993.

Coats, A.W. (1968a), 'Political economy and the tariff reform campaign of 1903', *Journal of Law and Economics*, 11, 181–229.

Coats, A.W. (1968b), 'The origins and early development of the Royal Economic Society', *Economic Journal*, 78, 349–71.

Coats, A.W. (1972), 'The appointment of Pigou as Marshall's successor: Comment', *Journal of Law and Economics*, 15, 487–95.

Cournot, A. (1838), *Recherches sur les Principes Mathématiques de la Théorie des Richesses*, Paris, L. Hachette. English translation *Researches into the Mathematical Principles of the Theory of Wealth*, 1897 edition prepared by I. Fisher, 1927, London, Macmillan.

Creedy, J. (1977–1998), *The History of Economic Analysis. Selected Essays by John Creedy*. 1998, Cheltenham, UK and Northampton, MA, USA, Edward Elgar. Contains, among others, as numbered chapters, the following articles: 2. 'Edgeworth's contribution to the theory of exchange', *Scottish Journal of Political Economy*, 26(2), June 1979, pp. 163–81; 3. 'Some recent interpretations of *Mathematical Psychics*', *History of Political Economy*, 12(2), Summer 1980, pp. 267–76; 4. 'The early use of Lagrange multipliers in economics', *Economic Journal*, 90, June 1980, pp. 371–6; 5. 'F.Y. Edgeworth, 1845–1926', in *Pioneers of Modern Economics in Britain*, D.P. O'Brien and J.R. Presley (eds), London, Macmillan, 1981, pp. 72–104; 6. 'F.Y. Edgeworth's mathematical training', *Journal of the Royal Statistical Society*, A, 146(2), 1983, pp. 158–62; 7. 'Edgeworth: utilitarianism and arbitration', *History of Political Economy*, 16(4), 1984, pp. 609–18; 10. 'Wicksell on Edgeworth's tax paradox', *Scandinavian Journal of Economics*, 90(1), 1988, pp. 101–12; 12. 'Marshall and Edgeworth', *Scottish Journal of Political Economy*, 37(1), February 1990, pp. 18–39; 20. 'Cournot on trade between regions and the transition from partial to general equilibrium modelling', *History of Economic Review*, 18, 1992, pp. 10–18; 22. 'Exchange equilibria: bargaining, utilitarian and competitive solutions', *Australian Economic Papers*, 33(62), June 1994, pp. 34–52.

Creedy, J. (1986), *Edgeworth and the Development of Neoclassical Economics*, Oxford, Basil Blackwell.

Creedy, J. (2006), 'The barter controversy', in T. Raffaeli, G. Becattini and M. Dardi, eds, in, *The Elgar Companion to Alfred Marshall*, Cheltenham, UK and Northampton, MA, USA, Edward Elgar.

Darnell, A., ed. (1991), *Early Mathematical Economics*, (6 volume), London, Pickering and Chatto. Vols. 2, 3, 4, 5 and 6 contain a selection of 21 articles and entries by F.Y. Edgeworth.

Dickens, Charles, Jr. (1888/1995) *Dickens's Dictionary of London*, Facsimile edition, Moretonhampstead, Old House Books.

Eatwell, J., M. Milgate, and P. Newman, eds (1987), *The New Palgrave Dictionary of Political Economy*, London, Macmillan.

Edgeworth, C.S. (1815), *Memoirs of the Abbé Edgeworth; Containing his Narrative of the Last Hours of Louis XVI*, London.

Edgeworth, F.A. (1867), *A Memoir of Maria Edgeworth: With Selections from her Letters*, Private edition of the 'Children of Mrs. Edgeworth', London, Augustus Hare and Edward Arnold. A public edition by Augustus Hare under the title *The Life and Letters of Maria Edgeworth* followed in 1894, London, Edward Arnold.

Edgeworth, F.Y. (1877), *New and Old Methods of Ethics*, Oxford and London, James Parker and Co., reprinted in Newman (2003).

Edgeworth, F.Y. (1881), *Mathematical Psychics*, London, Kegan Paul, and reprint 1967, New York, A.M. Kelley. Also reprinted in Newman (2003).

Edgeworth, F.Y. (1925), *Papers Relating to Political Economy*, New York, Burt Franklin for the Royal Economic Society.

Edgeworth, F.Y. (1882–1925/1994), *Edgeworth on Chance, Economic Hazard and Statistics*, edited by P. Mirowski, Lanham, Maryland, USA and London, Rowman and Littlefield (Works and reviews on probability and statistics).

Edgeworth, M. (1800/1994) *Castle Rackrent*, Ware, UK, Wordsworth Classics.

Edgeworth, R.L. (1820), *The Memoirs of Richard Lovell Edgeworth, Esq.: Begun by Himself and Concluded by His Daughter Maria Edgeworth*, 2 vols, printed for R. Hunter, London, 2nd edn, corrected,

1821; third edition 1844 abridged in one volume, London, Richard Bentley.

England, T.R. (1818), *Letters from the Abbé Edgeworth to his Friends, Written between the Years 1777–1807, with Memoirs of his Life*, Longman, London.

Espoz y Mina, F. (1852/1962), *Memorias del General D. Francisco Espoz y Mina*, Biblioteca de Autores Españoles (2 vols), Madrid, Ed. Atlas.

Feliu, G. (1975a), 'Ibáñez-Cuevas i de Valonga, Joaquim d'', in *Gran Enciclopèdia Catalana*, Barcelona, Enciclopèdia Catalana.

Feliu, G. (1975b), 'Malcontents, Guerra de', *Gran Enciclopèdia Catalana*, Barcelona, Enciclopèdia Catalana.

Fisher, I. (1892), 'Mathematical Investigations in the Theory of Value and Prices', *Transactions of the Connecticut Academy*, IX, July. Reprinted in 1925, New Haven, Yale University Press; reprinted in 1961, New York, Augustus M. Kelly.

Friendly, A. (1973), *Beaufort of the Admiralty*: *The Life of Sir Francis Beaufort*, New York, Random House.

Galli, F. (1831), *Campagne du Général Mina en Catalogne*, Paris, Chez Philippe.

Graves, A.P. (1930), *To Return to All That*, Dublin, Talbot Press.

Graves, R. (1929/1985), *Good bye to All That*, London and New York, Anchor Books, Doubleday.

Graves, R.P. (1882, 1885 and 1889), *Life of Sir William Rowan Hamilton*, Vol. I (1882), Vol. II (1885), Vol. III (1889), Dublin, Hodges and Figgis.

Gray, P. (1995), *The Irish Famine*, London, Thames and Hudson.

Groenewegen, P. (1995), *A Soaring Eagle: Alfred Marshall 1842–1924*, Cheltenham, UK and Northampton, MA, USA, Edward Elgar.

Guinan, J. (1909), 'Henry Essex Edgeworth', in *A Catholic Encyclopedia*, Vol. V, New York, Robert Appleton.

Hagen, K.H. (1844), *Die Nothwendigkeit der Handelsfreiheit für das Nationaleinkommen Mathematisch nachgeweisen*, Königsberg, Hagen.

Halsey, A.H., ed. (1976), *Traditions of Social Policy: Essays in Honour of Violet Butler*, Oxford, B. Blackwell.

Hamilton, C.J. (1904), *Notable Irishwomen*, Dublin, Sealy, Bryers and Co.

Hankins, T.L. (1980), *Sir William Rowan Hamilton*, Baltimore, Johns Hopkins University Press.

Hare, A.J.C. (1894/1971), *The Life and Letters of Maria Edgeworth*, Freeport, New York, Books for Libraries Press.

Harrod, R.F. (1951/1972), *John Maynard Keynes*, Harmondsworth, Middlesex, UK, Penguin Books.

Hicks, J.R. (1984a), *The Economics of John Hicks*, edited by D. Helm, Oxford, Blackwell.

Hicks, J.R. (1984b), 'Francis Ysidro Edgeworth', in A.E. Murphy, ed., *Economists and the Irish Economy: From the Eighteenth Century to the Present Day*, Dublin, Hermathena, Trinity College Dublin and Irish Academic Press, pp. 157–74.

Hicks, J.R. and R.G.D. Allen, (1934), 'A reconsideration of the theory of value, Parts I and II', *Economica*, 1, February and May, pp. 52–76 and 196–219.

Hildreth, C. (1968), 'Edgeworth, Francis Ysidro', in D.E. Sills, ed., *International Encyclopædia of the Social Sciences*, Vol. 4, New York, Macmillan and Free Press, pp. 506–9.

Hill, C. (1910), *Maria Edgeworth and her Circle in the Days of Buonaparte and Bourbon*, London and New York, John Lane.

Hollingworth, B. (1997), *Maria Edgeworth's Irish Writing*, London, Macmillan and St. Martin's Press.

Hughes-Hallett, P. (1990), '*My Dear Cassandra*: *Jane Austen*, London, Collins and Brown, Ltd.

Iglésies, J. (1974) *Estadístiques de Població de Catalunya: Primer Vicenni del s. XVIII*, Barcelona, Fundació Salvador Vives Casajoana.

Iglésies, J. (1981) *El Fogatge de 1553: Estudi i Transcripció.* Barcelona, Fundació Salvador Vives Casajoana.

Inglis-Jones, E. (1959), *The Great Maria*, London, Faber and Faber.

Instituto Luis de Salazar y Castro (1959), *Archivo General Militar de Segovia: Indice de Expedientes Personales*, Madrid, Ed. Hidalguía.

Jaffé, W. (1965), *Correspondence of Léon Walras and Related Papers*, 3 Volumes, Amsterdam, North-Holland Pub. Co.

Jaffé, W. (1973). 'Léon Walras's role in the "Marginal Revolution"', in R.D. Collison Black, A.W. Coats and C. D.W. Goodwin, eds, *The Marginal Revolution in Economics*, Durham, NC, USA, Duke University Press.

Jevons, H.W. and H.S. Jevons (1934), 'William Stanley Jevons: His Life', *Econometrica*, 2, 225–37.

Jevons, W.S. (1863), *A Serious Fall in the Value of Gold Ascertained, and its Social Effects Set Forth: with two diagrams*, London, Edward Standford. Included in Jevous, W.S. (1884), *Investigations in Currency and Finance*, ed. H.S. Foxwell, reprinted New York, Augustus Kelley, 1964.

Jevons, W.S. (1865), 'On the Variation of Prices and the Value of the Currency Since 1782', *Journal of the Statistical Society of London*, 192–240. Included in Jevous, W.S. (1884), *Investigations in Currency and Finance*, ed. H.S. Foxwell, reprinted New York, Augustus Kelley, 1964.

Jevons, W.S. (1871/1970), *The Theory of Political Economy*, Harmondsworth, Middlesex, UK, Penguin Books.

Jevons, W.S. (1886). *Letters and Journal of William Stanley Jevons*, edited by his wife Harriet A. Jevons, London, Macmillan.

Jevons, W.S. (1905). *The Principles of Economics, a Fragment of a Treatise on the Industrial Mechanism of Society and other Papers*, with a preface by Henry Higgs (editor), London, Macmillan.

Jones, T.W. (1978), 'The appointment of Pigou as Marshall's successor: the other side of the coin', *Journal of Law and Economics*, 21, 235–43.

Kadish, A. (1982), *The Oxford Economists in the late XIXth Century*, Oxford, Clarendon Press.

Kadish, A. (1989), *Historians, Economists and Economic History*, London and New York, Routledge.

Kavanagh, J. (1862), *English Women of Letters*, London, Hurst and Blackett.

Kendall, M.G. (1968), 'Francis Ysidro Edgeworth', *Biometrika*, 55:2, 269–75, reprinted in Pearson and Kendall (1970).

Keynes, J.M. (1921), *A Treatise on Probability*, London, Macmillan. Reprinted in Vol. VIII of Keynes (1971–1989).

Keynes, J.M. (1924), 'Alfred Marshall', *Economic Journal*, September 1924. Reprinted in G. Keynes, ed., *Essays in Biography*, 1951, London, Rupert Hart-Davis. Reprinted also in Vol. X of Keynes (1971–1989).

Keynes, J.M. (1926), 'Francis Ysidro Edgeworth (1845–1926)', *Economic Journal*, March 1926, reprinted as listed under Keynes (1924).

Keynes, J.M. (1930), *A Treatise on Money*, 2 Vols, London, Macmillan, reprinted in Vols V and VI of Keynes (1971–1989).

Keynes, J.M., (1936a), 'William Stanley Jevons', *Journal of the Royal Statistical Society*, Part III, reprinted as listed under Keynes (1924).

Keynes, J.M., (1936b), 'Herbert Somerton Foxwell', *Economic Journal*, December 1936, reprinted as listed under Keynes (1924).

Keynes, J.M. (1971–1989), *The Collected Writings of John Maynard Keynes*, 30 vols, ed. D.E. Moggridge and E.S. Johnston, London, Macmillan and Cambridge University Press.

Kowaleski-Wallace, E. (1991), *Their Fathers' Daughters: Hannah More, Maria Edgeworth and Patriarchal Complicity*, Oxford, Oxford University Press.

Lawless, E. (1904), *Maria Edgeworth*, London, Macmillan.

Levy, P. (1981), *Moore: G.E. Moore and the Cambridge Apostles*, Oxford, Oxford University Press.

Llorens, V. (1968), *Liberales y Románticos*, Madrid, Ed. Castalia.

Lopez, E., J. Peruga and C. Tudel (1988), *L'Andorra del segle XIX*, Andorra la vella, Grafinter.

Lyons, Tony (2003), *The Education Work of Richard Lovell Edgeworth: Irish Educator and Inventor, 1744–1817*, Lampeter, The Edwin Mellen Press.

MacDonald, Edgar E. (1970), *The American Edgeworths*, mimeo, Richmond, VA.

MacKenzie, N. and J. (1977), *The First Fabians*, London, Weidenfeld and Nicolson; reprinted 1979, London, Quartet Books Ltd.

MacKenzie, N. and J., ed. (1982), *The Diary of Beatrice Webb, 1873–1943*, Vol 1: 1873–1892, *Glitter Around and Darkness Within*, Boston, The Belknap Press of Harvard University Press.

Mangoldt, H.C.E. von (1863), *Grundriss der volkswirthschaftslehre*, Berlin, Verlag für Sprach- und Handelswissenschaft (2nd edn, 1871, F. Kleinwächter).

Marshall, A. (1868–1924) *The Correspondence of Alfred Marshall, Economist* (1868–1890, 1891–1902, 1903–1924, 3 vols), edited by J.K. Whitaker, Cambridge, Cambridge University Press.

Marshall, A. (1890/1972), *Principles of Economics*, London, Macmillan.

McCann, R.C. Jr. (1996), *F.Y. Edgeworth: Writings in Probability, Statistics and Economics*, 3 vols, Cheltenham, UK and Brookfield, USA, Edward Elgar.

Miller, J.P. (1967), 'Irving Fisher of Yale', in W. Fellner (ed.), *Ten Economic Studies in the Tradition of Irving Fisher*, New York, John Wiley and Sons.

Mirowski, P. (1990), 'Problems in the paternity of econometrics: Henry Ludwell Moore', *History of Political Economy*, 22:4, 587–609.

Mirowski, P. ed. (1994), *Edgeworth on Chance, Economic Hazard and Statistics*, Lanham, MA, USA, Rowman and Littlefield Pub., Inc. Contains Edgeworth's *Metretike* and a selection of his papers and reviews on statistics.

Moggridge, D.E. (1992). *Maynard Keynes: An Economist's Biography*, London and New York, Routledge.

Mozley, T. (1882), *Reminiscences, Chiefly of Oriel College and the Oxford Movement*, London, Longmans, Green and Co.

Murphy, A.E., ed. (1984), *Economists and the Irish Economy: From the Eighteenth Century to the Present Day*, Dublin, Hermathena, Trinity College, Dublin, and Irish Academic Press.

Newman, P. (1987), 'Francis Ysidro Edgeworth', in J. Eatwell, M. Milgate, and P. Newman, eds, *The New Palgrave*, Macmillan. Reprinted in *Utility and Probability*. The New Palgrave. London, Macmillan, pp. 38–69.

Newman, P., ed. (2003), *F.Y. Edgeworth's Mathematical Psychics and Further Papers on Political Economy*, Oxford, Oxford University Press. Contains Edgeworth's *Mathematical Psychics*, *New and Old Methods of Ethics* and a selection of his papers, contributions to the Palgrave Dictionary and book reviews on Political Economy.

O'Brien, D.P., ed. (1999), *The History of Taxation*, (8 volumes), London, Pickering and Chatto. Contains in Vol. 8 F.Y. Edgeworth's 'The Pure Theory of Taxation' (1897).

O'Donoghue, D.J. (1912), *The Poets of Ireland: Irish Writers of English Verse*, Dublin, Hodges, Figgis and Co.

Oliver, G. (1882), *A Study of Maria Edgeworth*, Boston, A. Williams and Co.

Pakenham, V. (2000), *The Big House in Ireland,* London, Cassell and Co.

Palgrave, R.H. (1894–1899), *Palgrave's Dictionary of Political Economy*, London, Macmillan.

Palgrave, R.H. (1925–26), *Palgrave's Dictionary of Political Economy*, revised edition, Reprinted 1963, New York, A.M. Kelley.

Pareto, V. (1896 and 1897), *Cours d'Économie Politique*, Reprinted 1964, Geneva, Librairie Droz.

Pareto, V. (1898/1966), 'Comment se pose le problème de l'économie pure', in *Œuvres Complètes*, Vol. IX, Geneva, Librairie Droz.

Pareto, V. (1900) 'Sunto di alcuni capitoli di un nuovo trattato di economia pura', *Giornale degli Economisti*, 2, 216–35 (Part I) and 511–49 (Part II).

Pareto, V. (1906/1909/1966), *Manuel d'Économie Politique*, Geneva, Librairie Droz. 1909, French translation of *Manuale d'Economia Politica* with some of Pareto's additions.

Pareto, V. (1964–84), *Oeuvres complètes de Vilfredo Pareto*, 28 vols, ed. G. Busino, Geneve, Librairie Droz.

Paterson, A. (1914), *The Edgeworths: A Study of Late Eighteenth Century Education*, London, University Tutorial Press, Ltd.

Pearson, E.S. and M.G. Kendall eds, (1970), *Studies in the History of Statistics and Probability*, London, Griffin.

Petridis, A. (1987), 'Price, Langford Lovell Frederick Rice (1862–1950)' in J. Eatwell, M. Milgate, and P. Newman, eds, *The New Palgrave Dictionary of Political Economy*, London, Macmillan.

Pigou, A.C. (1912), *Wealth and Welfare*, London, Macmillan.

Pigou, A.C. (1925), *Memorials of Alfred Marshall*, London, Macmillan Includes 'Reminiscences' by F.Y.Edgeworth, p. 66–73. Reprinted 1966, New York, A.M. Kelley.

Potier, J.-P. (1987/1991), *Piero Sraffa: Unorthodox Economist (1898–1983)*, London and New York, Routledge.

Price, L.L. (1926), 'Francis Ysidro Edgeworth', *Journal of the Royal Statistical Society*, 89, March, 371–7.

Price, L.L. (1946), 'Memoirs and Notes on British Economists 1881–1946', MSS, Brotherton Library, University of Leeds.

Reisman, D. (1990), *Alfred Marshall's Mission*, London, Macmillan.

Roncaglia, A. (2001), *La Ricchezza delle Idee: Storia del Pensiero Economico*, Rome-Bari, GLF Editori Laterza. 2005 English version by Cambridge University Press.

Saint Amand, I. de (1918), *La Dernière Année de Marie-Antoinette*, Paris, P. Lethielleux.

Samuelson, P.A. (1974), 'Complementarity: An Essay on the 40th Anniversary of the Hicks-Allen Revolution in Demand Theory', *Journal of Economic Literature*, 12, 1255–89.

Sánchez de la Campa, J.M. (1851), *El Valle de Andorra*, Barcelona, Imprenta el Barcelonés.

Schmidt, T. and C.E. Weber (2008), 'On the origins of ordinal utility: Andreas Heinrich Voigt and the mathematicians', *History of Political Economy*, 40:3, 481–510.

Schofield, R.E. (1963), *The Lunar Society of Birmingham*, Oxford, Clarendon Press.

Schumpeter, J.A. (1954), *History of Economic Analysis*, Oxford, Oxford University Press.

Seligman, B.B. (1962), *Main Currents in Modern Economics*, New York, Free Press of Glencoe, Macmillan.

Shubik, M. (1959), 'Edgeworth Market Games', in A.W. Tucker and R.D. Luce, eds, *Contributions to the Theory of Games*, Vol. 4, Trenton, NJ, USA, Princeton University Press.

Sidgwick, H. (1874), *Methods of Ethics*, London, Macmillan.

Skidelsky, R. (1989, 1992) *John Maynard Keynes.* Vol. I (1983) *Hopes Betrayed, 1883–1920*, Vol. II (1992) *The Economist as Saviour, 1920–1937*, London, Macmillan.

Slade, B.C. (1937), *Maria Edgeworth 1767–1849: A Bibliographical Tribute*, London, Constable Publishers.

Sraffa, P. and M. Dobb (1962 and 1965), *The Works and Correspondence of David Ricardo*, Vols IX and X. Cambridge, Cambridge University Press.

Stephen, L. and S. Lee (eds) (1908), *Dictionary of National Biography*, Oxford, Oxford University Press.

Stigler, S. (1978), 'Francis Ysidro Edgeworth, statistician', *Journal of the Royal Statistical Society*, Series A 141:3, 287–322.

Strachey, L. (1922), *Books and Characters: French and English*, London, Chatto and Windus.

Sully, J. (1877), *Pessimism: A History and a Criticism*, London, Henry S. King and Co.

Sully, J. (1918), *My Life and Friends: A Psychologist's Memories*, London, T. Fisher Unwin.

Uglow, J. (2002), *The Lunar Men*, London, Faber and Faber.

Voigt, A. (1893), 'Zahl und Mass in der Ökonomik. Eine Kritische Untersuchung der Mathematischen Methode und der Mathematischen Preistheorie', *Zeitschrift für die gesamte Staatswissenschaft*, 577–609.

Walras, L. (1874–1877), *Elements d'Économie Politique Pure*, Lausanne, Corbaz et Cie.

Webb, B.P. (1873–1943), *The Diary of Beatrice Webb*, see MacKenzie (1982).

Webb, B.P. (1926), *My Apprenticeship*, London, Longmans, Green and Co. Reprinted 1971, Harmondsworth, Penguin.

Weber, C.E. (2005), 'Edgeworth on complementarity, or Edgeworth, Auspitz-Lieben, and Pareto de-homogenized', *History of Political Economy*, 37:2, 293–307.

Whitaker, J.K., ed. (1996), *The Correspondence of Alfred Marshall, Economist*, (1868–1890, 1891–1902, 1903–1924, 3 vols), Cambridge University Press.

White, M.V. (2001), 'Indeterminacy in exchange: disinterring Jevons' trading bodies', *The Manchester School*, 69:2, 208–26.

Wicksell, K. (1936/1962), *Interest and Prices*, reprinted New York, A.M. Kelley.

Index